Synergetic Phenomena in Active Lattices

Springer
Berlin
Heidelberg
New York
Barcelona
Hong Kong
London
Milan
Paris
Tokyo

Springer Series in Synergetics

http://www.springer.de/phys/books/sssyn

An ever increasing number of scientific disciplines deal with complex systems. These are systems that are composed of many parts which interact with one another in a more or less complicated manner. One of the most striking features of many such systems is their ability to spontaneously form spatial or temporal structures. A great variety of these structures are found, in both the inanimate and the living world. In the inanimate world of physics and chemistry, examples include the growth of crystals, coherent oscillations of laser light, and the spiral structures formed in fluids and chemical reactions. In biology we encounter the growth of plants and animals (morphogenesis) and the evolution of species. In medicine we observe, for instance, the electromagnetic activity of the brain with its pronounced spatio-temporal structures. Psychology deals with characteristic features of human behavior ranging from simple pattern recognition tasks to complex patterns of social behavior. Examples from sociology include the formation of public opinion and cooperation or competition between social groups.

In recent decades, it has become increasingly evident that all these seemingly quite different kinds of structure formation have a number of important features in common. The task of studying analogies as well as differences between structure formation in these different fields has proved to be an ambitious but highly rewarding endeavor. The Springer Series in Synergetics provides a forum for interdisciplinary research and discussions on this fascinating new scientific challenge. It deals with both experimental and theoretical aspects. The scientific community and the interested layman are becoming ever more conscious of concepts such as self-organization, instabilities, deterministic chaos, nonlinearity, dynamical systems, stochastic processes, and complexity. All of these concepts are facets of a field that tackles complex systems, namely synergetics. Students, research workers, university teachers, and interested laymen can find the details and latest developments in the Springer Series in Synergetics, which publishes textbooks, monographs and, occasionally, proceedings. As witnessed by the previously published volumes, this series has always been at the forefront of modern research in the above mentioned fields. It includes textbooks on all aspects of this rapidly growing field, books which provide a sound basis for the study of complex systems.

Vladimir I. Nekorkin
Manuel G. Velarde

Synergetic Phenomena in Active Lattices

Patterns, Waves, Solitons, Chaos

With 207 Figures

Springer

Professor Vladimir I. Nekorkin
Radiophysical Department
Nizhny Novgorod State University
23 Gagarin Avenue
Nizhny Novgorod 603600, Russia

Professor Manuel G. Velarde
Universidad Complutense de Madrid
Instituto Pluridisciplinar (USM)
Paseo Juan XXIII, No. 1
28040 Madrid, Spain

Die Deutsche Bibliothek - CIP-Einheitsaufnahme
Nekorkin, Vladimir I.:
Synergetic phenomena in active lattices : patterns, waves, solitons, chaos /
Vladimir I. Nekorkin ; Manuel G. Velarde. - Berlin ; Heidelberg ; New York ;
Barcelona ; Hong Kong ; London ; Milan ; Paris ; Tokyo : Springer, 2002
(Springer series in synergetics)
(Physics and astronomy library)

ISSN 0172-7389
ISBN 978-3-642-62725-5 ISBN 978-3-642-56053-8 (eBook)
DOI 10.1007/978-3-642-56053-8

http://www.springer.de

Originally published by Springer-Verlag Berlin Heidelberg New York in 2002
Softcover reprint of the hardcover 1st edition 2002

Typesetting: Dataconversion by LE-TeX Jelonek, Schmidt & Vöckler GbR, Leipzig
Cover design: *design & production*, Heidelberg
Printed on acid-free paper SPIN: 10854087 55/3141/yl - 5 4 3 2 1 0

For Tatiyana Alekseevna and Maria del Pilar

Preface

In recent years there has been growing interest in the study of the nonlinear spatio-temporal dynamics of problems appearing in various fields of science and engineering. In a wide class of such systems an important place is occupied by active lattice dynamical systems. Active lattice systems are, e.g., networks of identical or almost identical interacting units ordered in space. The activity of lattices is provided by the activity of units in them that possess energy or matter sources. In real (1D, 2D or 3D) space, processes develop by means of various types of connections, the simplest being diffusion. The uniqueness of lattice systems is that they represent spatially extended systems while having a finite-dimensional phase space. Therefore, active lattice systems are of interest for the study of multidimensional dynamical systems and the theory of nonlinear waves and dissipative structures of extended systems as well. The theory of nonlinear waves and dissipative structures of spatially distributed systems demands using theoretical methods and approaches of the qualitative theory of dynamical systems, bifurcation theory, and numerical methods or computer experiments. In other words, the investigation of spatio-temporal dynamics in active lattice systems demands a multitool, synergetic approach, which we shall use in this book.

In this book we deal with basic concepts and models, with methodologies for studying the existence and stability of motions; for understanding mechanisms of formation of patterns and waves, their propagation and interactions in active lattice systems; and concerning how much cooperation or competition between order and chaos is crucial for synergetic behavior and evolution. The results described in the book have both inter- and transdisciplinary features and fundamental character.

When writing this monograph we had in mind graduate students and researchers from the nonlinear sciences including physics, biophysics, biomathematics, bioengineering, neurodynamics, electrical and electronic engineering, mathematical economics and computer sciences. We have done our best to make the material, and hence the methodologies developed, reasonably self-contained and accessible to the reader. Accordingly, we have provided numerous illustrations and computer results supporting the theory. We have also provided, as much as we were able, heuristic arguments and detailed discussions of time and/or space scales involved in the problems treated.

We expect that our book will prove to be useful as a textbook for graduate courses in the above-mentioned disciplines. As, to a large extent, chapters are independent units and relatively self-contained, the book is also expected to be useful as supplementary reading to graduate courses in dynamical systems, differential equations, applied mathematics and nonlinear science, in general.

As an honest warning, we ought to alert the potential reader that in order to really profit from our book he or she must be acquainted with the basics of dynamical systems theory, (linear and nonlinear) ordinary differential equations, (Lyapunov) stability theory, and wave equations (Afraimovich, Nekorkin, Osipov and Shalfeev, 1994; Shilnikov, 1997; Shilnikov, Shilnikov, Turaev and Chua, 1998; Thompson and Stewart, 1986; Andronov, Leontovich, Gordon and Maier, 1973; Andronov, Vitt and Chaikin, 1966; Arecchi, 1996; Hirsch and Smale, 1974). It is our understanding that such background material is currently taught either in senior-level courses (fourth year in the curriculum before graduation or licence diploma) or in the first year of most graduate schools.

Following a succinct introduction in Chap. 1, to help the reader place our book in perspective and to introduce the type of model problems to be discussed later on, we proceed to six chapters in which a substantial part of the research effort by the authors in the past decade is covered. Chapter 8 includes a few speculative ideas and suggestions for further work.

We will outline here the noticeable results and methodological developments to be found in this book. In Chap. 2, a detailed study is provided of the (1D) Boussinesq–Korteweg–de Vries (BKdV) equation augmented with dissipation and an input–output energy balance. The latter is capable of exciting and supporting solitons due to the (local) balance between nonlinearity and dispersion. The BKdV equation (Boussinesq, 1877; Korteweg and de Vries, 1895; Drazin and Johnson, 1989; Christov, Maugin and Velarde, 1996; Remoissenet, 1996; Scott, Chu and McLaughlin, 1973; Bishop, Krumhansl and Trullinger, 1980; Scott, 1999) is the paradigmatic model equation for the study of solitons and nonlinear (conservative) integrable systems. We show that contrary to common belief dissipation does not spoil solitons. Rather if energy input, leading to instability, helps past threshold to dynamically balance (and hence overcome) dissipation, solitons can survive very much like steady convective structures (Nicolis and Prigogine, 1977; Haken 1975, 1983 a, b, 2000; Colinet, Legros and Velarde, 2001; Nepomnyashchy, Velarde and Colinet, 2002; Velarde, Nepomnyashchy and Hennenberg, 2000; Scott, 1999; Koschmieder, 1993; Nicolis, 1995; Velarde and Normand, 1980; Gaponov-Gekhov and Rabinovich, 1992; Rabinovich, Ezersky and Weidmann, 2000). Indeed solitary waves, periodic wave trains, solitons and soliton bound states, if traveling with constant velocity, are steady (localized) structures in the corresponding moving frame. The input–output energy balance first selects the velocity and then helps maintain it while the energy supply lasts.

The results about the dissipation-modified BKdV equation are obtained by using a methodology developed for (3D) dynamical systems, which is a methodology used once and again throughout the book. For this reason we provide details of the methodology allowing the construction of stable and unstable manifolds and solutions (homoclinic and heteroclinic orbits). Single- or multiloop/humps waves, soliton-bound states, chaotic wave trains, etc., are constructed first by identifying their domain of existence and then by tracking them in what can be considered as numerical experiments.

In order to provide further methodological details and to help train the reader in its use, Chap. 3 is devoted to a succinct discussion of another paradigmatic equation in nonlinear science, the (1D) perturbed sine-Gordon (PSG) equation for long Josephson junctions (see, e.g., Ustinov, 1998) which appear widely in various branches of science and engineering.

Another building block of nonlinear science and engineering is the Chua circuit (Madan, 1993; Mira, 1997), whose rich panoply of dynamic behaviors (bistability, chaotic oscillations, etc.) make it particularly attractive for applied oriented uses of cellular neural networks or, with greater generality, cellular nonlinear networks (Chua, 1998; Haken, 1996; Manganaro, Arena and Fortuna, 1999). Accordingly, Chua's circuit is taken as a model unit, neuron-like element, and Chap. 4 is devoted to a thorough, albeit nonexhaustive, study of its dynamics and its role in (linear and circular) arrays and (2D) lattices, where many such units appear coupled together by resistors (diffusion) or by inductions. The emergence of (steady) patterns and waves of various types (pulses, solitons, and spiral, periodic and chaotic waves) is analysed in great detail using the methodologies presented in Chaps. 2 and 3 about homoclinic and heteroclinic orbits of the underlying dynamical system. In Chap. 4 these methodologies are augmented with further subtle and sophisticated details to help the reader master the subject. Analytical results are once more checked with numerical experiments.

A bistable unit, eventually excitable and hence an active element, appears as one of the most universal and versatile building blocks in a complex, synergetic system, be it a portion of the brain or of a man-made machine (Changeux, 1983; Hoppensteadt, 1986; Izhikevich, 2000; Llinás, 1988, 2001; Milton, 1996; Scott, 1995; Shepherd, 1991, 1998; Tass, 1999). For this reason, we have paid a great deal of attention to it in Chaps. 5 and 6. Particular care is taken with regard to coupled nonisochronous bistable oscillators and their role in lattices where (amplitude, frequency or phase) synchronization and clustering may occur. Thus, again the emergence of patterns and waves is the subject of our study, using once more the methodologies given in Chaps. 2 and 3. For mathematical simplicity we have made ample use of the FitzHugh–Nagumo–Schlögl (FNS) model functional law which accounts for the salient features, e.g., of the Hodgkin–Huxley axonal neuron dynamics (see, e.g., Cronin, 1987)

Linear or circular arrays and (2D and 3D) lattices with bistable units and diffusive bonds/couplings mimick well reaction–diffusion systems (Fife, 1979). Chapter 5 provides a wealth of information about synergetic reaction–diffusion features in various lattice geometries with bistable units.

Our greatest achievements are presented in Chap. 6. We refer to the question of replication of form (patterns or waves) and image and information transfer with a controllable degree of fidelity. This was to us a fascinating subject with potential applicability in the understanding of disparate phenomena ranging from replication processes in prebiotic evolution to designing computers in the (near) future (Feistel and Ebeling, 1989). Accordingly, great attention is paid in Chap. 6 to the emergence and replication of patterns and waves in 3D (multilayered) architectures. Many pictures obtained in computer experiments, carried out by V.B. Kazantsev, illustrate the use and results found with the methodologies based on the material presented in Chaps. 2, 3 and 4. Computer experiments have also permitted us to obtain new results that qualitative study and analytical theory cannot permit due to the computational (apparent though not real) "complexity" of the synchronization dynamics involved in the replication process and dynamic competition between patterns or waves in lattices with a high number of coupled active units.

In Chaps. 4 to 6 we deal with problems where space is discrete (lattices) but time proceeds as a continuous variable. Chapter 7 introduces time also as a discrete quantity, and hence it is devoted to a generalization to coupled map lattices (CMLs) of results found earlier in the book. CMLs appear to be useful in modeling dynamical events obtained from discrete time series, which after all is always the case in experiments (Chaté and Courbage, 1997; Kaneko and Tsuda, 2001; Kaneko, 1993). Once more, use is made of the methodologies based on the qualitative study of dynamical systems. In particular, we make use of the discrete version of the FHNS model. In a constructive and detailed process we provide methods for finding steady and dynamical hence evolving structures in CMLs of infinite dimension.

Finally, in Chap. 8 we provide a few concluding and speculative remarks, not really a conclusion, about avenues for future research and the potential applicability of methodologies and results earlier presented in the book.

Last but not least, we wish to express our appreciation first to Prof. Dr. Hermann Haken for his illuminating influence over decades. He was instrumental in the scientific survival of one of us (M.G.V.), in difficult times in Spain, thanks to a cooperative Grant he helped obtaining from the Stiftung Volkswagenwerk. Furthermore, the attendance of one of us (M.G.V.) at several of the Synergetics Workshops held at the Schloss Elmau (near Munich) decided much of his scientific interests and even largely shaped his approach to science.

The development of the research leading to this book was greatly influenced by Prof. Valentin S. Afraimovich, Prof. Gregoire Nicolis and Prof. John Ross.

Prof. Leon O. Chua was a source of inspiration and drive to both authors.

We also gratefully thank Prof. Rodolfo R. Llinas and Prof. Antonio Fernandez de Molina for taking time and patience in helping us understand brain physiology and neurodynamics, a realm of science which we have recently entered using the methodology developed in this book.

Our progress in the study of the problems treated in this book was possible thanks to the collaboration with Dr. Mederic Argentina, Dr. Jean Bragard, Prof. Christo I. Christov, Dr. Pierre Colinet, Dr. Ezequiel del Rio, Prof. Victor Fairen, Dr. Antonino Giaquinta, Dr. Andrej G. Maksimov, Prof. Alex A. Nepomnyashchy, Dr. Alex Ye. Rednikov, Dr. Nikolai F. Rulkov and, particularly, Dr. Viktor B. Kazantsev and Dr. Valeri A. Makarov.

The collaboration between the two coauthors was the result of a suggestion made by Prof. Mikhail I. Rabinovich back in 1993. He was right in foreseeing how much we could achieve working together, and hence how easy it would be, and how fruitful this collaboration could be. For this and also for his enlightening inspiration in our research on nonlinear problems, we express our gratitude to him.

In the preparation of the compuscript (text and figures) we benefited from the technical skills of Viktor Kazantsev, Valeri Makarov and Iouliia Makarova. The former performed for us many hours of painful typing and retyping. Our sincere thanks for their help and friendship.

This book and, indeed, the research behind it was possible thanks to various institutional grants from the Ministries of Education and Culture and of Science and Technology (Spain) under Grant PB96-599, from INTAS, and from the BCH Foundation (Spain). Visiting positions at the Instituto Pluridisciplinar (Madrid) were held through the years by one of us (V.I.N.) with support from the BBV Foundation (Spain), from the Ministry of Education and Culture (Spain) and from Universidad Complutense de Madrid (Spain). To all these institutions we express our gratitude.

July, 2001
Nizhny-Novgorod and Madrid

V.I. NEKORKIN and M.G. VELARDE

Contents

1. Introduction: Synergetics and Models of Continuous and Discrete Active Media. Steady States and Basic Motions (Waves, Dissipative Solitons, etc.)

1.1 Basic Concepts, Phenomena and Context

In modern science, "synergetics" owes its development to the extraordinary research output provided by H. Haken coupled with the many workshops he has organized in the past three decades. In short it deals with the emergence of dynamical and/or evolutionary features in complex systems where the total is not the mere addition of parts. Hence an emergent property belongs to a higher level of description than that of the underlying elements. Emergent properties have genuine laws of their own as is customary in biology, ecology, sociology, etc., and indeed engineering, chemistry and physics. Already thermodynamics is a science of emergence (of cooperative properties) in underlying atoms, and statistical mechanics is the bridge between the micro and macro (or even meso) levels. Equilibrium critical phenomena and phase transitions provide paradigmatic examples of transitions and cooperativity in nonevolving systems, and they define an area of physics rather well understood both at the emergent level and in the way emergent, macroscopic and phenomenological properties originate from the, microscopic, lower level. It is one of the areas where from first principles we know the recipe to account for emergent, synergetic, properties. The reductionism from the emergent to the lower level is far from trivial and clearly indicates that, although the basics are to be found at the microlevel, understanding the emergence of cooperative properties demands a great deal of imagination and, on occasion, computer power. It is sufficient to recall the breakthroughs provided, on the one hand, by L. Onsager in the 1940s, when he gave the exact solution of the 2D Ising model with no external field, and, on the other hand, by K.G. Wilson, who proposed the renormalization group approach to the same problem in arbitrary dimensions [1.9].

Following Haken [1.11, 1.12] we shall limit consideration to the synergetics referring to evolving systems either in nonequilibrium thermodynamics/physics or in any other science dealing with dynamically evolving structures. The common link is the fact that, when explicitly written, evolution equations are nonlinear ordinary, partial or functional differential equations, maps, etc., for one or a mere few slaving modes, also called order parameters following the terminology introduced by L.D. Landau. Nonlinear evolving systems may experience self-organization and various types of transitions called bifurcation phenomena. Haken's efforts were followed by many scientists, and today we have a wealth of disparate phenomena well embodied under the single, universal framework of synergetics.

To realize stable steady states or stable oscillatory states or waves, synergetic systems require a continuous influx of matter, energy or information. This allows the onset and development of spatio-temporal or functional (dissipative) structures or patterns in nonequilibrium systems due to their intrinsic dynamics more than to any specific external influence (the concept of dissipative structure was introduced around 1964 by I. Prigogine; see, e.g. [1.18, 1.19]). Hence, the characteristics of these dissipative structures are mostly defined by the properties and parameters of the system itself and within some limits do not depend on initial and/or boundary conditions.

A classical example of self-organization in physical systems is the formation of Bénard cells [1.4, 1.5, 1.7, 1.15, 1.23]. When a liquid layer (take, for instance, silicone oil) is heated from below, then, depending on the temperature difference, ΔT, across the layer, various types of flow behavior are possible, as experimentally observed a century ago by H. Bénard. When ΔT is small enough, the liquid stays at rest and heat is transferred by diffusion, hence solely by molecular thermal conductivity. For values of ΔT higher than some critical value the system produces a honeycomb pattern of hexagonal cells (Fig. 1.1). Whether the pattern is due to buoyancy, i.e. unstable stratification, or to surface stresses due to the variation of surface tension along the open surface (Marangoni effect), or a combination of both agents, its salient characteristics are practically independent of the initial conditions and for the basic elements may not depend on boundary conditions. Another case of self-organization in fluid layers is illustrated in Fig. 1.2. In this case the state of rest becomes unstable to waves and, eventually, solitons when, rather than heating the liquid from below, the heating is done from the air side, hence from above. Similar phenomena have been observed when the liquid absorbs enough of another lighter surface active vapor, e.g. pentane vapor being adsorbed (at the surface) and subsequently absorbed (in the bulk) by liquid toluene. Stable stratification due to gravity and surface stresses excite surface waves, which may in turn yield to internal waves. These waves or periodic wavetrains exhibit collisions crossing each other with no apparent deformation or regular and anomalous wall reflection (Fig. 1.2b) and hence exhibit properties very much like these of solitary waves, wavetrains, solitons and shocks, studied long ago by J.S. Russell and E. Mach [1.8, 1.16, 1.20, 1.21]. We return to this question in Chap. 2. These examples show the major features of self-organization, including the nonequilibrium property of the medium, the presence of a threshold relative to the external influence and, finally, emergent cooperativity from the molecular level. Mathematical models possessing such properties must be nonlinear and, for a nonuniform pattern to be formed, spatially distributed. Therefore, the investigation of self-organization processes deals with the spatio-temporal dynamics of nonlinear distributed systems.

Fig. 1.1. Pattern forming system. Synergetic behavior steady in time. The onset of steady patterned convection (Bénard cells) in a shallow silicone oil layer (about 1 mm deep) heated from below (the solid support is made of copper and the liquid is open to ambient air). First, due to the cylindrical boundary there is an axisymmetric circular roll-excitation (**a**–**c**), but, subsequently, the (Marangoni) surface stresses dominate leading to the honeycomb structure (**d**,**e**). (**f**) Close-up of Bénard cells (hexagons) showing streamlines due to suitably long exposure time. Flow is ascending in the cell centers, where the surface is depressed, and descending along their periphery. Velocities are about 10^{-2} cm/s, and wavelength is in the mm depth range for a thermal gradient of a few degrees per cm

Then, one of the major problems in the investigation is how a bifurcation or transition may occur between possible states of the system. Indeed, the values of parameters at bifurcation delineate those critical or threshold conditions on the system parameters which are responsible for the appearance, destruction and reorganization of spatio-temporal patterns, hence the self-organization of the medium. In connection, great attention in this book is given to the study of bifurcation sets of nonlinear, distributed systems.

In the above examples we have illustrated self-organization in a physical system. Self-organization is typical not only for hydrodynamics and physics, but it represents a widespread feature of nature. In particular, self-

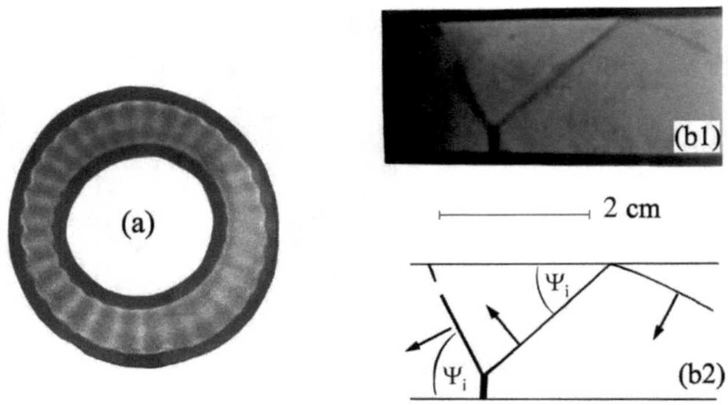

Fig. 1.2. Synergetic behavior leading to waves and solitons in Benard layers. (**a**) Traveling wavetrain in experiments of pentane vapor absorption on liquid toluene in an annular circular cylindrical container (outer diameter, 57 mm; inner diameter, 37 mm; liquid depth, 4 mm). (**b**) Soliton-wave reflection pattern (layer depth, 2 mm). (**b1**) Shadow picture of two reflections with and without Mach–Russell third wave or stem, corresponding to *large* and *small* enough incident angles, respectively ($\Psi_i \gtrsim 45$ and $\Psi_i \lesssim 45$). Similar phenomena have been observed in heat-transfer experiments with thermal gradients about 100 K/cm. Wave velocities are in the cm/s range. (**b2**) The diagram reconstructs the actually observed reflection pattern as shown in (**b1**)

organization occurs in a wide variety of biological systems. Several introductory texts provide a wealth of information, and we refer the reader to them [1.1, 1.11, 1.12, 1.18, 1.19]. Just because of its apparent relation to one of the problems treated in great detail in our book – replication of patterns with a controllable degree of fidelity – we consider Fig. 1.3. It shows the behavior of the tropical flatfish (*Bothus ocellatus*) located in a medium with various background textures. According to the texture, the animal skin mimics and hence replicates the pattern of the environment. To change the pattern it requires only from 2 to 8 s.

In our book we consider self-organization processes in nonequilibrium media containing spatially distributed energy or matter sources with an emphasis on *active* media. This means that local points, elements, cells or units are active, autonomous by themselves and linked to each other by transport processes, although the specific type of connection between elements may be very diverse. It is defined by the nature of the system. For example, the linkage can be by means of heat or particle diffusion, acoustic or luminous flux, electric current and so on. If the energy and/or matter sources are distributed in space "uniformly", the medium is continuous e.g. most chemical systems (Belousov–Zhabotinsky reaction, burning processes and so on), excitable membranes, long Josephson junctions and optically transparent monocristallic magnetic films. Mathematically such media are modeled by partial differential equations (PDE) [1.10].

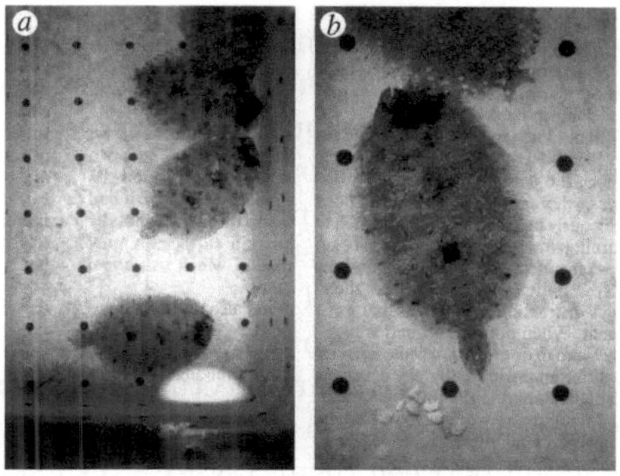

Fig. 1.3. Synergetic mimetism and replication of form in animals. Spatial patterns formed on the skin of a tropical flatfish (*Bothus ocellatus*) matching the floor background (From Ramachandran, V. S., Tyler, C. W., Gregory, R. L., Rogers-Ramachandran, D., Duensing, S., Pillsbury, C., and Ramachandran, C., "Rapid adaptive camouflage in tropical flounders", *Nature* **379** (1996) 815–818)

Another class of *active* media is formed by systems with nonuniformly distributed energy sources, i.e. located in local spatial "points" at sites with significant interspacing. Take, for instance, arrays of coupled lasers or distributed Josephson junctions, arrays of electronic oscillators, neural networks, myelinated nerve fibers and so on. Figure 1.4 illustrates salient features of this class of active media, including the localization of activity sources and the discreteness of spatial coordinates [1.22]. Since the major part of our book is devoted to self-organization processes in such systems, let us consider their properties in more details. Mathematical models of active systems with discrete spatial coordinates are provided by lattice dynamical systems. They can be of two types, represented by (i) coupled ordinary differential equations and (ii) coupled map lattices (CML). Type (i) systems have discrete spatial coordinates, continuous time and continuous states. At variance, type (ii) systems (CLMs) have discrete time as well as coordinates, but the state of the system is continuous. On the one hand, the CML models appear in the description of

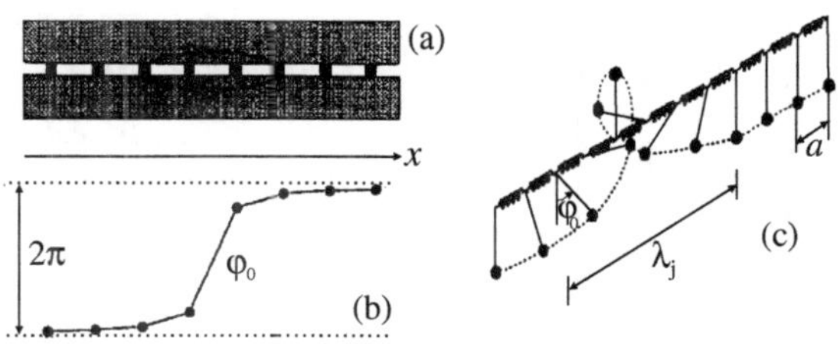

Fig. 1.4. Josephson junctions. (**a**) 1D array. (**b**) Phase difference, φ_j, of wave functions of the superconductors in the jth spatial site for a fluxon located inside the array. (**c**) Mechanical analog made using a chain of coupled mechanical pendula

the self-organization processes in systems whose states can be defined only at some discrete instants of time. They naturally come, for example, from some ecological problems where population quantity is measured discretely in time. Another example is a network of impulse-controlled automatic regulation. On the other hand, CML have been proposed as paradigmatic models for studying spatio-temporal chaos and, eventually, strongly dissipative turbulence [1.14]. The application aspect of CML models is very important and intensively studied at present. It is clear from the above-mentioned properties of lattice systems that the geometric architecture plays a major role in their collective dynamics, together with the dynamics of the single elements and the type of the interaction between them. Figure 1.5 shows diagrammatically the main geometrical configurations of lattice systems considered in this book. They vary from simple 1D arrays (chains) to complex multilayers (a layer is a 2D lattice). The black dots in the figure represent the active elements, and the lines the connection (interaction or bond) between them. The *activity* of the lattice is provided by the local activity of the elements, units or cells. The development of processes in space occurs due to the connectivity. Also, in general, all the systems in Fig. 1.5 have a local type of connection. We shall also consider the more complex case of global coupling, in which each element, at once, has a connection with all other elements.

In this book the basic characteristics of the dynamics of the elements are their capability for auto-oscillations and bistability. Let us explain the meaning of these terms. Although the work of other authors may be consid-

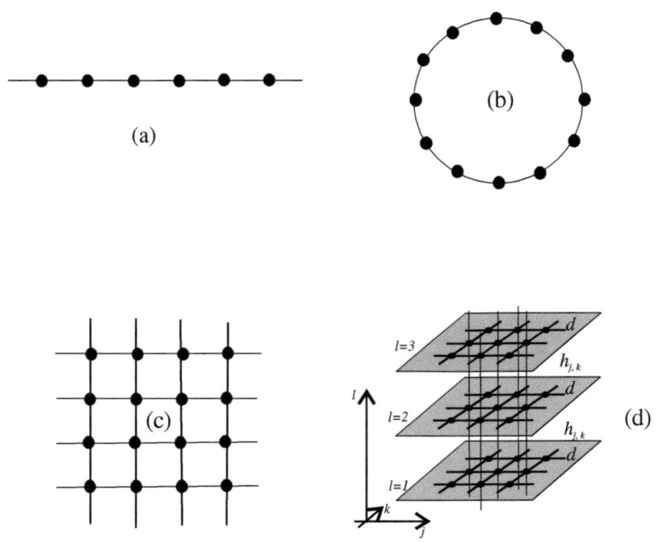

Fig. 1.5. Schematic representation of lattice systems with a local type of bond. (**a**) Fiber or (linear) array, chain, fiber or axon. (**b**) Circular chain or ring array. (**c**) 2D lattice. (**d**) Multilayer/lattice (3D) architecture or anatomy

ered as precursory, we can safely say that the concept of the "auto-oscillating system" was introduced by A.A. Andronov in first third of the twentieth century [1.2, 1.3]. This is a class of systems capable of generating self-sustained, periodic oscillations which are independent, within some bounds, of initial conditions. In auto-oscillating systems the energy dissipation is compensated by constant energy pumping Andronov found that the mathematical image of the auto-oscillations in systems with one degree of freedom is a stable limit cycle (a term introduced by H. Poincaré) on the corresponding phase plane. It is a closed trajectory attracting all neighboring trajectories asymptotically. If the trajectories from the neighborhood tend to the cycle with $t \to +\infty$, then the cycle is called asymptotically stable. At present the term auto-oscillation is interpreted in a wider sense. A dissipative system is called auto-oscillating if it is capable of intrinsically generating and sustaining oscillations not really dependent on external specific influences other than the input–output energy balance. From the point of view of the system phase space it means the existence of a nontrivial attractor, i.e. an isolated closed limit set of trajectories attracting other trajectories from the neighborhood. Bistability means the existence of two (multistability, more than two) attractors for the same parameter values. Figure 1.6 illustrates the phase spaces of the elements considered in this book. The element shown in Fig. 1.6a possesses auto-oscillations and bistability, simultaneously. Here the bistability is provided by a steady state and a limit cycle, and the auto-oscillations by the presence of a stable limit cycle. Therefore, depending on initial conditions the element either stays at rest (initial condition within the region bounded by the unstable limit cycle) or produces periodic oscillations (initial conditions outside this region). The simplest physical system with such a behavior is a generator with a hard mode of excitation. The element capable of chaotic oscillations is shown in Fig. 1.6b. An example of a CML element with discrete time is illustrated in Fig. 1.6c.

Having energy sources extended in space, active media can exhibit different structures and functions with synergetic behavior: homogeneous and inhomogeneous oscillations, nonlinear waves and solitons, steady or unsteady patterns, etc. Once patterns, waves, etc., have been generated in one lattice

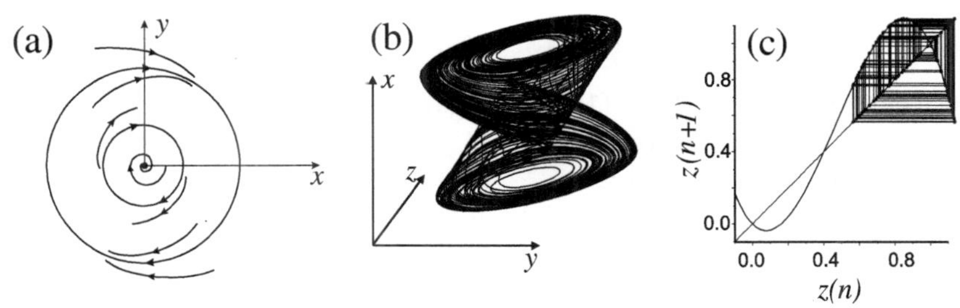

Fig. 1.6. Phase-space representation of the dynamics of elements or units. (**a**) Bistable auto-oscillating element. (**b**) chaotic element. (**c**) CML element

layer or in a chain or fiber, we can replicate them in a multilayer system or architecture.

In spite of such a variety of forms of self-organization in our book we follow a unified approach inspired by the work of both Andronov and Haken. Biased towards the former's view our approach uses the qualitative methods of the modern theory of dynamical systems. It rests on the detailed investigation of the structure of phase-space fibering and the available bifurcation sets. Needless to say that evolution equations are reduced to those of the leading or slaving modes in the corresponding system. We shall first discuss how patterns or waves appear in a system and how they can be transferred, copied or replicated in another or in many others in a 3D architecture.

In the remainder of this chapter we proceed to a succinct description of the type of (generic) models we shall be dealing with in this book.

1.2 Continuous Models

In one dimension (1D), the evolution of a synergetic, continuous system can be expressed by the nonlinear PDE

$$\mathbf{u}_t(x,t) = \mathbf{F}(\mathbf{u}, \mathbf{u}_x, \mathbf{u}_{xx}, \dots, \mu), \tag{1.1}$$

with

$$t > 0, \quad -\infty < x < \infty,$$

$$\mathbf{u}(x,t) = \{u_i(x,t)\} = u_1(x,t), u_2(x,t), \dots, u_p(x,t),$$

where $\mathbf{u}(x,t)$ is state vector, $\mathbf{F}$ is a vector of dimensionality p, x is a spatial coordinate, and μ is a control parameter which could be a set of parameters $\mu = \mu_1, \mu_2, \dots, \mu_k$. The right-hand side of (1.1) may depend on gradients and higher spatial derivatives of $\mathbf{u}(x,t)$.

Among the great diversity of motions and states described by (1.1), there are waves which propagate without changing their shape or some other forms of traveling waves, e.g.

$$\mathbf{u}(x,t) = \mathbf{v}(x - ct), \quad c = \text{const.} \tag{1.2}$$

A traveling wave of the form (1.2) propagates with the velocity c ($c > 0$) from left to right. Studying traveling waves in system (1.1) involves addressing two issues: existence and stability.

The solutions of the system (1.1) in the form of (1.2) satisfy the system

$$\dot{\mathbf{v}}(x,t) = \mathbf{F}(\mathbf{v}, \dot{\mathbf{v}}, \ddot{\mathbf{v}}, \dots, \mu), \tag{1.3}$$

where the dot denotes differentiation with respect to the coordinate $\xi = x - ct$ in the moving frame. System (1.3) is an autonomous dynamical system and has finite-dimensional phase space which we denote by G. Any bounded solution from G corresponds to some traveling wave existing in the extended system (1.1). Note that the wave velocity c is a parameter which increases the dimension of the parameter space by one unit. Let us denote by D the parameter space of (1.3). We shall be considering various types of traveling waves such as

- periodic waves (wave trains);
- wave fronts, shocks or kinks;
- pulses and dissipative solitons;
- waves of spatially chaotic profile.

Periodic Waves

In the phase space, G, such waves are hyperbolic limit cycles. Therefore, traveling periodic waves exist for the parameters taken from open regions in the parameter space D, and for fixed μ these waves have a one-parameter family or continuum of possible propagation velocities c.

Wave Fonts, Shocks or Kinks

These wave motions correspond to the heteroclinic trajectories of (1.3), i.e. to the trajectories joining two disjoint invariant sets. In particular, heteroclinic trajectories can be formed by equilibrium points (We shall also be using, indistinctly and according to context, the terms critical or fixed point, or steady state). The requirement of stability of wave fronts implies that at least one of the equilibrium points must be of saddle type with a 1D unstable manifold. If the second point is stable, then the fibering of trajectories in the phase space is structurally stable (see Fig. 1.7a). Therefore, the wave fronts corresponding to the heteroclinic orbits exist for values of the parameters taken from open regions in D and for fixed μ have a continuum set of possible propagation velocities c. However, in a strong sense these wave fronts are unstable. They can be formed in a medium only from small sets of initial distributions. A typical example of such fronts is the propagating wave in the Kolmogorov–Petrovsky–Piskunov–Fisher equation [1.10]. It is known that an initial distribution of the "step" form (Heaviside jump) evolves to a wave front propagating with minimum (from the whole available range) velocity. At the same time such a front is unstable to large scale perturbations. When a heteroclinic trajectory is formed by two saddle-type equilibrium points, the values of the parameters of (1.3) are located on some surface (of co-dimension 1) in the parameter space D (Fig. 1.7b). Therefore, in this case wave fronts exist only on surfaces of co-dimension 1 and for fixed μ we have a discrete spectrum of possible propagation velocities c. Note, that in this case the front is apparently identical to the first example, although, of course, it is a completely different solution. Such wave fronts are better suited to persist.

A number of examples of "reaction–diffusion" systems show exactly that such fronts are stable to small perturbations and are realized from a very wide class of initial distributions. This fact has a simple explanation. In such systems a saddle-type equilibrium point with 1D unstable separatrices of (1.3) corresponds to a stable spatially homogeneous state of (1.1). Therefore, a propagating wave front "switches" the system (1.1) from one locally stable homogeneous state to another also locally stable state. Thus, the main part of the fronts is "localized" near these stable states and then the fronts "hereditarily" accept their stability properties.

Pulses and Dissipative Solitons

These waves appear in the phase space G as homoclinic trajectories, i.e. trajectories formed by saddle-type equilibrium points of (1.3) (see Fig. 1.7c1). In dissipative systems a homoclinic trajectory exists for the parameter values located on surfaces of co-dimension 1 in the space D. Thus the spectrum of possible velocities of such waves is discrete. The term traveling "pulse", or earlier "auto wave", is used in "reaction–diffusion" systems (quasilinear equations of parabolic type). The case of "dissipative solitons" belongs to nonlinear systems of hyperbolic type albeit (super) parabolically perturbed in a singular way [1.6, 1.17]. All these wave motions are waves in the presence of variable dissipation and appropriate energy supply. Typical systems exhibiting dissipative soliton solutions that propagate with practically permanent form are the dissipatively PSG equation and the dissipative BKdV equation, about which we shall say a great deal in this book.

Pulse and dissipative solitons are to a large extent independent of initial conditions and mostly defined by the properties of the active medium; they crucially depend on an appropriate input–output energy balance, as we shall

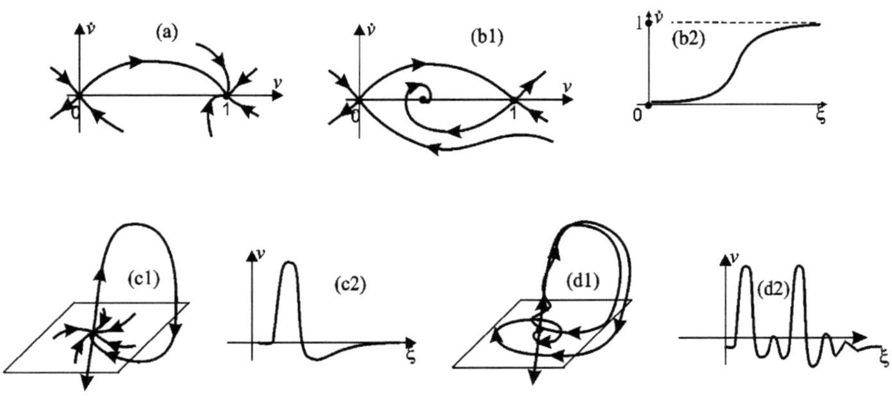

Fig. 1.7. Typical heteroclinic trajectories (**a,b1**) leading to a shock, kink, bore or hydraulic jump (**b2**) and homoclinic orbits (**c1,d1**), with corresponding traveling waves (**c2,d2**) of a dynamical system

see below. They can be evolutionary stable, i.e. they may last for a long time with a permanent form (long lived metastable states). Upon collision they may be annihilated or they may cross each other without changing much their form. Reaching the boundary of the medium, dissipative nonlinear wave pulses and dissipative solitons may disappear or may be reflected from the boundary. Thus pulses or dissipative solitons may experience elastic or inelastic collisions.

Waves of Chaotic Profile

Such waves can be realized only in media where (1.3) has a phase space G of dimensionality higher than two. Typical images of chaotic waves in G are multiloop homoclinic trajectories or other trajectories located in nonwandering sets lying near the homoclinic loop of a saddle-focus, near a Poincaré homoclinic trajectory (a trajectory which is homoclinic to a periodic orbit) or trajectories of chaotic attractors [1.17].

If (1.3) is 3D and has an equilibrium point of saddle-focus type with 1D unstable and 2D stable manifolds, then we can assume that the parameter space D has a surface Π of co-dimension 1, which corresponds to the homoclinic loop Γ of a saddle-focus. The behavior of trajectories in the neighborhood of G when Γ is disappearing may be determined from a theorem developed by Shil'nikov. According to this theorem, if the real eigenvalue has larger magnitude than the real part of complex eigenvalues, then the Poincaré map, on a cross-section transverse to the loop Γ, has an infinite number of Smale's horseshoes, a large number of which is preserved when Γ disappears. It means that the structure of nonwandering sets of trajectories lying near the homoclinic loop is very complex. There exist multiloop homoclinic trajectories making an arbitrary number of loops, countable sets of periodic orbits of all periods, etc. Figure 1.7d1 illustrates a two-loop homoclinic orbit. Therefore, (1.1) can display the propagation of pulses or dissipative solitons and periodic waves of very complex appearance (Fig. 1.7d2) and even spatially chaotic form. Such waves can be interpreted as wave trains consisting of nearly identical "blocks" linked by a finite number of oscillations of a smaller amplitude. The waves can be formed by an arbitrary number of such blocks. In the case of pulses or soliton-trains of chaotic profile the elementary block is a solitary pulse, but in the case of chaotic periodic waves it represents a wave train. Note that the number of chaotic waves is countable. The spectrum of propagation velocities c (for fixed μ) for chaotic pulses or solitons is discrete and consists of a countable number of elements, but chaotic periodic waves have a continuum set of propagation velocities. Thus, (1.1) can exhibit chaos in the form of traveling waves.

A similar situation occurs when a homoclinic trajectory is formed by means of the transverse intersection of the stable and unstable manifolds of a periodic orbit. In this case a Smale's horseshoe arises in a Poincaré cross-section. Thus a countable number of periodic and aperiodic waves of chaotic profile may propagate in (1.1). In the case of homoclinicity with respect to a periodic orbit, aperiodic waves have oscillatory "tails" and look like nearly periodic waves with an arbitrary distribution of intermediate splashes. Due to the roughness of the homoclinic structure most of these traveling waves also exist with little changes in parameter values. Hence, for fixed μ there is a continuum set of possible wave velocities c.

Another case giving rise to traveling waves of a chaotic profile is the appearance of a chaotic attractor in (1.3). Then (1.1) has chaos of waves with profiles formed by the trajectories of the chaotic attractor. The spectrum of possible propagation velocities c can also be a continuum set.

Thus, the problem of existence of traveling waves and their geometric properties in (1.1) is reduced to an issue in the qualitative theory of ordinary differential equations. Usually, this is a problem of global analysis of mostly nonlinear multidimensional systems and can be very difficult. In Chaps. 2–4, for three basic models of active media, we provide the methodology to solve this problem (and waves of chaotic profile) by qualitative methods.

Another no less complex problem is assessing the stability of the traveling waves. Presently, the analytical study of the stability has been carried out only for waves of rather simple form such as periodic waves and monotonic fronts, and this for just one-component systems. The stability of waves having more complex form has been studied numerically. However, even such an approach has great difficulties, due to the large spatial scale of the chaotic waves. These difficulties can be overcome for small-scale chaotic waves.

1.3 Chain and Lattice Models with Continuous Time

From the mathematical viewpoint, chain and lattice models represent interacting subsystems obeying identical differential equations constrained by some boundary conditions. The subsystems joined in the chains or lattices have intrinsic dynamics that can be complex and eventually chaotic (Fig. 1.5b). Such models in general form can be written as follows:

$$\dot{u}_j = \mathbf{f}(\{u_j\}^s), \tag{1.4}$$

where the dot denotes differentiation with respect to time; $u_j \in \mathbf{R}^p$, $\mathbf{R}^p$ is a p-dimensional Euclidean space, $j \in \mathbf{Z}^d$, and $\mathbf{Z}^d$ is the lattice in $\mathbf{R}^d$, i.e. the set of all vectors $j = (j_1, j_2, \dots, j_d)$ with integer coordinates; $\{u_j\}^s = \{u_j \mid |i-j| < s\}$, $s \geq 1$ is an integer. The integers p, s and d are interpreted as follows: p defines the number of components of the model, d its spatial dimensionality [for example, if $d = 1$, (1.4) describes a chain model, but if $d = 2$, it describes a 2D lattice model] and s the "distance" of spatial connection between the units in the lattice, i.e. s is the radius of the sphere of spatial interactions [if $s > 1$, (1.4) has nonlocal connection]. We will consider bounded $(1 \leq j_1, j_2, \dots, j_d \leq N)$ chains and lattice systems, i.e. systems such as (1.4) with appropriate boundary conditions. Note that, on the one hand, (1.4) represents a standard dynamical system with phase space of finite dimensionality. On the other hand, (1.4) can be interpreted as an ensemble of interacting active units located in the junctions or sites of the discrete lattice $\mathbf{Z}^d$ in space. In this case the indices j play the role of spatial coordinates. Therefore, (1.4) can be considered as an extended system and its dynamics has features evolving in both space and time.

As already said we are interested in studying pattern formation and traveling wave propagation. The correspondence between these phenomena and the trajectories of (1.4) can be easily seen using a 2D ($d = 2$), one-component ($p = 1$) system of reaction–diffusion type with local (nearest-neighbor) interactions, using the FitzHugh–Nagumo–Schlögl (FNS) model [1.13]

$$\begin{aligned} \dot{u}_{j,k} &= f(u_{j,k}) + \kappa(\Delta u)_{j,k} \,, \\ j,k &= 1,2,\, \dots \,, N \,, \end{aligned} \tag{1.5}$$

where κ is a diffusion coefficient, $f(u) = u(u - a)(1 - u)$, the parameter a satisfies the condition $0 < a < 1$ and $(\Delta u)_{j,k}$ is the discrete Laplace operator

$$\begin{aligned} (\Delta u)_{j,k} &= (\Delta u)_j + (\Delta u)_k - 4u_{j,k} \,, \\ (\Delta u)_j &= u_{j+1,k} + u_{j-1,k} \,, \\ (\Delta u)_k &= u_{j,k+1} + u_{j,k-1} \,. \end{aligned}$$

Stationary patterns correspond to the steady states of (1.5). The coordinates of these states are determined by the system of nonlinear equations

$$\begin{aligned} f(u_{j,k}) + (\Delta u_{j,k}) &= 0 \,, \\ j,k &= 1,2,\, \dots \,, N \,. \end{aligned}$$

In a 3D space $\{\mathbf{Z}^2, \mathbf{R}\}$, stable steady states correspond to stationary patterns (Fig. 1.8). Note that the simplest steady states which do not depend on the indices j, k determine spatially homogeneous states of the medium described by (1.5).

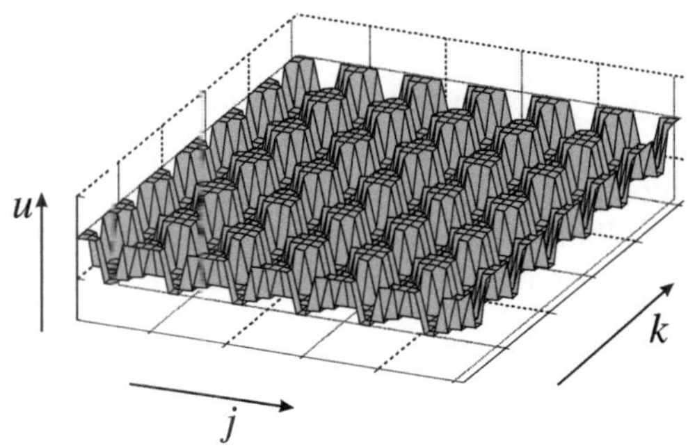

Fig. 1.8. Stationary pattern in a 2D lattice

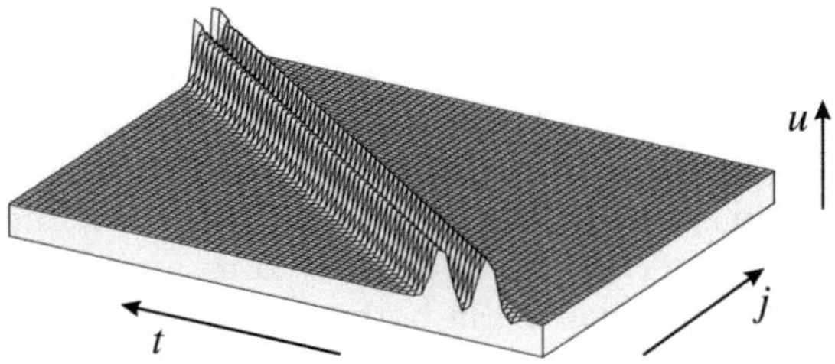

Fig. 1.9. Space (lattice sites)–time plot of a two-hump pulse propagating in a (linear) chain or fiber

Spatially homogeneous oscillations correspond to trajectories of (1.4) which are independent of the spatial coordinates j, k. In the space $\{\mathbf{Z}^2, \mathbf{R}\}$, such motions determine oscillations changing synchronously in time. In other words, when such oscillations are realized the medium "breathes" like a single "organism". These spatially homogeneous oscillations are governed by a dynamical system which coincides with the system describing a local unit of the lattice. For example, for (1.5) such oscillations must satisfy the equation

$$\dot{u} = f(u)\,.$$

In this equation only steady states exist and, therefore, (1.5) does not have spatially homogeneous oscillations.

Traveling waves correspond to solutions of (1.4) of the following type:

$$u_{j,k} = \Psi(j\cos\Theta + k\sin\Theta + ct)\,,$$

where $\Theta \in \mathbf{R}$ is given, and both Ψ and $c \in \mathbf{R}$ are to be determined. The function $\Psi(\cdot)$ satisfies the functional differential equation

$$\begin{aligned} c\Psi'(\xi) = f(\Psi) + d[\Psi(\xi + \cos\Theta) + \Psi(\xi - \cos\Theta) \\ + \Psi(\xi + \sin\Theta) + \Psi(\xi - \sin\Theta) - 4\Psi(\xi)]\,, \end{aligned} \tag{1.6}$$

where $\xi = j\cos\Theta + k\sin\Theta + ct$ is a "moving coordinate" and the prime denotes the differentiation with respect to ξ. Any bounded solution of (1.6) corresponds to a traveling wave in (1.4). The propagation of the wave of pulse-form is shown in Fig. 1.9. Note that the problem of searching the bounded solution in (1.6) can be very difficult. As far as we know, at present there is no literature discussing this topic giving a strict analytical solution of strongly nonlinear functional differential equations of such a type. The problem can be investigated using numerical methods, but in some particular cases when (1.4) can be approximated by PDEs (the so-called quasicontinuum approximation), it admits analytical solution.

Note also that the discrete character of the spatial coordinates j causes the wave motions in (1.4) to be formed by means of definite "phase" shifts between the oscillations in neighboring subsystems of the lattice.

1.4 Chain and Lattice Models with Discrete Time

Chains and lattice models with discrete time are chains and lattices of coupled maps. We focus our attention on models of unbounded media. Such models are of major importance for the description of spatial chaos, wave propagation and so on. Chains and lattice models of nonequilibrium unbounded media are also taken as dynamical systems. Such an approach demands introducing a phase space and the operator of evolution in this space. It is easy to define these terms for the bounded media, but in the case of unbounded media, this is a difficult task.

Chain Models of Unbounded Media

Consider the set $\mathbf{B}$ of an infinite sequence

$$\mathbf{B} = \{\mathbf{u} = (\dots, u_{-1}, u_0, u_1, \dots, u_j, \dots)\},$$

with $u_j \in \mathbf{R}^p$; $\mathbf{R}^p$ is p-dimensional Euclidean space with the standard inner product $(\cdot,\cdot)$ and the norm $\|\cdot\| = \sqrt{(\cdot,\cdot)}$. Taking $\mathbf{B}$ as a linear space (with coordinate addition and multiplication by a number) we define a scalar product $\langle\cdot,\cdot\rangle$ as follows:

$$\langle \mathbf{u}, \mathbf{v}\rangle = \sum_{j=-\infty}^{+\infty} \frac{(u_j, v_j)}{q^{|j|}}, \tag{1.7}$$

where $q > 1$ is some fixed integer. $\mathbf{B}$ is a Hilbert space. The norm in $\mathbf{B}$ is defined in the standard way, $|\cdot\| = \sqrt{\langle\cdot,\cdot\rangle}$.

Consider the map $F : \mathbf{B} \to \mathbf{B}$:

$$\big(F(\mathbf{u})\big)_j = f(u_{j-s}, \dots, u_j, \dots, u_{j+s}), \tag{1.8}$$

where $s \geq 1$ is some natural number, $j \in \mathbf{Z}$, $f : (\mathbf{R}^p)^{2s+1} \to \mathbf{R}^p$ is a differentiable map of class C^2 such as

$$\|f(u_{j-s}, \dots, u_j, \dots, u_{j+s})\| \leq c \max_{|i-j|\leq s} \{|u_j|\}, \tag{1.9}$$

and hence $F(\mathbf{B}) \subset \mathbf{B}$. From the physical viewpoint only bounded motions are of interest. Accordingly, (1.9) can be considered as fulfilled for models of concrete physical systems. Moreover, since only bounded solutions are of interest to us the behavior of the function f at infinity can be assumed to be arbitrary but satisfying (1.9). We assume that all first and higher partial derivative components of the vector function $\mathbf{f}$ vanish outside the full-sphere (in $\mathbf{R}^{(2l+1)p}$) of radius $r \gg 1$. These conditions ensure the existence of the constant M, which determines the inequality

$$\left|\frac{\partial f}{\partial u_i}\right| \le M\,, \ \forall (u_{-k}, \dots, u_k)\,, \tag{1.10}$$

where i is an arbitrary index. If (1.10) is fulfilled, then the operator $\mathbf{F}$ satisfies the Lipschitz condition, i.e.

$$\|\mathbf{F}(\mathbf{u}') - \mathbf{F}(\mathbf{u})\| \le M(2s+1)^{3/2} q^{s/2} \|\mathbf{u}' - \mathbf{u}\|\,, \tag{1.11}$$

and thus a chain is defined as a dynamical system.

The dynamical system $\{\mathbf{F}^n; \mathbf{B}\}$, $n \in Z_+$ *is a chain with discrete time*:

$$\mathbf{u}(n+1) = \mathbf{F}\big(\mathbf{u}(n)\big) \tag{1.12}$$

or

$$u_j(n+1) = f\big(u_{j-s}(n), \dots, u_j(n), \dots, u_{j+s}(n)\big)\,, \quad j \in Z\,.$$

Note that in the norm (1.7) the "damping" of disturbances imposed to the motion investigated in (1.12) implies that all components of the disturbances tend to zero for any fixed index j. This decay can be nonuniform with respect to the spatial coordinates j. In other words, in any finite domain of the acceptable space, the disturbances tend to zero, but at "infinity" they are bounded only or increase slower than $q^{|j|}$. From the physical viewpoint such stability seems to be acceptable.

Lattice Models of Unbounded Media

Consider the set $\mathbf{B} = \{\mathbf{u} = \{u_j\}, j \in \mathbf{Z}\}$, where $u_j \in \mathbf{R}^p$, $\mathbf{R}^p$ is a p-dimensional Euclidean space with the standard inner product $(\cdot,\cdot)$ and the norm $\|\cdot\| = \sqrt{(\cdot,\cdot)}$. For example, for $d = 2$ and $p = 1$, $\mathbf{B}$ is a set of infinite matrices, and in this case u_j is a state variable of a 2D discrete medium. Having considered $\mathbf{B}$ as a linear space, we introduce the scalar product according to (1.7), where $q > 1$ is some fixed integer and

$$|j| = \max\{|j_1|, |j_2|, \dots, |j_d|\}\,.$$

$\mathbf{B}$ is a Hilbert space with the norm defined in the standard way.

Consider the map $\mathbf{F} : \mathbf{B} \to \mathbf{B}$:

$$\big(\mathbf{F}(\mathbf{u})\big)_j = f(\{u_j\}^s)\,, \tag{1.13}$$

$$\{u_j\}^s = \{u_j \mid |i-j| \le s\}\,,$$

where $s \ge 1$ is a some natural number and

$$f : (\mathbf{R}^p)^{2s+1^d} \to \mathbf{R}^p$$

is a differentiable map of class C^2 such that

$$|f(\{u_j\}^s)| \leq c \max_{|i-j|\leq s} \{|u_i|\}. \tag{1.14}$$

Similar to the chain models, the inclusion $\mathbf{F}(\mathbf{B}) \subset \mathbf{B}$ is also valid for the lattice models [it follows from (1.14)], and the operator $\mathbf{F}$ is Lipschitz continuous for the bounded motions. Therefore, a lattice model can be studied by means of a dynamical system.

The dynamical system $\{\mathbf{F}^n; \mathbf{B}\}$, $n \in Z_+$ *is a lattice model with discrete time*:

$$\mathbf{u}(n+1) = \mathbf{F}\big(\mathbf{u}(n)\big) \tag{1.15}$$

or

$$u_j(n+1) = f(\{u_j(n)\}^s), \quad j \in Z.$$

We here study three types of motions in lattice dynamical systems: steady solutions which describe stationary patterns in a medium, spatially homogeneous solutions and traveling waves. We illustrate these motions using the discrete version of one-component nonlinear diffusive equations (FNS):

$$u_j(n+1) = u_j(n) + f\big(u_j(n)\big) + \kappa\big(u_{j-1}(n) - 2u_j(n) + u_{j+1}(n)\big), \tag{1.16}$$

where $u_j \in \mathbf{R}$, $j \in \mathbf{Z}$ and $n \in \mathbf{Z}_+$; the nonlinearity f satisfies (1.10) and κ is a diffusion coefficient.

Steady Solutions

Steady solutions are solutions which do not depend on time n. The steady solutions of (1.16), $u_j(n) = \Psi_j$, satisfy

$$k(\Psi_{j+1} - 2\Psi_j + \Psi_{j-1}) + f(\Psi_j) = 0. \tag{1.17}$$

This equation can be rewritten in the following equivalent form:

$$x_{j+1} = y_j, \quad y_{j+1} = 2y_j - \frac{1}{k} f(y_j) - x_j. \tag{1.18}$$

Any bounded trajectory of (1.18) corresponds to a bounded stationary state of (1.16).

Spatially homogeneous solutions do not depend on the spatial coordinate j, and each of these solutions, $u_j(n) = \Psi(n)$, satisfies

$$\Psi(n+1) = \Psi(n) + f\big(\Psi(n)\big). \tag{1.19}$$

Traveling Waves

By analogy with the continuous case, we define a stationary traveling wave as the solution of (1.16) of the following type:

$$u_j(n) = \Psi(lj + mn), \quad l, m \in \mathbf{Z}_+,$$

which must satisfy

$$\begin{aligned}\Psi(lj+mn+m) &= \Psi(lj+mn) + f\big(\Psi(lj+mn)\big) \\ &+k\big(\Psi(lj+mn-l) - 2\Psi(lj+mn) + \Psi(lj+mn+l)\big)\,.\end{aligned} \tag{1.20}$$

Let us introduce the "moving coordinate" as $i = lj + mn - l$. Then (1.20) becomes

$$\Psi(i+m+l) = \Psi(i+l) + f\big(\Psi(i+l)\big) + k\big(\Psi(i) - 2\Psi(i+l) + \Psi(i+2l)\big)\,. \tag{1.21}$$

Let $m > l+1$. Then, with the convention

$$\Psi(i) = x_i^{(1)}, \ldots, \Psi(i+l) = x_i^{(l+1)}, \ldots, \Psi(i+2l) = x_i^{(2l+1)}$$
$$\Psi(i+m+l-1) = x_i^{(l+m)},$$

(1.21) can be rewritten as follows:

$$\begin{aligned}x_{i+1}^{(1)} = x_i^{(2)}, \quad x_{i+1}^{(2)} = x_i^{(3)}, \ldots, x_{i+1}^{(l+m-1)} &= x_i^{(l+m)}\,, \\ x_{i+1}^{(l+m)} = x_i^{(l+1)} + f\left(x_i^{(l+1)}\right) + k\left(x_i^{(1)} - 2x_i^{(l+1)} + x_i^{(2l+1)}\right)&.\end{aligned} \tag{1.22}$$

Equations (1.22) define a dynamical system with a $(l+m)$-dimensional phase space. The time for this system runs with the discrete index i. Any bounded trajectory of (1.22) determines a bounded solution of (1.21), and, consequently, traveling waves for (1.16). Note that the dimension of the phase space of (1.22) is defined by the wave velocity m/l.

We focus attention on the fact that the existence of solutions is defined by the finite-dimensional dynamical system. We aim to study the behavior of these solutions in the phase space of the initiating system and determine the conditions of their stability.

2. Solitary Waves, Bound Soliton States and Chaotic Soliton Trains in a Dissipative Boussinesq–Korteweg–de Vries Equation

2.1 Introduction and Motivation

A solitary wave that, eventually, becomes a soliton represents a unique and very attractive object in modern nonlinear science of spatially extended systems. The term "soliton" was coined by N.J. Zabusky and M.D. Kruskal [2.30] to characterize solitary waves in nonlinear systems whose properties (energy etc.) are mostly localized in a bounded region of space at any instant of time such that upon collision they act like particles (e.g. electron, proton, etc.) and, in the case they studied, elastically, crossing each other or interchanging places with no significant change of form. As such, solitons were for long time a concept in the realm of conservative (and integrable) systems and hence solutions of dissipationless nonlinear equations [2.2, 2.9, 2.10, 2.12, 2.21, 2.24]. Recently, this concept has been extended to driven-dissipative systems, where an appropriate input–output energy balance exists and helps sustain the particle-like traveling localized structure. The work of Zabusky and Kruskal referred to the Korteweg–de Vries (1895) equation, an equation earlier explicitly derived by Boussinesq (1872, 1877) (Fig. 2.1) [2.4–6,2.8,2.14]. Accordingly, in this book we shall refer to it as the Boussinesq–Korteweg–de Vries (BKdV) equation (Fig. 2.1). The localized structure and hence the solitary wave solution of the BKdV equation result from the appropriate (local) balance between nonlinearity (velocity depends on amplitude) and dispersion (velocity depends on color or wavelength), as Boussinesq and Lord Rayleigh [2.22] clearly understood (see, however, the work by Ursell [2.27]). Since the pioneering work by Zabusky and Kruskal, intensive and highly fruitful research has developed on solitons and soliton-bearing equations and related matters. In fact a new chapter in applied mathematics and a new research line with broad scope in physics and other sciences has appeared. Examples are the phenomena of self-focusing and self-induced transmittancy in optics, the description of developed turbulence in plasma, the interpretation of solitons as particles in field theory, the exploration of statistical properties of some models in solid-state physics and hydrodynamics, solitons in Josephson junctions, action potentials in neurodynamics, and so on.

On the one hand, many numerical computations have been done on soliton-bearing equations. On the other hand, new analytical tools for solving nonlinear PDEs have been developed. As a result of these investigations the fundamental role of solitons in nonlinear science has, indeed, been established. It has been found that they are stable to finite perturbations and capable of self-restoration after interaction with one another. Moreover a major finding is that they significantly define the character of evolution of wave perturbations in many nonlinear spatially extended systems. In other words, solitons are evolutionarily stable for a wide class of initial field distributions. As particular solutions of nonlinear field equations, solitons can be treated as elementary objects whose role in nonlinear systems is either like particles in (classical) mechanics or like harmonic waves in linear systems, i.e. they appear as building blocks for nonlinear dynamics and a generalized (nonlinear) Fourier description (and spectral decomposition) of nonlinear signals.

This chapter is devoted to the study of solitary waves, chaotic soliton-trains, bound soliton states in a dissipation-modified Boussinesq–Korteweg–de Vries (BKdV) equation and hence wavy dissipative structures. That equation embraces both the original BKdV equation and the Burgers equation, and it is the simplest equation, so far known, exhibiting dissipative solitons. Note that the BKdV equation is integrable and that the Burgers equation can be reduced to the diffusion equation using the Cole–Hopf transformation [2.10]. The dissipation-modified BKdV equation, discussed below, is not integrable, although some exact solutions have been found, mostly for physically unrealistic parameter values. Besides dispersion, the dissipation-modified BKdV equation possesses the basic ingredients of nonlinear nonequilibrium media: nonlinearity, dissipation and instability, and hence an input–output energy balance. We shall show that the typical result of self-organization of such a system is the appearance of solitary waves of very diverse shapes. The existence of solitary waves, soliton-trains or bound soliton states will be established through studying the homoclinic orbits of the appropriate (third-order) nonlinear dynamical system underlying the dissipation-modified BKdV equation. We can safely say that our conclusions are qualitatively valid for other models.

The work by Boussinesq, Lord Rayleigh, Korteweg and de Vries originated in the serendipitous finding but subsequent systematic investigation of a ship-building engineer, J.S. Russell, around 1830–1840. The discovery happened at Union Canal (near Turning Point) in the outskirts of Edinburgh (not far from the present-day quarters of Heriott–Watt University). Russell's 1842 [2.25, 2.26] report on waves triggered theoretical work by the mathematician Airy, who proved that waves of permanent form were not permitted in the realm of (nonlinear) hydrodynamics. Airy's conclusion (not

$$(283\ bis) \qquad \frac{dh'}{dt} + \omega_0 \frac{d}{ds}\left[h' + \frac{k''}{2}\left(\frac{2+k}{2}\frac{h'^2}{\mathrm{H}} + \frac{k'\mathrm{H}^2}{3}\frac{d^2h'}{ds^2}\right)\right] = 0,$$

Fig. 2.1. The so-called Korteweg–de Vries equation (1895) as written in 1877 by Boussinesq wrong but besides the point) was due to his neglect of the significant role played by linear wave dispersion [2.10]. As earlier said, Boussinesq and later Lord Rayleigh, who gave credit to the earlier work by Boussinesq, clearly understood how nonlinearity and (linear) dispersion can lead, when in balance, to solitary waves capable of traveling with no deformation or accomodating geometrical constants. Their work either was misunderstood or just passed unnoticed (Boussinesq's 1877 report is over six-hundred pages long in French and not easy to read) until a publication by Korteweg and de Vries (1895) and, subsequently, until the middle of twentieth century.

The work of Russell deserves some comment, for he carefully described the major, albeit in part overlocked, properties of solitons. First, he characterized wave velocity and wave evolution; then, he embarked on characterizing wave reflection (see, in particular, plate VII in his celebrated 1842 report on waves). Similar experimental results were later found around 1865 by H. Bazin (in collaboration with his father-in-law Darcy) with experiments carried out using a "rigole" parallel to the Bourgogne canal in the outskirts of Dijon (France) [2.1, 2.3]. This work was done presumably at the suggestion of Boussinesq. Several other researchers have provided further support to Russell's discoveries and to the reality of solitons. For our purpose here, let us mention the work of H. Linde around 1965 in Berlin and, recently, in Madrid (with A. Wierschem), which, for surface tension gradient-driven dissipative systems, provided a wealth of experimental results encompassing Russell's discoveries on water waves and E. Mach's findings on shocks. Russell's and Bazin's findings on wave reflections share, indeed, common features with shocks as illustrated by the pioneering discoveries, around 1875, by Mach and collaborators, on shocks and also on hydraulic jumps in water waves [2.15, 2.18]. In fact 2D hydraulics has been used as a testing ground for shock-wave phenomena. Note that Russell's solitary wave is related to the conservative BKdV equation, while shocks are described by the dissipative Burgers equation.

One particular kinematic feature of solitary waves, hydraulic jumps and (compressible) shocks is the appearance of a third wave in collisions or wave reflections with a wide enough incoming angle [2.15–18,2.25,2.29]. This third wave, called the Mach stem, travels phase locked with the outgoing or post-collision waves and will be referred to in this book as the Mach–Russell stem or third wave, as Russell's discovery (he spoke about "lateral accumulation") predated Mach's. Such a third wave, if reinterpreted as a particle, appears as a phenomenon of soliton or particle creation.

2.2 Model Equation

We take as a model problem for qualitative analysis the 1D dissipation-modified BKdV equation [2.11, 2.20]:

$$u_t + uu_x + \nu u_{xx} + \beta u_{xxx} + \gamma u_{xxxx} + \alpha(uu_x)_x = 0\,, \tag{2.1}$$

where ν, β, γ and α are constants and, for instance, u denotes surface deformation in the oscillatory instability of Bénard layers (Fig. 1.1). The generalization of (2.1) to a 3D geometry exists, but we need not consider such a case here, on the one hand, in view of the limited scope of our book and, on the other hand, because there are few analytical results about solutions and numerical experiments. Suffice to mention the theoretical support for the above-mentioned Mach–Russell stem or third wave in collisions [2.20].

Equation (2.1) has the family of spatially homogeneous steady states $u = u_0 = \text{const}$. Substituting $u = A\exp(ikx + \omega t)$ into the linearized version of (2.1), one obtains the linear dispersion relation

$$\omega = ik(k^2\beta - u_0) + k^2(\nu + \alpha u_0 - \gamma k^2)\,.$$

The constants γ and α characterize stability or instability (self-excitation) of the state u_0, γ being dissipation and β dispersion. Let, for instance, $u_0 = 0$, i.e. we consider the spatio-temporal evolution of the variable $u(x,t)$ of (2.1) around the zero level. Let us assume that the parameters α, β and γ are positive, but the parameter ν can take either positive or negative values. Note that for $\nu < 0$ the steady state $u = 0$ is linearly stable, but for $\nu > 0$ it is unstable to small amplitude sinusoidal waves. It follows from the dispersion relation that these waves grow linearly for long wavelengths, while their growth is damped for short wavelengths. The maximum growth rate occurs at the wave number $k_{max} = \sqrt{\nu/\gamma}$. The conditions which we impose on the parameters of (2.1) are applicable, in particular, to surface waves in shallow Bénard fluid layers, where buoyancy is negligible. As a matter of fact, in such a case, in the absence of buoyancy, γ and α are strictly positive; β is positive or negative according to the value of the air–liquid surface tension, as in the original or standard BKdV equation. The coefficient ν is negative below the instability threshold and positive otherwise, when excitation develops and thus accounts for the supercritical distance to threshold and for energy pumping from, say, the imposed thermal gradient. Thus, the parameter ν is an experimentally tunable quantity which controls energy input at long wavelengths; γ controls the extent of energy dissipation on shorter wavelengths; and α accounts for the nonlinear feedback provided by the surface stresses at the open surface, thus allowing a nonlinear correction to the long wavelength energy input. Further dissipation at, say, the bottom boundary of the liquid layer may be incorporated by adding a term proportional to u or even to $u|u|$. If, however, proper account of the bottom boundary is taken and its corresponding friction is incorporated into the formulation of the problem, then (2.1) becomes integrodifferential. We shall not dwell on this case here. With appropriate parameter values, (2.1) reduces either to the standard BKdV equation ($\nu = \gamma = \alpha = 0$), the Burgers equation ($\beta = \gamma = \alpha = 0$), a combination of both ($\gamma = \alpha = 0$), the Kuramoto–Sivashinsky (KS) equation ($\beta = \alpha = 0$), or an extended KS equation ($\beta = 0$).

Variations on (2.1) arising in various contexts have been studied by several authors (see, e.g., [2.20]).

At variance with the standard BKdV equation, which has a one-parameter family of (solitary) wave solutions with amplitudes and phase velocities originating in *initial conditions* with subsequent energy conservation, in a dissipative system such as that described by (2.1), past the instability threshold dissipation helps in the selection of a *single* value for the wave velocity with a corresponding single wave amplitude. Recall that the standard BKdV equation is integrable, and besides the energy possesses an infinite number of conservation laws. Thus to survive the solitary wave of (2.1) demands an (input–output, pumping–dissipation) energy balance as well as the usual (local) balance between (inertial) nonlinearity and dispersion. The same applies to cnoidal wave trains of (2.1), but in such case both amplitude and velocity accommodate the selected wavelength. This is typical of all dissipative structures far from thermodynamic equilibrium. Indeed, multiplying (2.1) by u, integrating with respect to x from minus infinity to plus infinity (or over a wave period) and suitably integrating by parts, we find that the energy associated with a solitary wave at a *given* value of an external constraint such as a thermal gradient, i.e. the energy

$$E = \int_{-\infty}^{\infty} u^2 \, \mathrm{d}x \,, \tag{2.2}$$

is governed by the balance equation

$$\frac{1}{2}\frac{\partial}{\partial t}\int_{-\infty}^{\infty} u^2 \, \mathrm{d}x = \nu \int_{-\infty}^{\infty} u_x^2 \, \mathrm{d}x - \gamma \int_{-\infty}^{\infty} u_{xx}^2 \, \mathrm{d}x + \alpha \int_{-\infty}^{\infty} u u_x^2 \, \mathrm{d}x \,, \tag{2.3}$$

which is zero at steady state. The balance equation (2.3) shows what was earlier pointed out: the first term on the right-hand side of (2.3) accounts for the energy input at long wavelengths excited by the external energy input; the second term corresponds to energy dissipation at shorter wavelengths; and the third term provides feedback, i.e. a nonlinear correction to the energy balance also maintained by the external energy input. The uu_x and βu_{xxx} terms separately conserve energy and vanish identically in the standard BKdV equation.

2.3 Traveling Waves

Let us describe (2.1) in a moving frame where for convenience the new space coordinate is $\xi = x + ct$, with c being the (phase) velocity of the expected wave traveling right to left. A similar analysis applies to right-moving waves ($\xi = x - ct$). Upon integration of (2.1) from $-\infty$ to the traveling ξ-coordinate, we obtain the following system of three nonlinear coupled ordinary differential equations (ODE):

$$\dot{u} = y\,, \tag{2.4}$$
$$\dot{y} = z\,, \tag{2.5}$$
$$\gamma\dot{z} = -\beta z - \alpha u y - \nu y - F(u)\,, \tag{2.6}$$

where the dot denotes the derivative with respect to ξ, $F(u) \equiv cu + u^2/2$. For illustration, we consider the system (2.4–6) for $\nu > 0$. The investigation of this system for $\nu < 0$ is analogous and is not given here. Equations (2.4–6) define the underlying (nonlinear) dynamical system to the PDE (2.1) [2.19].

2.3.1 Steady States

The steady solutions or fixed points of (2.4–6) are $O_1 \equiv (-2c, 0, 0)$ and $O_2 \equiv (0,0,0)$ in the space $G : \{u, y, z\}$. O_1 is a *saddle–focus* when $\gamma > \gamma_S$ with $\lambda_3 > 0$, $\mathrm{Re}\,\lambda_i < 0$, $\mathrm{Im}\,\lambda_i \neq 0$ $(i = 1,2)$, whereas it is a *saddle* when $\gamma > \gamma_S$ with $\lambda_3 > 0$, $\mathrm{Re}\,\lambda_i < 0$, $\mathrm{Im}\,\lambda_i = 0$ $(i = 1,2)$. We have used

$$\gamma_S = \frac{\{-(\nu - 2\alpha c)[9\beta c + 2(\nu - 2\alpha c)^2] + 2[(\nu^2 - 2\alpha c)^2 + 3\beta c]^{3/2}\}}{27c^2}\,. \tag{2.7}$$

λ_i denote the roots of the characteristic equation for O_1, i.e. the solution of

$$\lambda^3 + \frac{\beta}{\gamma}\lambda^2 + \frac{(\nu - 2\alpha c)}{\gamma}\lambda - \frac{c}{\gamma} = 0\,. \tag{2.8}$$

There are two invariant manifolds passing through the fixed point: the 2D manifold $W^s(O_1)$ formed by all incoming trajectories, and the 1D manifold $W^u(O_1)$ formed by two outgoing orbits W_1^u and W_u^2. Let G denote the phase space of the system (2.4–6). Analyzing the linearized system corresponding to (2.4–6) we establish the following result: The trajectories W_1^u, W_2^u in G originate from the point O_1 and are tangent to the line

$$u + 2c = \frac{y}{\lambda_3} = \frac{z}{\lambda_3^2}\,. \tag{2.9}$$

It follows from (2.9) that one of the separatrices, W_1^u, goes to the half-space $y > 0$, and the other, W_2^u, goes to the half-space $y < 0$ (Fig. 2.2). The plane tangent at the point, O_1, to the manifold W^s is defined by

$$\frac{c}{\lambda_3\gamma}(u + 2c) + \left(\lambda_3 + \frac{\beta}{\gamma}\right) y + z = 0\,. \tag{2.10}$$

The second point, O_2, is asymptotically stable when $\gamma < \beta\nu/c$ and characteristic roots $\kappa_3 < 0$, $\mathrm{Re}\,\kappa_i < 0$ $(i = 1,2)$, and unstable when $\gamma > \beta\nu/c$ and $\kappa_3 < 0$, $Re\kappa_i > 0$ $(i = 1,2)$. As in this chapter we are only interested in homoclinic orbits, we note that the point O_2 becomes unstable through the Andronov–Hopf bifurcation and provides the appearance of a limit cycle.

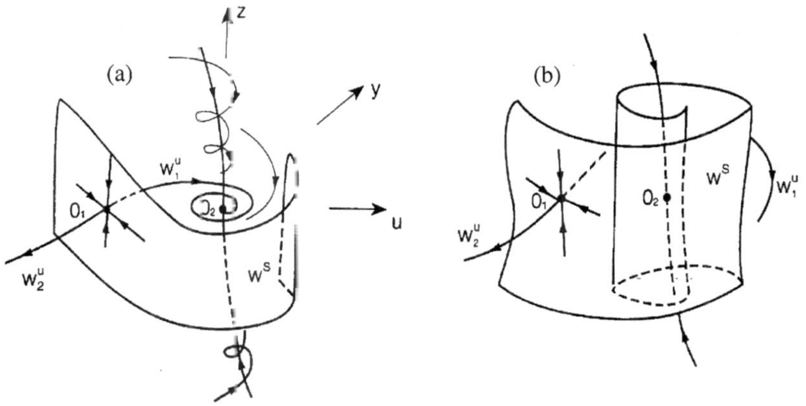

Fig. 2.2. The two fixed points of (2.4–6) and their stable and unstable manifolds: (**a**) $\alpha\beta - \gamma > 0$, $\nu - \alpha c > 0$; (**b**) $\alpha\beta - \gamma < 0$, $\nu - \alpha c < 0$

2.3.2 Lyapunov Functions

To assess the global stability of the stationary solutions it is useful to consider a suitable Lyapunov function or potential. Take

$$\begin{aligned} V_1 = (\alpha\beta - \gamma)[\gamma z + \beta y + \alpha F(u)]^2/2 + \gamma[\gamma z + \alpha F(u)]^2/2 \\ + \alpha\beta\gamma(\nu - \alpha c)y^2/2 + \beta[\alpha\beta - \gamma + \alpha^2(\nu - \alpha c)] \int_0^u F(u)\,\mathrm{d}u\,, \end{aligned} \tag{2.11}$$

which according to (2.4–6) has a "time" derivative

$$\dot{V}_1 = -\beta(\alpha\beta - \gamma)(\nu - \alpha c)y^2 - \beta[\gamma z + \alpha F(u)]^2\,. \tag{2.12}$$

Then in the region $d_1 = \{\alpha\beta - \gamma > 0, \nu - \alpha c > 0\}$ we have $\dot{V}_1 \leq 0$. Thus V_1 is the Lyapunov potential. Having analyzed the location of the level surfaces of the function V_1 in G, it can be shown that the manifold W^s of O_1 splits the phase space G into the regions Q^+ and Q^- such that all trajectories of the system, including the W_1^u, tend to O_2 as $\xi \to \infty$ in the region Q^+, while all trajectories in the region Q^-, including the trajectory W_2^u, go to infinity as $\xi \to +\infty$ (Fig. 2.2a).

Analogously, we can construct the second Lyapunov function:

$$\begin{aligned} V_2 = (\gamma - \alpha\beta)[\gamma z + \beta y + \alpha F(u)]^2/2 - \gamma[\gamma z + \alpha F(u)]^2/2 \\ + \alpha\beta\gamma(\alpha c - \nu)y^2/2 + \beta[\gamma - \alpha\beta + \alpha^2(\alpha c - \nu)] \int_0^u F(u)\,\mathrm{d}u\,, \end{aligned} \tag{2.13}$$

whose "time" derivative is

$$\dot{V}_2 = \beta(\gamma - \alpha\beta)(\alpha c - \nu)y^2 + \beta[\gamma z + \alpha F(u)]^2\,. \tag{2.14}$$

Since for the points in the region $d_2 = \{\gamma - \alpha\beta > 0, \alpha c - \nu > 0\}$ we have $\dot{V}_2 > 0$, all trajectories (2.4–6), with the exception of those on the manifold W^s and two trajectories extending to O_2, go to infinity as $\xi \to \infty$ (Fig. 2.2b).

2.4 Homoclinic Orbits. Phase-Space Analysis

Let us construct the invariant domains in the phase space G, allowing us to control the location of the manifolds W^s and W^u. Then we shall prove the existence of the homoclinic orbit, formed by the trajectory W_1^u of the fixed point O_1.

2.4.1 Invariant Subspaces

For convenience we change from z to v by introducing

$$\gamma z = \gamma v + \mu y - \alpha F(u) + \nu L(u + 2c)\,, \tag{2.15}$$

with

$$L \equiv \alpha \frac{(\beta + \mu)}{\gamma} - 1 \tag{2.16}$$

and

$$\mu \equiv -\frac{(\beta + \alpha\nu)}{2} + \left(\frac{(\beta - \alpha\nu)^2}{4} + \alpha\gamma c\right)^{1/2}\,. \tag{2.17}$$

Then (2.4–6) become

$$\dot{u} = y\,, \tag{2.18}$$

$$\dot{y} = (\mu/\gamma)y - (\alpha/\gamma)(u/2 - \nu L/\alpha) + (u + 2c) + v\,, \tag{2.19}$$

$$\gamma\dot{v} = -(\beta + \mu)v + L[u/2 - \nu(\beta + \mu)/\gamma](u + 2c)\,. \tag{2.20}$$

Thus setting $L = 0$ yields

$$\gamma = \alpha\beta - \alpha^2(\nu - \alpha c)\,. \tag{2.21}$$

The system (2.18–20) has a stable integral plane, $v = 0$, and the motion is reduced to

$$\begin{aligned} \dot{u} &= y\,, \\ \dot{y} &= (\mu/\gamma)y - (\alpha/2\gamma)u(u + 2c)\,. \end{aligned} \tag{2.22}$$

Illustrated in Fig. 2.3 are the possible phase portraits of (2.22) according to the value taken by μ. Clearly, (2.22) possesses a *homoclinic* orbit formed by the trajectory W_1^u if (2.21) is fulfilled and $\mu = 0$, i.e. when $\gamma = \alpha\beta$ and $\nu = \alpha c$.

Consider now the parameter region of (2.18–20) where $L > 0$, i.e. when

$$\gamma < \alpha\beta - \alpha^2(\nu - \alpha c)\,. \tag{2.23}$$

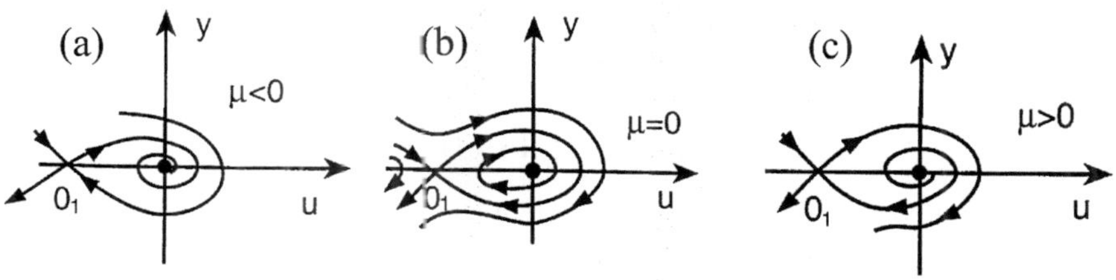

Fig. 2.3. The two fixed points of (2.4–6) and possible phase portraits of (2.18–20) according to values of μ: (**a**) μ positive, (**b**) $\mu = 0$ (with homoclinic trajectory), and (**c**) μ negative

Then from (2.20) we get $\dot{v} < 0$ when $v < -h^-$, with

$$h^- \equiv [L/2(\beta+\mu)][c+\nu(\beta+\mu)/\gamma]^2 \,. \tag{2.24}$$

Thus the set of singular trajectories of (2.18–20) lies completely in the half-space G^+ determined by the inequality $v > -h^-$. Analogously, for $L < 0$ there is a second set of singular trajectories lying in G^+ corresponding to the inequality $v < g^+$, with

$$g^+ \equiv [|L|/2(\beta+\mu)][c+\nu(\beta+\mu)/\gamma]^2 \,. \tag{2.25}$$

Thus, either

$$v(\xi) \geq -h^- \quad \text{for} \quad L > 0 \tag{2.26}$$

or

$$v(\xi) < g^+ \quad \text{for} \quad L < 0 \tag{2.27}$$

are obeyed by any trajectory $\{u(\xi), y(\xi), v(\xi)\}$ of (2.18–20) with the initial conditions in the half-space G^+.

2.4.2 Auxiliary Systems

Let us start with $L \geq 0$. For convenience we introduce the *auxiliary* systems

$$\dot{u} = y \,, \tag{2.28}$$
$$\dot{y} = -h^- + (\mu/\gamma)y - (\alpha/\gamma)(u/2 - \nu L/\alpha)(u+2c) \tag{2.29}$$

and

$$\dot{u} = y \,, \tag{2.30}$$
$$\dot{y} = h^+ - (\mu/\gamma)y - (\alpha/\gamma)(u/2 - \nu L/\alpha)(u+2c) \,, \tag{2.31}$$

with $h^+ \equiv \alpha(L)^{1/2}/2\gamma$. As (2.28–29) do not depend on v, all trajectories lie on cylindrical surfaces in G^+. Note that (2.28–29) and (2.30–31) define systems of the form of (2.22). It follows that

$$\gamma > \frac{4\alpha^3\nu^2 c + 2\alpha\nu(\beta - \alpha\nu)\left[\nu + (\nu^2 + 4\alpha^2c^2)^{1/2}\right]}{\left[\nu + (\nu^2 + 4\alpha^2c^2)^{1/2}\right]^2} . \qquad (2.32)$$

Thus their phase portraits are equivalent to one of the three depicted in Fig. 2.3. The fixed point coordinates will be different, in general. Here for our purpose it suffices that the *qualitative* behaviors of the trajectories are similar.

Consider now the relative position of the saddle separatrices of (2.28–29) and (2.30–31) in the (u, y) plane for v = const. Let us denote the saddles of (2.28–29) and (2.30–31) by $O_1^-(u_1^-, 0)$ and $O_1^+(u_1^+, 0)$, respectively. It appears that

$$u_1^+ < u < u_1^- , \qquad (2.33)$$

where $u_1 = -2c$ is one of the coordinates of the saddle point O_1 of the 3D system (2.18).

Compare now the vector fields of (2.28–29, 30–31):

$$R = \left(\frac{\mathrm{d}y}{\mathrm{d}u}\right)_{(2.30-31)} - \left(\frac{\mathrm{d}y}{\mathrm{d}u}\right)_{(2.28-29)} = \frac{h^+ + h^-}{y} , \qquad (2.34)$$

from which it follows that $Ry > 0$. This together with (2.33) results in the fact that the separatrices of the saddle of (2.28–29) and (2.30–31) do not intersect on the (u, y) plane. The results of the investigation of the dynamics of (2.22) shows that when $\mu = 0$, the separatrix of the saddles O_1^+, O_2^- are like those depicted in Fig. 2.4b, where the solid and dotted lines refer to (2.30) and (2.28), respectively. Note that for $L = 0$ the two systems (2.28–29) and (2.30–31) coincide with (2.22) and their phase portraits are those depicted in Fig. 2.3. Consequently, for parameter values leading to L close to zero, the separatrices of (2.28–29) and (2.30–31) are like those illustrated in Fig. 2.4a (Fig. 2.4c) for negative μ (positive μ).

Using now ν and γ as control parameters (all other parameters being held fixed), from the previous results we can infer the existence of curves Γ_1 and Γ_2 tangent to $\gamma = \beta\nu/c$ at $\mu = 0$ ($\gamma = \alpha\beta, \nu = \alpha c$, Fig. 2.5). Figure 2.6a and b illustrate the relative positions of the saddle separatrix of (2.28–29) and (2.30–31) corresponding to Γ_1 and Γ_2, respectively. Note that these curves delineate in the (ν, γ) plane of Fig. 2.5 the regions d_3 and d_4 for the points corresponding to saddle separatrices of (2.28–29) and (2.30–31) as illustrated in Figs. 2.4a and c, respectively.

2.4.3 Construction of Regions Confining the Unstable and Stable Manifolds W^u and W^s

We shall now find the orientation of the vector field on the cylindrical surfaces formed by the separatrices of (2.28–29) and (2.30–31). To do this we need

$$R^- = \left(\frac{\mathrm{d}y}{\mathrm{d}u}\right)_{(2\,18-20)} - \left(\frac{\mathrm{d}y}{\mathrm{d}u}\right)_{(2.28)} = \frac{v + h^-}{y}\,,$$

$$R^+ = \left(\frac{\mathrm{d}y}{\mathrm{d}u}\right)_{(2.18-20)} - \left(\frac{\mathrm{d}y}{\mathrm{d}u}\right)_{(2.30)} = \frac{v - h^+}{y}\,.$$

From 2.26) it follows that $R^- y > 0$ in the region G^+. Consequently, the trajectories of (2.18–20) intersect the cylindrical surfaces formed by the saddle separatrices of 2.30) such that the y-coordinate increases (see arrows in Fig. 2.4). As $R^+ y < 0$ when $v < h^+$, the portions of cylindrical surfaces formed by the saddle separatrices of (2.30–31) when $v < h^+$ are intersected by the trajectories of (2.18–20) such that y decreases (arrows in Fig. 2.4). In the construction of the tunnels or regions where the manifolds W^u and W^s are confined, we shall use the plane $v = h^+$. The inequality $\dot{v} < 0$ must be satisfied on the portion of the plane $v = h^+$ that is used in the construction. Equation (2.20) shows that this is so if $u \in (u^-, u^+)$, with

$$u^{-,+} \equiv -c + \frac{\nu(L+1)}{\alpha} \pm \sqrt{\left(c + \frac{\nu(L+1)}{\alpha}\right)^2 + \frac{\nu(L+1)}{\sqrt{L}}}. \qquad (2.35)$$

We shall consider the values of the parameters of the system at which the cylindrical surfaces enclosed between the planes $u = u_1^+$ and u^s (respectively, $u = u^0$ or $u = u^u$ in Fig. 2.4) lie in the zone between the planes $u = u^-$

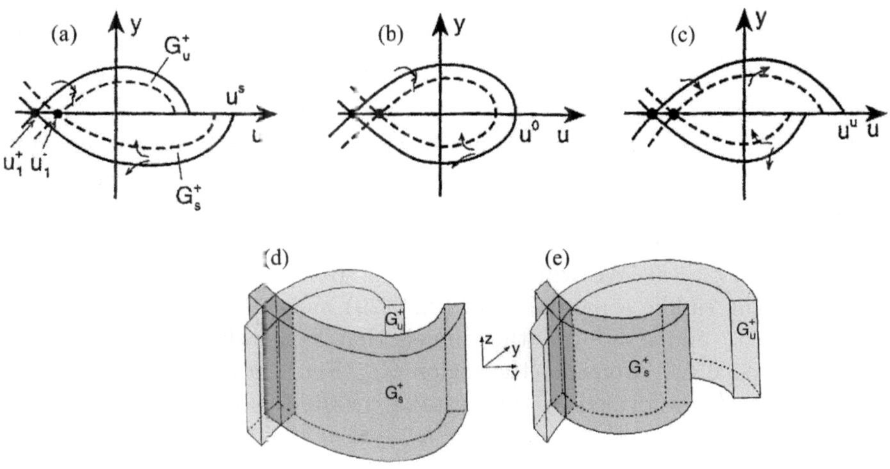

Fig. 2.4. Different phase portraits for the *auxiliary* systems (2.28–29) and (2.30–31). The "tunnels" are suitably chosen to enclose the trajectories depicted in Fig. 2.3, stable and unstable separatrices of saddle points according to values of μ: **(a)** μ positive, **(b)** $\mu = 0$ (with homoclinic trajectory) and **(c)** μ negative. **(d)** and **(e)** correspond, respectively, the 3D unfolding of **(a)** and **(c)**

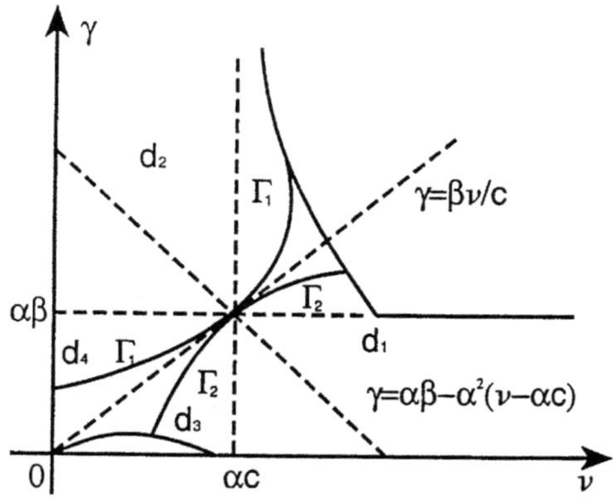

Fig. 2.5. The (ν, γ) plane. Relative to the (inertial) nonlinearity, the variables ν and γ control energy input and dissipation, respectively

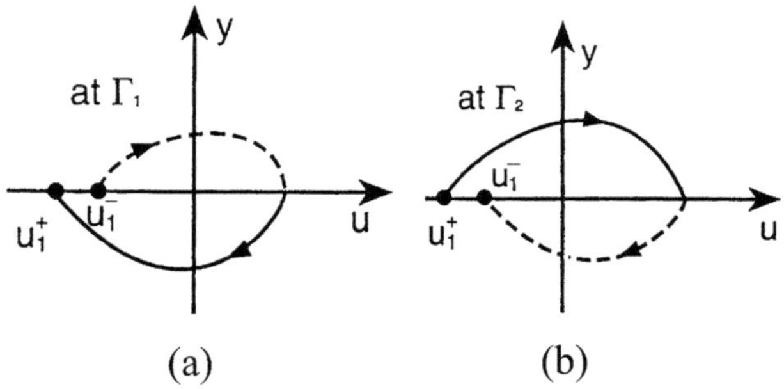

Fig. 2.6. Relative positions of the portions (taken from Fig. 2.4) of the saddle separatrices of (2.28–29) and (2.30–31) corresponding to (**a**) Γ_1 and (**b**) Γ_2, respectively, of Fig. 2.5

and $u = u^+$. These parameter values border rather closely on the straight line $L = 0$ because $|u^{-,+}| \to \infty$ as $L \to 0$. Therefore, by the regions d_3 and d_4 we shall understand either the regions as a whole, if the requirement specified above is fulfilled at all its points, or the portions of these regions bordering on $L = 0$. We will now construct from the cylindrical surfaces formed by the saddle separatrices of (2.28–29) and (2.30–31) and from the planes $v = h^+$ and $v = -h^-$, the tunnels that locate these separatrices in the phase space G^+. Consider the region G_u^+ (Fig. 2.4d,e), bounded by the cylindrical surfaces formed by the unstable saddle separatrices of (2.28–29) and (2.30–31) and by the planes $v = h^+$ and $v = -h^-$. As earlier found, the vector field of (2.18–20) at the boundary of G_u^+ is oriented inwards to G_u^+ and the saddle point O_1 is located inside G_u^+. Therefore the trajectory W_1^u of (2.18–20) in its motion "along" G_u^+ approaches the plane $y = 0$ and intersects it at the point $M^u \in G_u^+ \cap \{y = 0\}$ (Fig. 2.4d,e). Consider now the region G_s^+ (Fig. 2.4d,e) that is bounded by the cylindrical surfaces formed by the stable saddle separatrices of systems (2.28–29) and (2.30–31) and by the planes $v = h^+$ and $v = -h^-$. This region also contains the point O_1 and is intersected by the trajectories of the system in a definite manner: inwards to G_s^+ on the planes and outside this region on cylindrical surfaces. Thus the 2D manifold W^S is located between the cylindrical surfaces and intersects the planes $v = h^+$ and $v = h^-$, i.e. G_s^+ is the sought tunnel for W^S. Taking into account that the tunnel G_s^+ intersects the plane $y = 0$, we find that the 2D separatrix W^S intersects this plane along the curve I^S that connects the planes $v = h^+$ and $v = -h^-$. We will now pass over to the space of the parameter values of (2.18–20) from the points in region d_3 to region d_4. In such a transition, on the secant plane $y = 0$, there always exist "traces" of 1D and 2D manifolds: the point M^u and the line I^S. Because G_u^+ and G_s^+ exchange places, M^u and I^S change their relative positions too. Accordingly, between regions d_3 and d_4 there exists at least one line, Π, which corresponds to the situation when the point M^u lies on the curve I^S and in the phase space G^+ there exists a homoclinic orbit Γ formed by the trajectory W_1^u.

The tunnels are constructed analogously ($L < 0$) using 2D systems of the form

$$\dot{u} = y\,, \quad \dot{y} = -g^- + \frac{\mu y}{\gamma} - \frac{\alpha}{\gamma}\left(\frac{u}{2} + \frac{\nu|L|}{\alpha}\right) + (u + 2c)\,, \tag{2.36}$$

$$\dot{u} = y\,, \quad \dot{y} = g^+ + \frac{\mu y}{\gamma} - \frac{\alpha}{\gamma}\left(\frac{u}{2} + \frac{\nu|L|}{\alpha}\right) + (u + 2c)\,, \tag{2.37}$$

where $g^- \equiv 3\alpha/8\gamma c - (\nu|L|)/\alpha)]^2$. As before, these tunnels allow us to separate the regions d_3 and d_4 when

$$L < 0\,, \quad \gamma < \frac{\alpha\nu[\alpha^2 c\nu + (\beta = \alpha\nu)(\nu - \alpha c)]}{(\nu - \alpha c)^2}\,, \tag{2.38}$$

and to prove the existence of line Π corresponding to the homoclinic trajectory Γ.

2.5 Multiloop Homoclinic Orbits and Soliton-Bound States

2.5.1 Existence of Multiloop Homoclinic Orbits

The behavior of the trajectories in the neighborhood of a homoclinic orbit that breaks at the intersection with Π is nearly always determined by the sign of the saddle value

$$\sigma = \lambda_3 + \max\{\mathrm{Re}\,\lambda_1, \mathrm{Re}\,\lambda_2\}\,. \tag{2.39}$$

If $\sigma < 0$ the situation is rather simple: a single stable periodic motion exists and no other homoclinic orbits appear in the neighborhood of the breaking loop Γ for both saddle and saddle-focus O_1. If $\sigma > 0$, the behavior of the trajectories is different for a saddle and for a saddle-focus. In the case of saddle, a saddle periodic motion appears in the neighborhood of the breaking Γ and no other homoclinic trajectories exist. In the case of a saddle-focus, a countable set of multiloop homoclinic orbits and a nontrivial hyperbolic set containing a countable set of saddle periodic motions exist in the neighborhood of the loop breaking on either side. The multiloop homoclinic structure passes its broadened neighborhood in several sections. Hence multiloop homoclinic orbits exist when the parameters of the system belong to a countable set of bifurcation surfaces located in the neighborhood of Π. Consequently, the existence of multiloop homoclinic orbits may be established by finding the parameter values of (2.18–20) at which O_1 is a saddle-focus and $\sigma > 0$. For this purpose the space of the parameter values must be partitioned into regions corresponding to the different signs of σ and to the different types of the equilibrium state O_1. Analysis of the location of the roots λ_i on the complex plane shows that, for constant α, β and γ, three partitioning types are possible (Fig. 2.7). The curves P and Q in Fig. 2.7 are specified by the following equations:

$$P : \{c(2\alpha\beta - \gamma) = \beta\nu\},$$

$$Q : \{c\gamma(\gamma + 2\alpha\beta) = \gamma\beta\nu + 2\beta^3\};$$

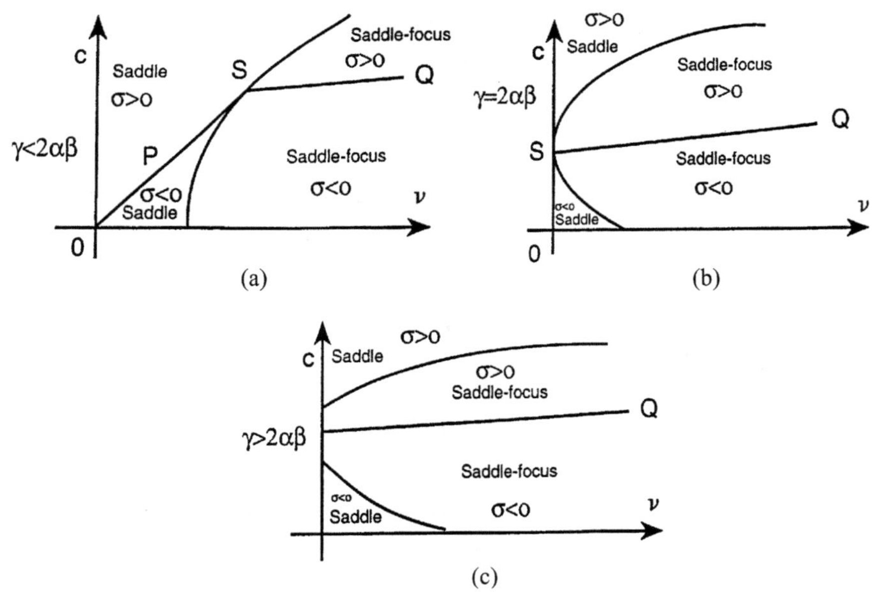

Fig. 2.7. The three possible partitions of the (c, γ) plane according to the relative values of γ depicting, for a given wave speed, the possible solution: (**a**) $\gamma < 2\alpha\beta$, (**b**) $\gamma = 2\alpha\beta$ and (**c**) $\gamma < 2\alpha\beta$

the point S has the coordinates

$$\nu = \frac{\beta^2}{\gamma^2}(2\alpha\beta - \gamma)\,, \quad c = \frac{\beta^3}{\gamma^2}\,. \tag{2.40}$$

Let us qualitatively find the location of the line Π on the (ν, c) plane using the knowledge of its boundaries. Π must be somewhere between the lines Γ_1 and Γ_2 (Fig. 2.5). Thus for $\gamma < \alpha\beta$ the plane Π can only be located in the region $c > \nu/\alpha$. The line $c = \nu/\alpha$, when $\gamma < \alpha\beta$, is located above the line Π, and the equilibrium state O_1 is a saddle with a positive saddle value $\sigma > 0$ for points of line Π. Consequently, no homoclinic trajectories different from Γ exist in its neighborhood. The line Π is shown in Fig. 2.8a for $\gamma < \alpha\beta$. When $\gamma = \beta\alpha$, the line Π becomes to the line $c = \nu/\alpha$. [As already noted, in this case (2.18–20) has a stable integral plane $v = 0$, and the trajectory Γ is formed by the separatrix of the 2D system (2.22) (see Fig. 2.8b).] Analogously, we find that for $\gamma > \alpha\beta$ there exists on line Π a portion Π_c (Fig. 2.8c) whose points correspond to the trajectory Γ formed by the saddle-focus having a positive saddle value. Consequently, there exists in the neighborhood of Π_c a countable set of bifurcation surfaces corresponding to multiloop homoclinic trajectories. Unfortunately, the general layout of the complete bifurcation picture of the neighborhood of Π_c is not known. Point 1 on line Π is specified by the condition $\sigma = 0$, and point 2 by the condition of existence of the saddle O_1 with equal negative characteristic roots and by $\sigma > 0$. Then there exists in the neighborhood of point 1 a countable set of curves $\{\Pi_n^2\}_{n=1}^{\infty}$, each corresponding to a double-loop homoclinic trajectory (Fig. 2.9d). The subscript n denotes the number of oscillations of the homoclinic trajectory in the neighborhood of point O_1 in the first loop. The family Π_n^2 is a set

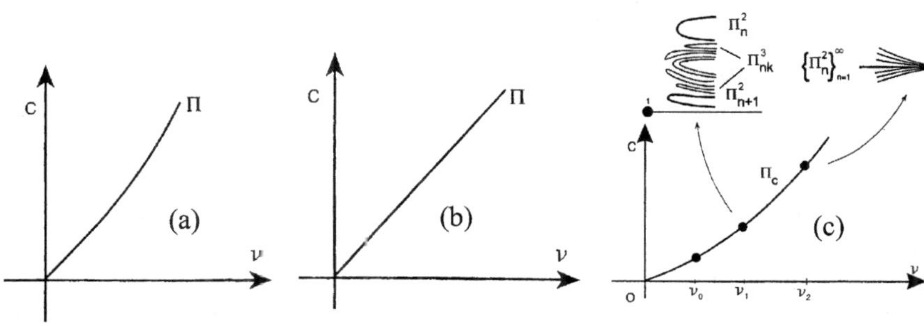

Fig. 2.8. Qualitative location of the line Π illustrating the possible soliton behavior as a function of energy input ν and the corresponding wave speed c. Recall that relative to the (inertial) nonlinearity, γ accounts for energy dissipation and β accounts for the (local) balance between dispersion and the (inertial) nonlinearity. α is the ratio of the genuine nonlinear effect of the Marangoni stress corresponding to a supercritical value of the thermal gradient to the (inertial) nonlinearity: (**a**) $\gamma < 2\alpha\beta$, (**b**) $\gamma = 2\alpha\beta$, and (**c**) $\gamma < 2\alpha\beta$. We see that for a given (inertial) nonlinearity the dispersion and the nonlinear feedback reinforce each other to sustain the solitary wave

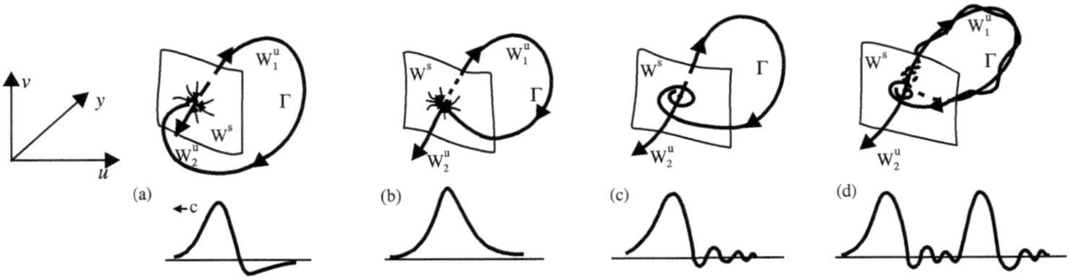

Fig. 2.9. Homoclinic trajectories and dissipative localized structures: (**a**)–(**c**) solitons, (**d**) bound state as a train of two solitons

of parabolic curves whose peaks accumulate at point 1 (Fig. 2.8c). A set of curves Π^3_{nk} corresponding to 3-loop homoclinic trajectories exists between each pair of curves Π^2_n, Π^2_{n+1}. The subscript k denotes the number of oscillations of the homoclinic trajectory in the neighborhood of point O_1 in the second loop. The number of Π^3_{nk} curves is finite for each fixed n and grows without bound as $n \to \infty$. Each of the Π^3_{nk} consists of two branches whose relative positions may be of two types. Only one pair of curves Π^2_n and Π^2_{n+1} and one set of curves Π^3_{nk} are depicted in Fig. 2.8c. A set of curves corresponding to 4-loop homoclinic trajectories exists between each pair Π^3_{nk} and Π^3_{nk+1}. Five-loop homoclinic trajectories exist between the 4-loop trajectories, and so on. Consider now the neighborhood of point 2. A bundle of curves $\{\Pi^2_n\}^\infty_{j=1}$ corresponding to 2-loop homoclinic trajectories (Fig. 2.8c) emerges from point 2. A bundle Π^3_{nk} emerges from point 2 between each pair Π^2_n and $\Pi^2_{n=1}$, and so on. Thus, there exists in the neighborhood of the component Π_c a countable set of bifurcation curves corresponding to homoclinic trajectories making an arbitrary number of loops.

2.5.2 Solitonic Waves, Soliton-Bound States and Chaotic Soliton-Trains

By generalizing the soliton concept used to describe solitary waves of the BKdV equation, the investigation of homoclinic trajectories allows us to establish the soliton features of model (2.1). A single soliton of a relatively simple form (Fig. 2.9a,b) whose velocity is uniquely specified by the parameter values of the model may propagate at $\gamma \leq \alpha\beta$ for each value of ν. The situation is drastically different if $\gamma > \alpha\beta$. When $\nu < \nu_1$ (Fig. 2.8c), a single value of velocity and a rather simple soliton form again correspond to each ν. The qualitative form of the soliton is depicted, for $\nu < \nu_0$, in Fig. 2.9a,b and for $\nu_0 < \nu < \nu_1$ in Fig. 2.9c. In the latter case the soliton has an oscillating tail as expected in view of the (linear, fourth-order) dissipative terms in (2.1). When $\nu_1 < \nu < \nu_2$, each ν corresponds to a countable set of values of the parameter c which specify the velocity of propagation of soliton-trains corresponding to multiloop homoclinic trajectories. Such groups of solitons may be treated as *bound states* of (2.1). A soliton-train of two "elementary" wave crests is illustrated in Fig. 2.9d. An arbitrary number of such elementary crests may take part in the formation of a soliton-train for there may exist trajectories with an arbitrary number of loops. Hence, depending on the initial conditions (i.c.), a bound state of a certain form propagating with a *definite* velocity is realized for $\nu_1 < \nu < \nu_2$. Model (2.1) is highly sensitive to initial conditions because there exists at each $\nu \in (\nu_1, \nu_2)$ a countable set of soliton-trains. Thus we can speak about chaotic soliton-trains for $\nu \in (\nu_1, \nu_2)$. When $\nu > \nu_2$, there are no bound states and solitons of simple form may propagate (Fig. 2.9a,b). Equation (2.1) also possesses periodic waves like the cnoidal waves of the BKdV equation, but we shall not consider them here [2.23].

2.5.3 Homoclinic Orbits and Soliton-Trains. Some Numerical Results

The qualitative study of the homoclinic orbits and solitary waves has guided us in the numerical exploration of the bifurcation sets corresponding to these orbits [2.7, 2.13, 2.28]. Analytical results also exist – although very limited in scope – for the stability of the solitary waves, but we shall limit consideration here to what can be considered as numerical experiments supporting our analytical, albeit qualitative, predictions. We shall carry this out in several steps. Some results for the cases $\nu > 0$ and $\nu < 0$ are given in the following:

The Case when $\nu > 0$. Figure 2.10 shows the results of numerical computation providing the bifurcation diagram, corresponding to Fig. 2.8c. The curves presented define the dependence of the velocity, c, of propagating solitary waves on the value of the parameter ν when all other parameters are fixed. For

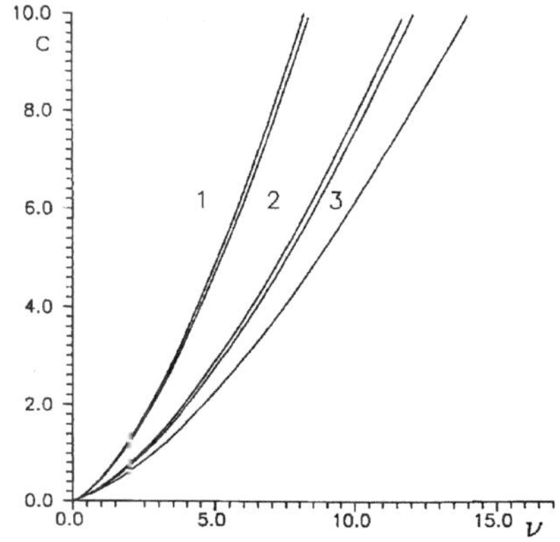

Fig. 2.10. Typical bifurcation diagram in the (ν, c) plane for multiloop homoclinic orbits (labelled 1, 2 and 3) for positive ν ($\alpha = 0.01, \beta = 1, \gamma = 10$). Curves 2 and 3 are parabolas

$\alpha = 0.01, \beta = 1$, and $\gamma = 10$ in the (ν, c) plane we have the following curves: the bifurcation curve Π-1, corresponding to the "basic" homoclinic orbit, one curve of the family Π_n^2-2, corresponding to the double-loop orbit, and one curve of the family Π_{nk}^3-3, corresponding to the three-loop homoclinic orbits of the system (2.4–6). For the parameter values mentioned above, the point ν_1 (Fig. 2.8c) in Fig. 2.10 is located very close to the origin, while the point ν_2 is very far from it. Then, the curves 2 and 3, with parabolic form, appear very "sharpened" near the origin. The bifurcation set located near point 1 in Fig. 2.8c is not shown in Fig. 2.10. In Fig. 2.10 the points of curve 1 are associated with isolated solitons, and the points of curves 2 and 3 with trains of two or three elementary solitons, respectively. For instance, the abscissa $\nu = 1.5$ provides in Fig. 2.10 five crossings (taken from top to bottom), hence five values of the velocity and thus the five waves displayed in Fig. 2.11a (curve 1: isolated soliton), Fig. 2.11b,c (curve 2: trains of two solitons) and Fig. 2.11d,e (curve 3: trains of three solitons). Figure 2.12 illustrates the space–time dynamics of the two-soliton train. For a 20% amplitude disturbance upon the expected homoclinic solution, Fig. 2.12 provides the evolution

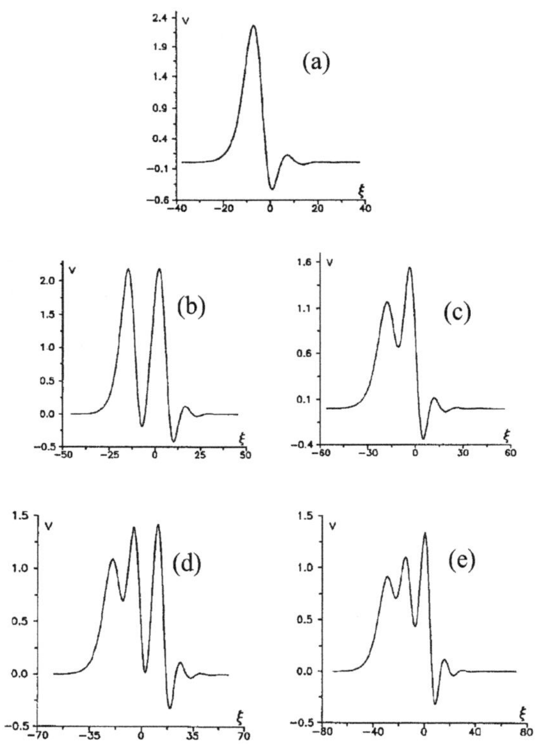

Fig. 2.11. Typical single- and multiple-hump solitary waves for $\nu = 1.5$ with velocities obtained from the crossings from top to bottom: (**a**) curve 1 of Fig. 2.10; (**b**) and (**c**) curve 2 of Fig. 2.10; (**d**) and (**e**) curve 3 of Fig. 2.10

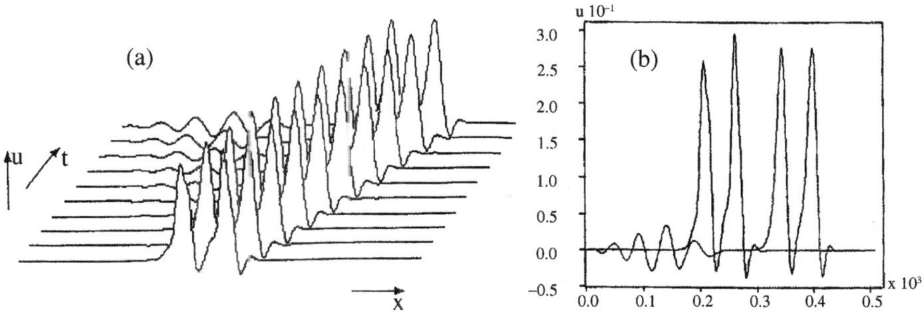

Fig. 2.12. (**a**) Time evolution of a 20% perturbed two-hump solitary wave for parameter values given in Fig. 2.10 (upper crossing of line 2) and $\nu = 0.33$; (**b**) Two superposed snapshots of the two-hump solitary wave corresponding to the initial time ($t = 0$) and "practical" infinity ($t = 1350$)

of such initial states The initial condition evolves towards a long-lived, two-hump, equal-amplitude bound wave, although a wavy tail (radiation) develops lagging drastically behind the main wave train. This radiation depends on the instability of the spatially homogeneous state $u = 0$ (see Sects. 2.1 and 2.6.1).

The Case when $\nu < 0$. When $\nu < 0$, in contrast with the previous case, there exist two essentially different types of diagram defining the form and the propagation-velocity dependence of the solitary waves from the parameter ν. In the first case there exists in the (γ, c) plane only one curve, while in the other there is an infinite number of curves. The first type of diagram is presented in Fig. 2.13a, while Fig. 2.13b illustrates a portion of the diagram of the second type. Figure 2.13a corresponds to the case of a single-hump

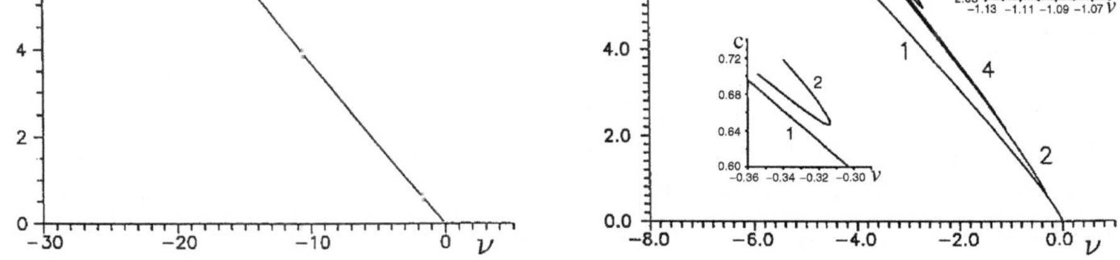

Fig. 2.13. Typical bifurcation set in the (ν, c) plane: (**a**) single-loop homoclinic orbit for *negative* ν with $\alpha = 2.5, \beta = 2, \gamma = 3.99$; (**b**) multiple-loop homoclinic orbits (labelled 1, 2, 3 and 4) for negative ν with $\alpha = 1$, $\beta = 2$, $\gamma = 3.99$

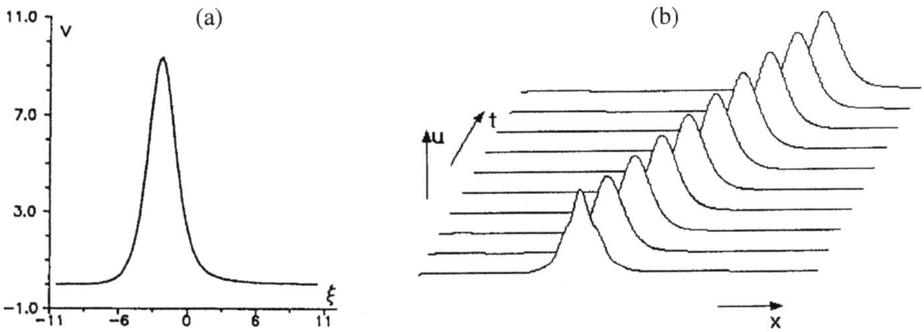

Fig. 2.14. (**a**) Case ν *negative.* Single-hump solitary wave for the case of Fig. 2.13a with $c = 3$ and $\nu = -8.142$. (**b**) Time evolution of a 20% perturbed single-hump solitary wave corresponding to the parameter values given in Fig. 2.13a and $\nu = -0.142$

wave as illustrated in Fig. 2.14. At variance with the (standard) BKdV equation ($\nu = \gamma = \alpha = 0$) for given $\{\alpha, \beta, \gamma\}$, once ν is fixed then a single wave velocity is selected and, moreover, the form of the hump is slightly asymmetric. Our numerical study shows that such solitary waves are very long lived (Fig. 2.14b). Figure 2.13b shows that for a given value of ν our analysis predicts the existence of single solitons and wave trains, containing two, three and four elementary solitons. We find that for a given value of the energy pumping, the nonlinear evolution equation (2.1) accommodates different wave solutions (displayed in Fig. 2.15) with indeed correspondingly different wave velocities. For instance, for $\nu = -4$, Fig. 2.15a depicts the soliton predicted with a wave velocity about 6 units, according to curve 1 in Fig. 2.13b. Figure 2.15b and c, also for $\nu = -4$, correspond to the waves with velocities selected from the crossings of the abscissa $\nu = -4$ with curve 2 (below and above, respectively), and so on.

Lacking the explicit analytic forms of these traveling localized dissipative structures or solitons, we have numerically explored their long-term persistence of *practical* stability on an infinitely extended 1D space support. Results are depicted in Figs. 2.12 and 2.14b. We must insist on the fact that (2.1) is not integrable and, in particular, energy is not conserved, but we have a continuous supply of energy. Thus the energy balance through energy pumping (nonvanishing ν) and nonlinear energy redistribution (nonvanishing α) helps compensate for the dissipation (nonvanishing γ leading to radiation etc.) due to viscosity and heat, hence maintaining the traveling localized structure, as, for example, an "aging" soliton. This is the crucial difference with the standard or ideal BKdV soliton just modified with viscous terms as in a BKdV–Burgers equation. Moreover, as already noted, this energy balance reduces the continuum of solutions of the BKdV equation ($\nu = \gamma = \alpha = 0$) to just one.

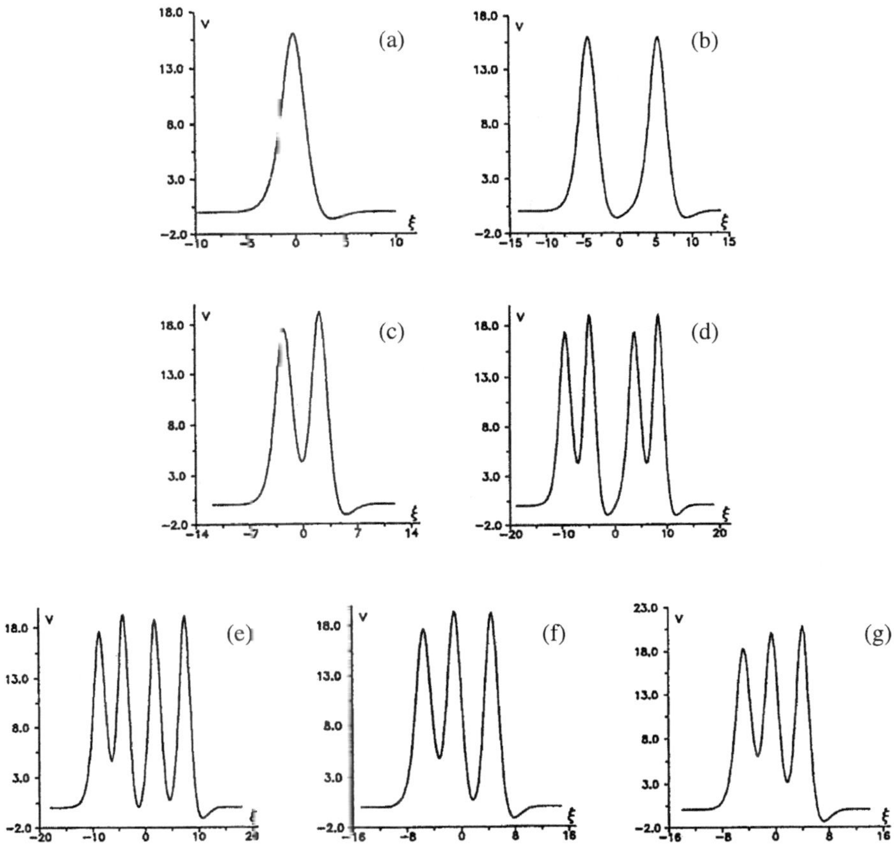

Fig. 2.15. Case ν *negative* (see Fig. 2.13b). Typical single- and multiple-hump solitary waves for $\nu = -4$ with velocities obtained from the crossings from bottom to top: (**a**) curve 1 of Fig. 2.13b; (**b**) and (**c**) curve 2 of Fig. 2.13b; (**d**) and (**e**) curve 4 of Fig. 2.13b; (**f**) and (**g**) curve 3 of Fig. 2.13b

2.6 Further Numerical Results and Computer Experiments

In Sects. 2.5.2 and 2.5.3 we have shown how slightly perturbed forms (up to, say, 20%) of the expected homoclinic solutions of the dynamical system (2.4–6) underlying (2.1) evolve to a (quasi)steady pattern (a single soliton, a two-bump soliton, a two-soliton bound state, etc.) and survives as a very long-lasting soliton-like structure. Let us now recall further numerical results carried out to emphasize the utility of the "aging" soliton concept, cross-check the time and space evolution of soliton solutions to (2.1) and study their collisions and formation of bound states when initial conditions depart from an expected solution of (2.1) [2.7, 2.13].

2.6.1 Evolutionary Features

As we have mentioned, soliton propagation in (2.1) occurs with radiation. Let us consider this phenomenon for $\nu > 0$ in further detail, following a recent report on the numerical integration of (2.1) for particular albeit significant values of its parameters [2.13]. We shall see that, as time proceeds, radiation packets move far away from the localized structure that can be assimilated to a true solitonic structure. Needless to be recalled, however, is that there is no energy conservation here, and hence radiation cannot spoil the "reality" of dissipative solutions of (2.1), for their very existence is due to the input–output energy balance much discussed above.

For convenience, we change the variables in (2.1) such that $x = (1/2a)\tilde{x}$, with positive parameter $a > 0$, and introduce the new variables B, D, Σ, and C as follows:

$$\nu = \varepsilon\frac{B}{4a^2}\,, \quad \gamma = \varepsilon\frac{D}{16a^4}\,, \quad \alpha = \frac{\varepsilon\Sigma}{4a^2}\,, \quad \beta = \frac{C}{8a^3}\,,$$

with $0 < \varepsilon \ll 1$. In this case we seek the solution of (2.1) with the help of an asymptotic procedure [2.13]. We take u depending upon the fast variable, η, and a slow time, T, such that

$$\eta_x = 1\,, \quad \eta_t = -V(T)\,, \quad T = \varepsilon t\,.$$

Then (2.1) becomes

$$Cu_{\eta\eta\eta} - Vu_\eta + 2a\; uu_\eta + \varepsilon[u_T + Bu_{\eta\eta} + Du_{\eta\eta\eta\eta} + \Sigma(uu_\eta)_\eta] = 0\,. \tag{2.41}$$

If the solution of (2.41) is sought in the form

$$u(\eta, T) = u_0(\eta, T) + \varepsilon u_1(\eta, T) + \ldots \tag{2.42}$$

and we are interested in studying localized solutions vanishing together with its derivatives at $|\eta| \to \infty$, then the leading-order solution is

$$u_0 = \frac{6C}{a}S^2\cosh^{-2}(S\eta)\,, \quad S = S(T)\,. \tag{2.43}$$

where $S(T)$ is a slowly varying function and $V = 4CS^2$. From the solvability condition to (2.41) at order ε it follows that

$$S'_T = \frac{8}{15}S^3(B + RS^2)\,, \tag{2.44}$$

where

$$R = \frac{4}{7}\left(6\frac{C}{a}\Sigma - 5D\right)\,. \tag{2.45}$$

The quantity $u_1(\eta, T)$ is a solution of a linear inhomogeneous ordinary differential equation (ODE); it will contain therefore free parameters depending on T, such as $S(T)$ in the leading-order problem. They are fixed when imposing the solvability condition in the next order problem. Higher-order approximations may be studied similarly.

Equation (2.45) allows us to find the variation of the amplitude and velocity of the soliton in time. The behavior of S depends on the signs of B and R and on the value of $S_0 \equiv S(T=0)$. Indeed, when both B and R are positive, S diverges, while when both are negative it will vanish. For $B < 0$, $R < 0$, the parameter S vanishes if $S_0 < \sqrt{-B/R}$, while it diverges if $S_0 > \sqrt{-B/R}$. In the case of divergence, the explicit expression of the blow-up time of the soliton is

$$T = \frac{4R}{15B^2} \log \left| \frac{S_0^2(B + Rb^2)}{S^2(B + RS_0^2)} \right| + \frac{4}{15BS_0^2}. \tag{2.46}$$

Note that in this case the initial assumption about the slow evolution of the parameters defining the soliton is violated: near blow-up, the soliton changes too rapidly.

If there is no blow-up, S vanishes when T tends to infinity. The most interesting case occurs when $B > 0, R < 0$, when S tends to $\sqrt{-B/R}$ independent of S_0. Relation (2.44) may be directly integrated, giving the implicit dependence of S on T:

$$T = \frac{4R}{15B^2} \log \left| \frac{S_0^2(B + RS^2)}{S^2(B + RS_0^2)} \right| - \frac{4(S_0^2 - S^2)}{15BS^2S_0^2}. \tag{2.47}$$

The interesting feature of (2.47) is that it provides an analytical description of the time-dependent process for the selection of the solitary wave (2.43) imposed by the input–output energy balance in (2.1) or (2.41). For sufficiently large disturbances the diffusion term $(uu_{\tilde{x}})_{\tilde{x}}$ becomes dominant and leads to blow-up. Equation (2.44) does not allow us to capture this subtle effect, and hence the asymptotic procedure should be carried further to describe the phenomenon.

If we additionally assume $C = 2aD/\Sigma$, the asymptotic dissipative soliton (2.43) will tend to the exact traveling solitary wave solution of (2.1). Noteworthy is that this cannot be attained from infinitesimal initial disturbances.

The direct numerical integration of (2.1) or (2.41) permits to cross-check the analytical results described above. Let $k_c = 1/2a\sqrt{\nu/2\gamma}$ be the wave number corresponding to the most unstable linear mode in (2.1). The corresponding wavelength is $\lambda_c = 2\pi/k_c$. For the numerical calculations the length of the spatial domain was chosen to be $256\lambda_c$, i.e. rather long relative to the (localized) soliton size. At the same time, the number of discretization points was chosen to be 4096, i.e. λ_c is covered by 16 points. The latter ensured fair resolution of all solutions computed. Periodic boundary conditions were used.

The pseudospectral technique was employed for the spatial discretization and the Runge–Kutta fourth-order scheme for the time evolution. The time step was chosen to be 0.01. Tests with smaller time steps and better resolution did not really improve the results.

One can use a general localized initial condition (Gaussian). Since ν, γ and α are small, at the first stage, the dynamics of the initial conditions is expected to be governed by the standard part of the BKdV equation, and the solitons will be formed on this preliminary stage as the (local) balance between nonlinearity and dissipation operates. The selection process, due to the energy balance, is expected to appear at later times, when solitons are already formed.

For the parameter values $\varepsilon = 0.1, a = 1, B = 1 > 0, C = 1, D = 6/5$, $\Sigma = -2$, and $R = -72/7 < 0$, the asymptotic value S_∞ of $S(T)$ at $T \to \infty$, obtained from (2.43) and (2.45), is $S_\infty = \sqrt{-B/R} = 0.312$. The asymptotic amplitude of the solitary wave is $6CS_\infty^2/a = 0.583$, and its corresponding asymptotic velocity is $V = 4CS_\infty^2 = 0.389$. Let us see how selection of the dissipative soliton occurs from "below" when the magnitude of an initial Gaussian pulse is smaller than that of the eventually selected solitons (Fig. 2.16a) and from "above" when the selected soliton amplitude is smaller than that of the initial pulse magnitude (Fig. 2.16b). To make the evolving structures distinguishable, only part of the long spatial domain is shown. At $t \sim 120$ an initial Gaussian pulse, with magnitude $0.3 < 0.583$ and width 36, breaks into a train of three localized pulses aligned in decreasing magnitude. Due

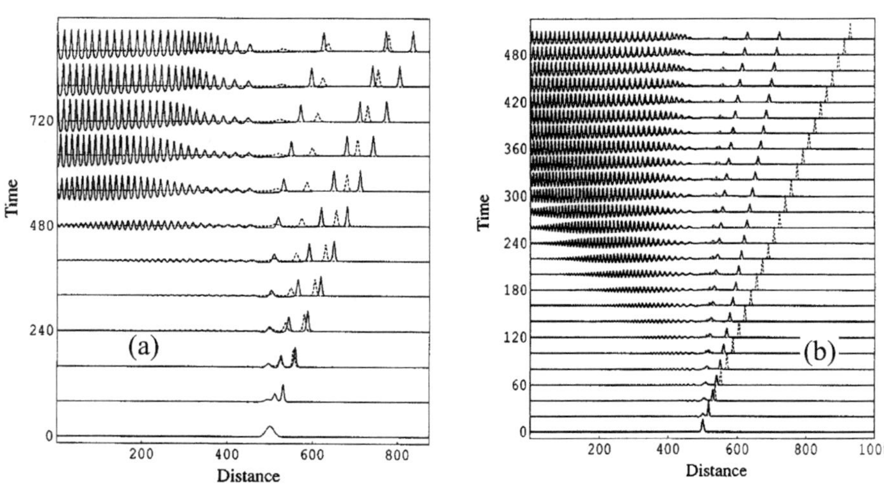

Fig. 2.16. (**a**) Selection of solitary wave from "below"; initial condition is a Gaussian profile with amplitude 0.3 and width 36 units; only a part of the long spatial domain is shown here. (**b**) Selection of solitary wave from "above"; initial condition is a Gaussian profile with amplitude 1 and width 12 units; only a part of the long spatial domain is shown here. The dotted (Sech-like) profiles indicate the expected evolution (and corresponding trajectories) according to the BKdV equation

to the smallness of ε, the influence of the dissipative terms in (2.1) is really small at this stage. This may be seen by comparing the solutions of (2.1), shown as solid lines, with the standard BKdV case, $B = D = \Sigma = 0$, depicted by dashed lines. At later stages, the initial pulse transforms into a train of solitons, all reaching the same amplitude (and velocity) with enough time. For nonzero ε, the amplitude and velocity of each solitary wave tend to the values 0.585 and 0.38, respectively, in agreement with the above-mentioned wave selection by the input–output energy balance. Each of three solitons of the standard BKdV equation continues propagation with its own amplitude and velocity. Unequal spacing between equally high crests reflects the original separation of the solitons in the standard BKdV stage when higher solitons travel faster. The radiation tail appears as a result of long-wave instability. These irregular waves cannot be described by the theory given above. Noteworthy, as evidence of the evolutionary stability of the dissipative solitons as traveling localized structures, and hence of their very long lifetime, is that the solitary waves have a higher velocity than the velocity of the growing wave packet. As a result, the solitons escape the destructive influence of the tail lagging behind.

The selection process realized from "above" is shown in Fig. 2.16b, when the initial Gaussian pulse has magnitude $1 > 0.583$ and width 12. Two equal solitary waves with common amplitude 0.585 and velocity 0.38 appear as a result of the *decrease* in magnitude of the initial pulse. The comparison with the standard BKdV case is shown by dashed lines. All features of the selection process are similar to the selection from "below" as, again, besides the fact that the two solitons have equal amplitude and velocity, it clearly appears that the distance between the (main) two-soliton bound state structure and the radiation packet considerably grows as time proceeds.

The localization of the disturbances in space has, indeed, a crucial role here. If we consider a conventional small-amplitude but large-band widespread noise as the initial condition, the growing long-wavelength disturbances will destroy the possible formation of the solitary waves. Finally, it is important to mention the significance of the periodic boundary conditions used in the computer experiment. After a sufficiently long time, solitons, moving faster than the irregular waves, will reach the end of the computational domain, will appear at the beginning of the domain, will further reach and collide with the irregular waves, and will ultimately be destroyed. This is the reason for the consideration of a long spatial interval, to allow the waves to evolve naturally and to separate, without collision thus forming a soliton bound state.

2.6.2 Numerical Collision Experiments

True collision and wall reflection events with the exception of overtaking collisions are only possible when dealing with a 3D extension of (2.1). Although such a 3D extension of the dissipation-modified BKdV equation has been provided [2.20] for surface tension gradient-driven instability and waves, here, however, we shall limit ourselves to a succinct account of overtaking collisions with various initial conditions [2.7]. Before embarking on such an account and just for the record, let us mention that one of the predictions [2.20] is that when two solitary waves move obliquely, forming an angle 2Ψ, their interaction leads to a mere shift in their trajectories whose value depends on Ψ. They also predicted that the sign of the deviation depends on whether $\Psi > \Psi^*$ or $\Psi < \Psi^*$, where Ψ^* is a value with which no change of trajectories occurs (neutral collisions). For $\Psi > \Psi^*$ there is the formation of a Mach–Russell third wave or stem moving phase locked with the postcollision waves. As a wall reflection can be thought of as the virtual collision of a wave with its mirror image, this prediction agrees well with the experiment (Fig. 1.2).

Restricting consideration to the simplest collision events, numerical experiments on soliton interactions or collisions, again for particular albeit significant values of the parameters in (2.1) and, now to reduce computer time, initial conditions not far from the expected solutions, were carried out by Christov and Velarde [2.7]. Take now

$$\nu = \frac{q}{4\alpha_1^3}, \quad \beta = \frac{1}{8\alpha_1^3}, \quad \gamma = \frac{q}{16\alpha_1^4}, \quad \alpha = 0, \tag{2.48}$$

where $\alpha_1 = 3$ and q is the control parameter. In Fig. 2.17 we present the overtaking collision of two soliton solutions of (2.1) with phase velocities $c = 10$ and $c = 5$, respectively. The result is very instructive in the sense that it confirms the expectation that on time intervals shorter than the dissipation time, q^{-1}, the interaction must be essentially similar to the standard BKdV equation. Indeed Fig. 2.17b shows the trajectories of the centers of traveling localized structures. The dotted lines are projected trajectories of stationary propagating bumps, here taken as hyperbolic secants (for short, called seches). As we are considering one-side traveling waves for the dissipation-modified BKdV equation [(2.1) with, however, $\alpha = 0$], this study is the natural extension to a nonconservative case of the pioneering work by Zabusky

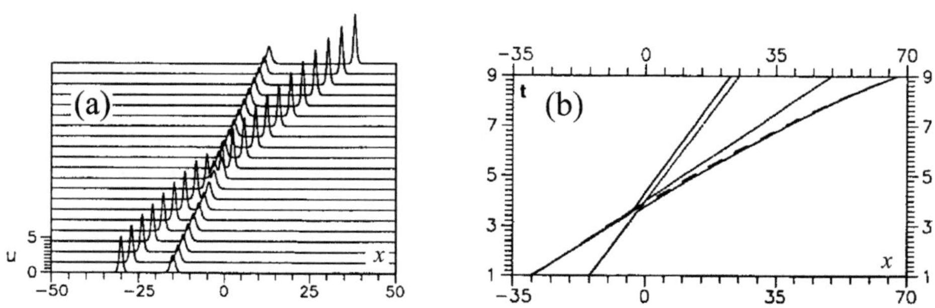

Fig. 2.17. Interaction of two seches with dimensionless velocities $c = 10$ and $c = 5$ in (2.1) with $q = 0.001$ (2.48). (**a**) Evolution of shape with time. (**b**) Trajectories of centers of solitons

and Kruskal on the standard BKdV equation [(2.1) with $\nu = \gamma = \alpha = 0$]. The dashed line is the trajectory of the accelerating sech. It is seen that the trajectory change or phase shift is essentially similar to the result found for the standard BKdV equation, and the smaller sech experiences the larger phase shift. The phase shift is "negative" in the sense that the structures are temporarily accelerated during the interaction and re-appear at positions farther ahead in the directions of their motion relative to the mass center than the position they would have reached were the interaction not to have taken place.

What is seen in Fig. 2.18 is the, earlier mentioned, slow "aging" of the soliton, decreasing in amplitude but keeping its sech-like shape. After long enough dimensionless time, $t \approx 8000$, the coherent structure approximately attains its terminal shape (Fig. 2.18a). This time interval is in fact almost an order of magnitude longer than the $q^{-1} = 1000$ scale and, hence, is practical infinity for experimental purposes. Just to be on the safe side, the calculations were carried out until $T = 15\,000$ (Fig. 2.18b) and the result was only marginally refined. Its convergence to a terminal state is beyond any doubt. It is interesting to note that the shape is virtually equal to a sech with phase velocity $c = 1.399$. The four digits are correct. Consequently, the amplitude is 0.69924, with the same accuracy of four digits. Thus, with the energy-balance mechanism being of order of $q = 0.001$, little quantitatively changes the shape of the solution, but it provides a qualitative effect on the spectrum of sech-like solutions. While for the standard BKdV equation the spectrum is continuous $c > 0$, and hence the equation has a one-parameter family of solitons, here we see again that one has a single value $c = 1.399$ imposed by the steady input–output energy balance.

In Fig. 2.19a–c we present the overtaking of a small soliton of initial phase velocity $c = 2$ by a bigger one with $c = 10$ for $q = 0.1$. Taking

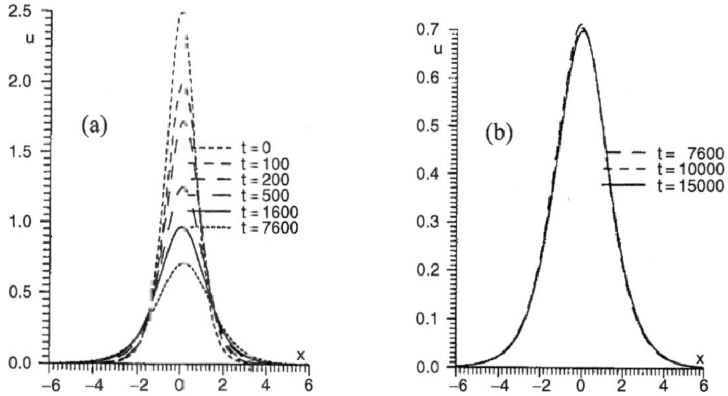

Fig. 2.18. "Aging" of the solitary wave with velocity $c = 5$ in (2.1) for $q = 0.001$ (2.48): (**a**) time evolution; (**b**) slight asymmetry of the "aged" soliton from the initially symmetric sech profile

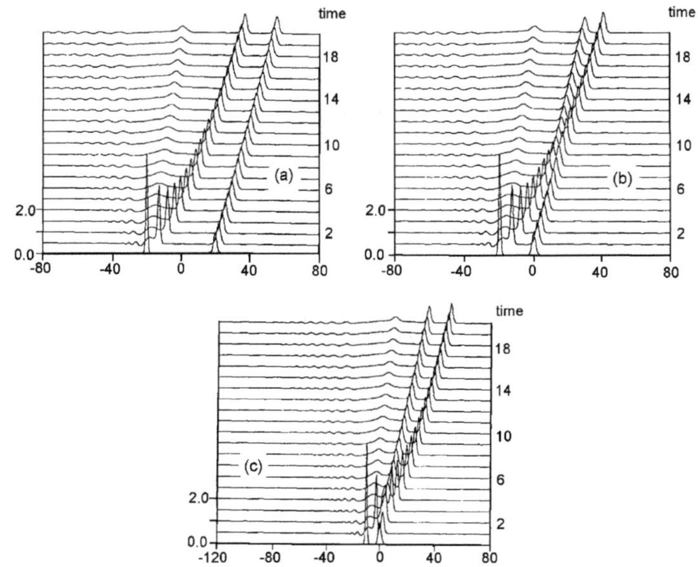

Fig. 2.19. Overtaking interactions of solitary waves with $c_1 = 10$ and $c_2 = 2$ in (2.1) for $q = 0.1$. (**a**) Large, (**b**) moderate, and (**c**) small initial separation

such a large value for the parameter q allows us to observe significant effects within a shorter dimensionless time (and still a reasonable computational time). In Fig. 2.19a the two solitons are far enough separated and within the characteristic time $q^{-1} = 0.1^{-1} = 10$ that they could not really collide, save for the formation of a two-soliton bound state. Because of their "aging", for the chosen parameter values they attain the permanent shape (and hence the terminal phase velocity) before they become close enough to interact. The amplitude of the large soliton diminishes by half for a dimensionless time equal to one! Approximately one half is the reduction of the phase velocity. It is not an effect of order $q = 0.1$, which is another facet of the singular asymptotic nature of the problem under consideration. Much less violent is the evolution of the smaller soliton, which is much closer to the stationary solution of (2.1) for this particular value of q.

Another computer experiment is presented in Fig. 2.19b, where the initial separation of the solitons is reduced by half relative to Fig. 2.19a. This distance is short enough, and even after the faster crest has lost half of its velocity, it is still able to get close enough to the slower one so that both feel each other. However, they cannot pass through each other. They reach a distance when the "repelling" part of the interaction potential becomes significant, and the underdog slows its motion even further due to the repulsive force from the wavy tail. The energy of the former is transferred to the latter, which grows and becomes somewhat faster. Then the two structures slowly separate to a distance compatible with one of the local minima of the potential of interaction and one structure "nests" in the local minimum of the other. By then, they are already of permanent shape and move precisely with the same phase velocity. Thus, as earlier discussed from the qualitative viewpoint, they form a bound state which is metastable in the sense that it can be broken only by a strong enough disturbance, but not by an infinitesimal one. Further reduction in the initial separation reveals a genuine clash of the two structures, during which they form a single bump (see the line corresponding to $t = 2$ in Fig. 2.19c). Hereafter the scenario is the same as the case in Fig. 2.19b. Keeping track of the evolution of the signal in Fig. 2.19c, one can see that the two large bumps eventually attain the same terminal shape and phase velocity and form a bound state. The third bump is growing on a much slower time scale than the one presented in the figure. It eventually reaches the same terminal velocity and shape after a longer time; it lags far behind the other two, and hence no triplet bound state is formed (Fig. 2.19c).

In the numerical experiments, one can see that the total amplitude of the compound signal never exceeds the amplitude of the larger sech. From the analytical two-soliton solution of the standard BKdV equation it follows that the amplitude of the compound signal is smaller than the sum of the amplitudes of the two initial solitons, but it is still considerably larger than the amplitude of the bigger soliton. Thus this inelastic facet of the interaction of solitons is significantly exaggerated by the presence of dissipative terms. In visual, mechanical, terms dissipative solitons appear much more like clay balls than hard spheres or billiard balls.

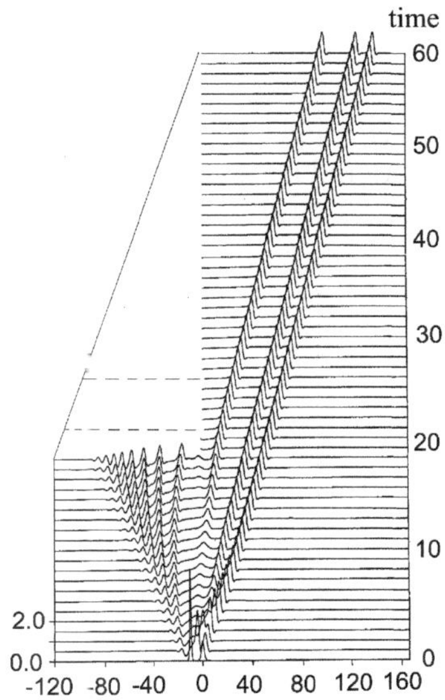

Fig. 2.20. A bound state formed with three crests of equal size following the collision of two rather different solitons ($q = 2$; 2.48)

After overtaking, different signals are excited, and part of them form bound states with different degrees of "tightness". Especially instructive is the portion of the signal shown in Fig. 2.20, where one can see that following the overtaking collision of two solitons of dimensionless velocities $c = 10$ and $c = 2$, respectively, a three-hump (triplet) bound state is formed.

2.7 Salient Features of Dissipative Solitons

In conclusion we can safely claim that the solitary wave solutions of the dissipation-modified BKdV equation (2.1) are real, albeit imperfect, solitons aging with time and experiencing collisions with variable degrees of inelasticity. Accordingly, we feel that our concept of dissipative (and "aging") solitons extends the Zabusky and Kruskal [2.7, 2.30] definition to waves that are not of rigorously permanent shape but that survive while aging with a long lifetime, traveling steadily, in accordance with the given input–output energy balance. There is nothing wrong in doing this, because even the proton is known to be a long-lived, though presumably unstable, particle. The only problem is the time scale, because physically speaking there is no "infinite" time for which the waves (or particles) could attain the truly permanent shapes. What we claim is that in times shorter than "practical infinity" (the latter formally represented by the inverse of the small parameter q^{-1}), the interactions of solitary waves (or wave crests in wave trains) is solitonic. A difference from the study of Zabusky and Kruskal [2.30] is that here a recurrence of the initial state is impossible due to the dissipative nature of (2.1). In experimental situations, both in the lab and in nature, the experimental time is never infinite but long enough for the observation of steadily traveling structures. Accordingly, long-lived structures of the above-discussed type whose "aging" is slow enough can be considered as (quasi)steady and imperfect, long lasting metastable structures [2.2, 2.7, 2.19]. After all, molecules when they obey the van der Waals equation of state are still called molecules, albeit imperfect ones. The van der Waals theory of imperfect molecules provides a better metaphor (or model) for reality than the perfect gas model.

3. Self-Organization in a Long Josephson Junction

3.1 Introduction and Motivation

The discovery of the Josephson effect stimulated the development of superconductor microelectronics with a huge number of both theoretical and experimental papers devoted to it. Such interest is caused by its great technological potential and the value of the Josephson effect in information processing, generation of ultra-high frequency oscillators, fast switching devices, mixers, detectors, amplifiers and so on (see, e.g., [3.1–3,3.7]).

In its simplest form the Josephson junction is a sandwich system of two superconducting films separated by a thin (about 10^{-7} cm) layer of insulator (Fig. 3.1). The Josephson effect is a quantum tunneling effect where two superconductors sufficiently close to each overcome the insulator barrier and current goes through. When the superconductors are located sufficiently close to each other, their wave functions overlap in the insulator region. Therefore, the supercurrent appearing in the system depends on the phase difference φ of the wave functions at the two sides of the insulating layer.

There is a significant difference between short (or small) and long Josephson junctions. The distinguishing measure of this classification is the so-called Josephson penetration depth, λ_J, connected with the phase difference, $\varphi \sim \exp(-x\lambda_J)$, where x is the spatial coordinate. If the geometric sizes of the junction is less than λ_J, then the contact is called small (in practical terms a point), otherwise it is long. We shall study a long Josephson junction

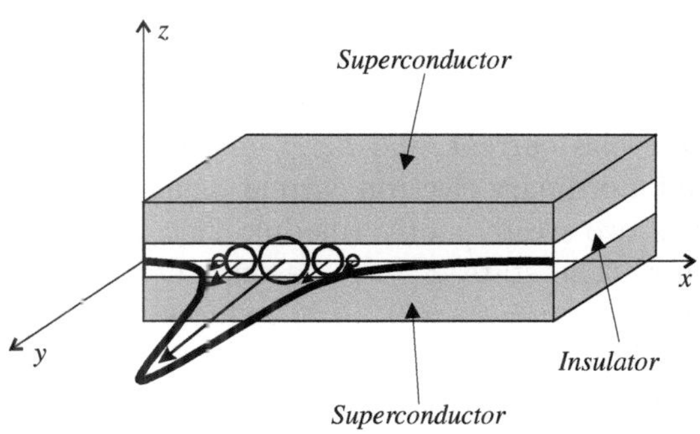

Fig. 3.1. Schematic of a long Josephson junction

(or transmission line). Let us summarize the physical processes occurring in such a junction. Figure 3.1 shows circulating Josephson supercurrents (the so-called Josephson vortices) and the corresponding distribution of magnetic field oriented along the y-axis and localized within the insulating layer. The magnitude of this field is $\Phi_0 \varphi_x$, with $\Phi_0 = \hbar/2l = 2.07 \times 10^{-15} Wb$ being the quantized magnetic flux. The electric field is oriented along the z-axis and its value is φ_t. If the phase difference φ is varying from 0 to 2π then a pulse or soliton (for φ_x) appears in the contact (Fig. 3.1). It can move along the contact, transferring a magnetic flux quantum. Accordingly, the soliton in a long Josephson junction is often called a "fluxon". Besides such one-soliton states, long Josephson junctions allow more complex multisoliton states or multihump solitons which are a form of soliton bound state. Such solitons may be characterized by topological charge $n = (2\pi)^{-1}[\varphi(t, +\infty) - \varphi(t, -\infty)]$. This quantity takes only integer values. In particular, an isolated soliton has $n = +1$, while a two-soliton state has $n = +2$ and so on.

In this chapter we consider the self-organization occurring in dissipative long Josephson junctions (LJJ). In particular, we consider the dynamics of kinks and solitons in the perturbed sine–Gordon (SG) equation modeling an LJJ. We analytically show the existence in the SG equation of kink solutions corresponding to the propagation of topological solitons with high (higher than unity) arbitrary topological charges. We also study the evolutionary stability of these dynamical structures. On the basis of these results we provide a qualitative explanation of possible current–voltage characteristics of LJJ.

3.2 The Perturbed Sine–Gordon Equation

Magnetic flux quanta (fluxons) in LJJ in one dimension can be described by the perturbed sine-Gordon (PSG)

$$\varphi_{xx} - \varphi_{tt} = \sin\varphi + \alpha\varphi_t - \beta\varphi_{xxt} - \gamma \,, \tag{3.1}$$

where $\varphi(x,t)$ is the quantum phase difference across the junction; x is distance going away from the junction, normalized to the Josephson penetration depth; t is the time normalized to the inverse of the Josephson plasma frequency; γ is the normalized bias current; and $\beta\varphi_{xxt}$ and $\alpha\varphi_t$ are dissipative components responsible for ordinary electron current along and across the junction. Equation (3.1) has two steady states (modulo 2π): $\varphi = \varphi_1$ and $\varphi = \pi - \varphi_1$, with $\varphi_1 = \arcsin\gamma$. The state $\varphi = \varphi_1$ is linearly stable but $\varphi = \pi - \varphi_1$ is unstable. When $\alpha = \beta = \gamma = 0$, (3.1) reduces to the standard form of the SG equation. The SG equation has solutions in the form of 2π kinks (or antikinks), topological solitons with $+1$ (-1) charge corresponding to jumps between the two stable states $\varphi = 0$ and $\varphi = 2\pi$ ($\varphi = 2\pi$ and $\varphi = 0$). Solitons with higher topological charges are not admitted by the SG equation. The situation differs in the case of the PSG equation. For (3.1), solutions of $2\pi n$-kink form ($n = 2, 3, \dots$), corresponding to transitions between $\varphi = \varphi_1$ and $\varphi = \varphi_1 + 2\pi n$ stable states, are possible. These solutions determine topological solitons with charges higher than unity. Such solitons have been studied numerically, experimentally and analytically by various authors [3.4–6]. Homoclinic trajectories of different types which exist in the phase space of this system have been studied. The evolution of multihump solitons (solitons with topological charge higher than unity) was then numerically investigated, providing quantitative values of the $I - V$ characteristics.

3.3 Bifurcation Diagram of Homoclinic Trajectories

The model equation (3.1) is analyzed qualitatively in Sect. 3.6 for a class of solutions of the form $\varphi(-x + ct)$. These solutions are described by the nonlinear dynamical system

$$\begin{gathered}\dot{\varphi} = y\,, \quad \dot{y} = z\,, \\ \beta c\dot{z} = -(1 - c^2)z + \alpha c y + \sin\varphi - \gamma\,,\end{gathered} \tag{3.2}$$

where the dot denotes differentiation with respect to $\xi = -x+ct$. In the phase space $\mathbf{R}^3$ of the system (3.2) there exist alternating saddle steady points of two types: O_1^n ($\varphi = \varphi_1 + 2\pi n$, $y = z = 0$) and O_2^n ($\varphi = \pi - \varphi_1 + 2\pi n$, $y = z = 0$), with $\varphi_1 \equiv \arcsin\gamma, n = 0, \pm 1, \pm 2, \dots$ The stationary points O_1^n are saddle-foci or saddles with a 1D unstable manifold $W^u(O_1^n)$ and a 2D stable manifold $W^s(O_1^n)$ (Fig. 3.2a). The manifold $W^u(O_1^n)$ consists of the point O_1^n and two outgoing trajectories W_1^u and W_2^u. The stationary points O_2^n may be either saddle-foci or saddles with 1D stable manifold $W^s(O_2^n)$ and 2D unstable manifold $W^u(O_2^n)$ (Fig. 3.2b). The manifold $W^s(O_2^n)$ consists of the point O_2^n and two stable trajectories W_1^s and W_2^s. Due to the cylindrical phase space of (3.2), a homoclinic trajectory may wind a cylinder an arbitrary number of times, n, before closing. For $n = 1$ (Fig. 3.2a,b) this trajectory

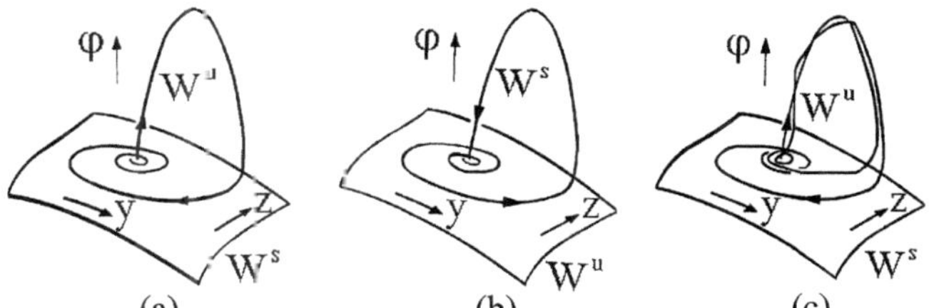

Fig. 3.2. Homoclinic trajectories (orbits) exhibiting stable and unstable manifolds: (**a**) single-loop orbit formed by the separatrix of the saddle point O_1^n, (**b**) single-loop orbit formed by the separatrix of the saddle point O_2^n, (**c**) double-loop orbit formed by the separatrix of the saddle point O_1^n

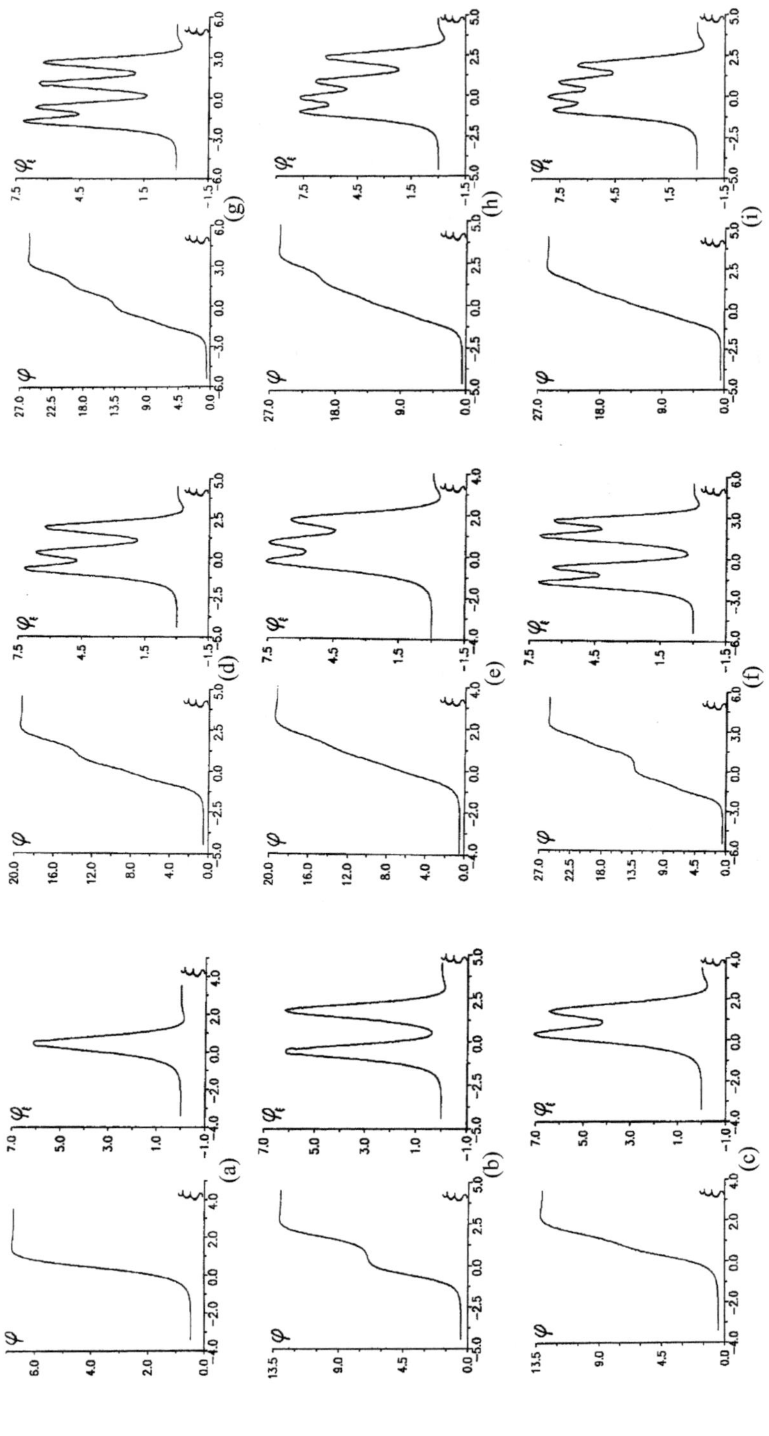

Fig. 3.3. Profiles of $2\pi n$ kinks and n-hump solitons

corresponds either to an ordinary jump of the φ profile or to a soliton for φ_t. The waves are monotonic in the case of a saddle state, or have an oscillatory tail if the steady state is a saddle-focus. For $n > 1$, we have solitons with topological charge n or an n-hump soliton (a train of solitons) ($n = 2$ in Figs. 3.2c and 3.2b).

We consider different types of stationary waves at constant α and β (e.g. $\beta \sim 10^{-2}$) and investigate the dependence of velocity, c, on the parameter γ. For this purpose it suffices to analytically construct, in the 3D phase space of (3.2), surfaces that are intersected by the trajectories of this system transversally in one direction. These surfaces permit us to distinguish in the phase space some regions ("tunnels") inside which the stable and unstable trajectories of the saddle point must be located. The analytical investigation of the location of these tunnels yields the existence of n-envelope homoclinic trajectories and, consequently, of n-hump solitons in (3.1) (see Sect. 3.5 for further details).

The analytical proof of the existence of n-hump solitons guided us in the computer plot of c_n versus γ. Results of this study are presented in Fig. 3.4, where lines of multihump solitons comprising a different number of peaks in the train are depicted on the (γ, c) plane. The lines designated by Π^n and P^n correspond to different types of homoclinic trajectories and, consequently, to different types of solitons. The homoclinic trajectories of (3.2) with a stable 2D manifold (Figs. 3.2a,c) correspond to the Π^n curves, while those with unstable 2D manifold (Fig. 3.2b) correspond to the P^n curves. Numerical integration of the initial system (3.1) has shown that the solitons corresponding to the P^n curves are evolutionarily unstable, and, therefore, they are not considered here. There is only one Π^1 curve on the (c, γ) plane (Fig. 3.4b) that corresponds to 1-hump solitons. The situation is quite different in the case of multihump solitons: A family of an infinite number of lines corresponds to each type of multihump solitons. For instance, Fig. 3.4b depicts two curves, Π^4 and Π^4_1, which belong to such a family and correspond

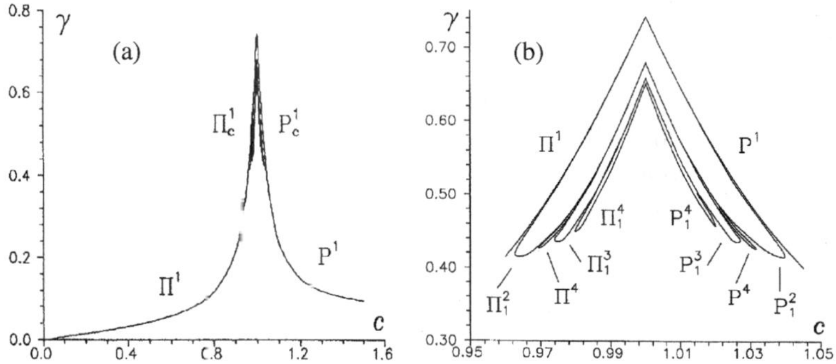

Fig. 3.4. (a) Elements of base bifurcation diagram ($\alpha = 0.05, \beta = 0.02$). (**b**) enlargement around the peak in (**a**)

to 4-hump solitons (Figs. 3.2f–i). They are parabolic curves between which the lines, each corresponding to a different number n (Fig. 3.4b), are located, with the Π_1^3 curve describing 3-hump solitons (Fig. 3.2d,e) seated between the Π^4 and Π_1^4 lines. The curves obtained for multihump solitons are given in Fig. 3.4. These curves are enlarged in Fig. 3.4b as they are too close to Π^1 (Fig. 3.4a). The kinks and soliton profiles corresponding to the curves depicted in Fig. 3.4 are presented in Fig. 3.2. Note that Fig. 3.4 illustrates the curves of the set Π only for $c \leq 1$ and the curves of the set P for $c \geq 1$. Actually, the curves of the set Π can "extend" into the region $c > 1$, and those of the set P into the region $c < 1$. However, on the (γ, c) plane they are very close to the line $c = 1$ and have nontrivial geometric structure. Therefore, we shall only discuss the behavior of the curves of the sets Π and P in the neighborhood of $c = 1$ (see Sect. 3.5).

3.4 Current–Voltage Characteristics of Long Josephson Junctions

The potential difference, $v(x,t)$, between the superconducting electrodes varies according to the value of $\varphi_t(x,t)$. If (3.1) has a solution in the form of a kink, then the quantity $\varphi_t(\xi)$ has a soliton form. Although the kink propagating through the junction gives rise to voltage pulsations, the time-averaged voltage is constant and is

$$v = \frac{c \cdot 2\pi n}{l}, \tag{3.3}$$

where l is the dimensionless length of the junction. From (3.3) it follows that to determine the $I - V$ characteristics of the junction, i.e. the dependence of mean voltage, v, on the bias current, γ, one needs to know the soliton velocity, c. For the solitons (kinks) to fix the $I - V$ characteristics of a junction they must be stable, in some sense. Under the term of stability we again consider here the evolutionary behavior of solitons with respect to a rather broad class of initial distributions $\varphi(0,x)$. As the basis for the initial distribution $\varphi(0,x)$ we take the solution $\varphi(\xi)$ found in the numerical simulation of the self-similar system (3.2) for the bifurcation values of parameters corresponding to the propagation of an n-hump soliton in an infinitely long domain. Various additional perturbations are imposed on the solution $\varphi(\xi)$ in order to find $\varphi(0,x)$. Numerical integration of (3.1) reveals that single-hump solitons corresponding to the Π^1 points are stable, while among multihump solitons only those corresponding to the "lower" portions of the parabolas are stable (Fig. 3.4b), in agreement with known results about soliton stability for almost zero α and β. As an example, the evolution of two initial distributions $\varphi(0,x)$ in the (x,t,φ) space is shown in Fig. 3.5. The initial distribution $\varphi(0,x)$ is a perturbation against the background of the

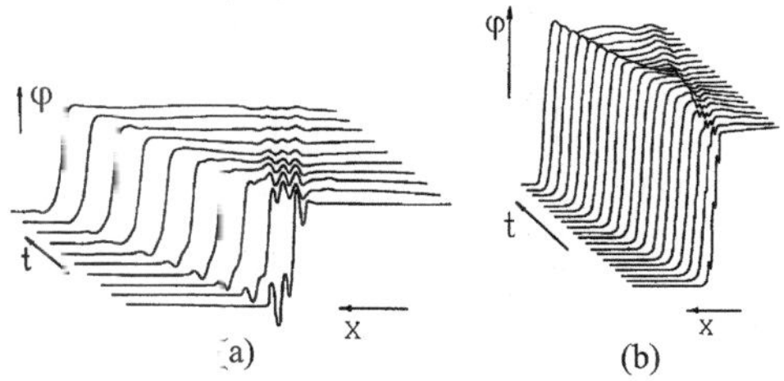

Fig. 3.5. Evolution of two initial distributions ($\alpha = 0.05, \beta = 0.02, \gamma = 12$). (**a**) $n = 1$; (**b**) $n = 3$

traveling front with $n = 1$ (Fig. 3.5a) and $n = 3$ (Fig. 3.5b). Clearly, these perturbations evolve to the profiles of the corresponding kinks. Numerical investigation shows that the kinks (here also termed solitons) are stable not only to small perturbations, like in the previous case, but also to large disturbances. For example, a 5-hump soliton is stable even to the perturbation shown in Fig. 3.6, which may be treated as a train of 2-hump and 3-hump solitons. The dynamics of such a perturbation evolving towards a 5-hump soliton greatly resembles that of perturbations in the dissipation-modified Boussinesq–Korteweg–de Vries equation, as shown in Chap. 2.

Results of the qualitative analysis of (3.2) that have been verified by computer experiment lead to the following conclusions about the $I - V$ characteristics of a LJJ: There are two different diagrams of $I - V$ characteristics. If the value of β is close to zero, then there is only one Π^1 curve on the (c, γ) plane and, consequently, the $I - V$ characteristics of the LJJ consist of one branch (Fig. 3.7a). The increase in the parameter β leads to the formation on the (c, γ) plane of a family of curves corresponding to multihump solitons (Fig. 3.4b). This is accompanied by the complication of the $I - V$ characteristics, which now contains an infinite number of branches. A fragment of $I - V$ characteristics consisting of four branches is depicted in Fig. 3.7b.

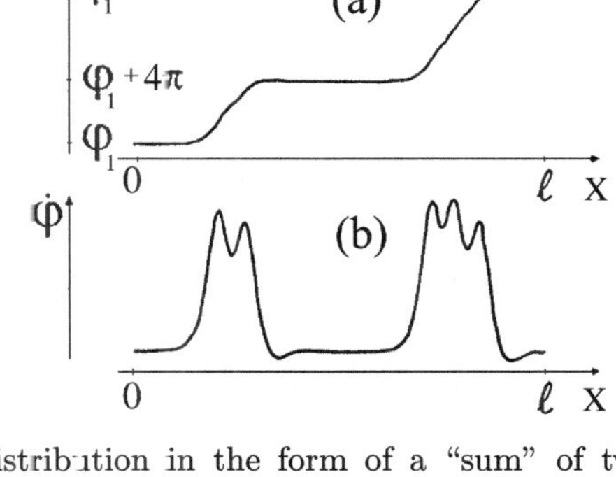

Fig. 3.6. Initial distribution in the form of a "sum" of two kinks ($\alpha = 0.05$, $\beta = 0.02, \gamma = 0.454$). (**a**) φ, (**b**) $\dot{\varphi}$

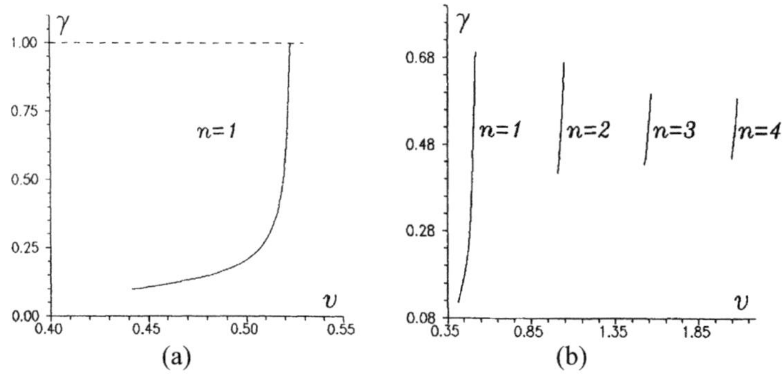

Fig. 3.7. Elements of $I - V$ characteristics of LJJ

Note that each of the voltage branches corresponding to multihump solitons ($n = 2, 3, 4$) comprises an infinite number of curves. However, because these bifurcation lines have very close values of c (Fig. 3.4a), the respective portions of the $I - V$ characteristics are so close to each other that they are almost indistinguishable in the diagram. For example, the $I - V$ branch formed by the Π^4 curve (Fig. 3.4b) is so close to the $I - V$ branch corresponding to the Π_1^4 curve that they almost coincide in Fig. 3.7b. Besides, the mean voltage, v, on the junction grows with increasing n but has an upper limit because only solitons comparable in size with the junction of finite length l are acceptable. At the same time, the range of variation of the bias current decreases as n increases. All these features of the $I - V$ characteristics agree with the $I - V$ characteristics of a real Josephson junction.

3.5 Bifurcation Diagram in the Neighborhood of $c = 1$

3.5.1 Spiral-Like Bifurcation Structures

Let us investigate the structure of the bifurcation set, which corresponds to the heteroclinic trajectories $O_1^0 \to O_1^n$ ($n = 1, 2, \dots$) in the region $c \sim 1$. For this purpose we numerically integrate (3.2) using an algorithm based on the analysis of the relative position of a 1D unstable separatrix and a family of contactless surfaces in the phase space $\mathbf{R}^3$. It appears that all elements of the bifurcation set in the (c, γ) plane, in the neighborhood $c = 1$, look like spirals near the foci at points A^j ($c = 1, \gamma = \gamma_j$) ($j = 1, 2, \dots$), (Figs. 3.8–11). In Fig. 3.4 it appears that in the (c, γ) plane all elements (except Π^1) outside the neighborhood $c = 1$ have a periodic shape. These elements initiate and terminate at points A^i and A^j, respectively. Moreover, there are two cases: (i) $i \neq j$ (for example, $\Pi_1^2, \Pi_1^3, \Pi_1^4$), and (ii) $i = j$ (Π^4, Π^2), as illustrated in Figs. 3.4 and 3.10.

The elements of the bifurcation set P in the (c, γ) plane have a similar behavior. They also initiate and terminate by the spiral-like parts with foci

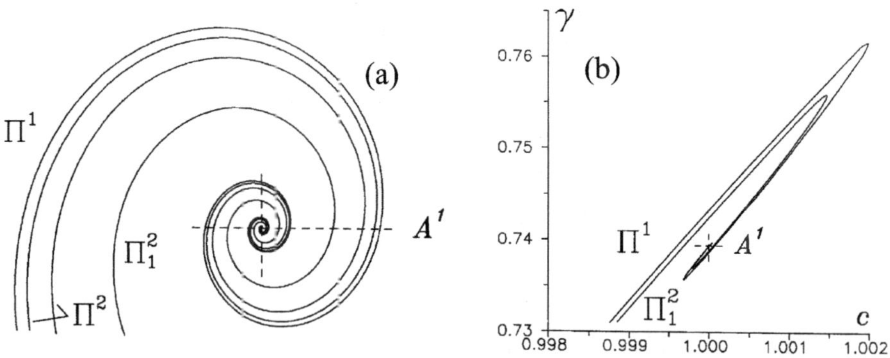

Fig. 3.8. Part of the bifurcation set Π in the neighborhood of A^1: (**a**) qualitative picture (**b**) numerical result ($\alpha = 0.05, \beta = 0.02$)

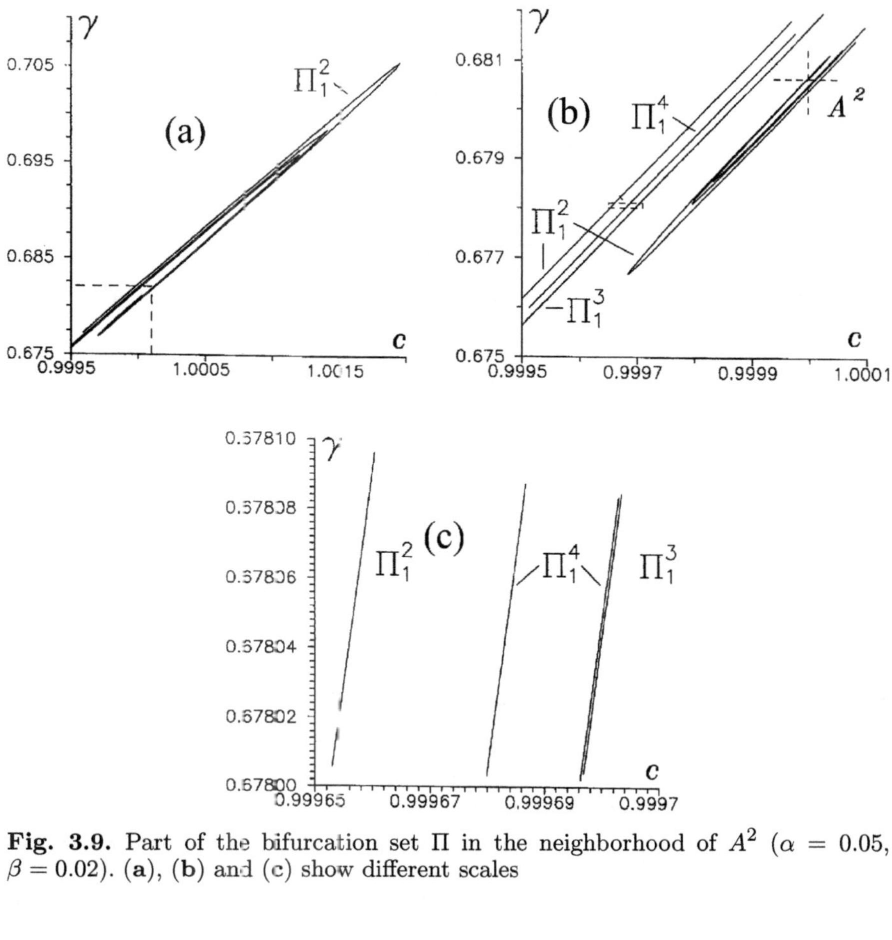

Fig. 3.9. Part of the bifurcation set Π in the neighborhood of A^2 ($\alpha = 0.05$, $\beta = 0.02$). (**a**), (**b**) and (**c**) show different scales

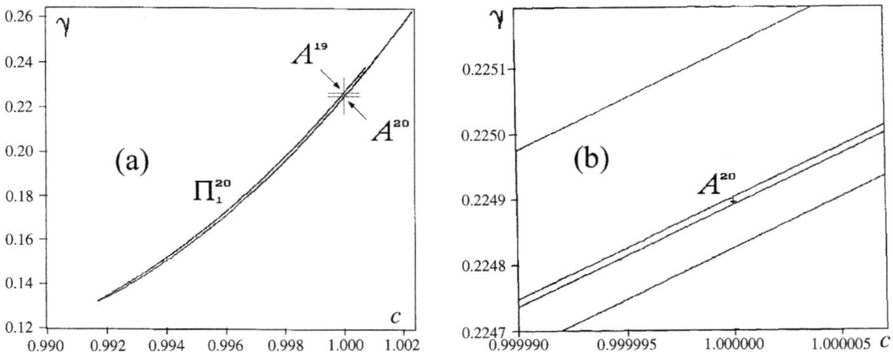

Fig. 3.10. The Π^{20} line. (**a**) Overview; (**b**) the part of the *lower* branch of Π_1^{20} in the neighborhood of A^{20} ($\alpha = 0.05, \beta = 0.02$)

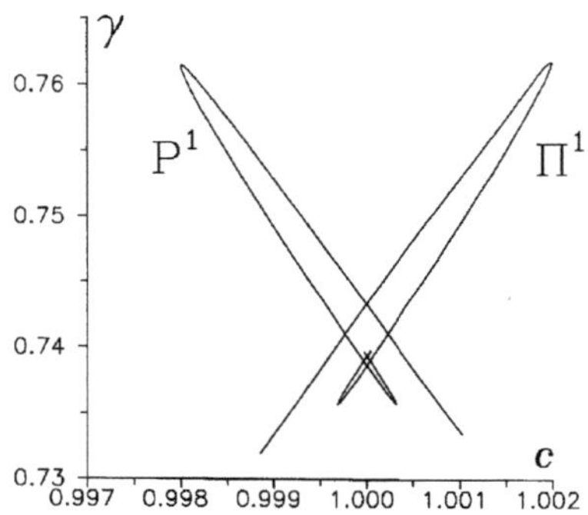

Fig. 3.11. Intersections of Π^1 and P^1 in the neighborhood of A^1

in the same point, A^j, as the corresponding elements from Π. This occurs as (3.2) is reversible when $c = 1$ (i.e. invariant under the transformation: $\varphi \to \varphi - \pi$, $\xi \to -\xi$, $z \to -z$). Consequently, if there exists a trajectory connecting the points O_1^0 and O_1^n, then, for the same values of the parameters, there also exists a trajectory connecting the points O_2^0 and O_2^{-n}.

Thus, according to the properties of the spirals, the corresponding bifurcation set elements (i.e. with the same indices) intersect on the (c, γ) plane at $c = 1$ an infinite number of times. For example, Fig. 3.11 shows the intersections of Π^1 and P^1 in the neighborhood of the point A^1.

3.5.2 Heteroclinic Contours

To understand why the bifurcation lines corresponding to n-loop homoclinic orbits with different n converge to a single point A^i (e.g. to A^1: $n = 1, 2$; to A^2: $n = 2, 3, 4$; to A^3: $n = 3, 4$; and so on), we consider the phase space $\mathbf{R}^3$ of (3.2) for the parameter values corresponding to the point A^i. In Fig. 3.12 for $i = 1$ the structure of $\mathbf{R}^3$ is shown. Here, and in what follows we denote,

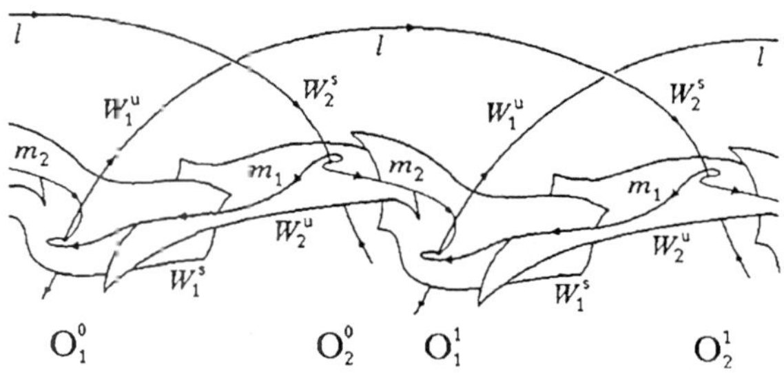

Fig. 3.12. Phase space G of (3.2) for the parameter values corresponding to A^1

for convenience, the 2D manifolds $W^s(O_1^u)$ and $W^u(O_2^n)$ by W_1^s and W_2^u and call them by 2D separatrices

The "intersection" of the unstable 1D trajectory W_1^u of the steady state O_1^0 and the stable 1D trajectory W_2^s of the steady state O_2^1 yields in $\mathbf{R}^3$ the heteroclinic trajectory 1: $O_1^0 \to O_2^1$ ($W_1^u \cap W_2^s = 1$). The "intersection" of the stable 2D separatrix W_1^s of the steady state O_1^1 and the unstable 2D separatrix W_2^u of O_2^1 yields a trajectory m_1: $O_2^1 \to O_1^1$ ($W_1^s \cap W_2^s = m_1$). Thus the intersection of the stable 2D separatrix W_1^s of O_1^2 and the unstable 2D separatrix W_2^u of O_2^1 yields a trajectory m_2: $O_2^1 \to O_1^2$ ($W_1^s \cap W_2^s = m_2$) (for simplicity no new index for W is introduced, because for a cylindrical phase space $\mathbf{R}^2 \times \mathbf{S}^1$ such an index is meaningless).

Let us consider the shapes of the 1-loop homoclinic orbit and the 2-loop homoclinic orbit corresponding to the parameters of the Π^1 and Π^2 lines, respectively, in the vicinity of A^1. In this case the single-loop homoclinic orbit first enters (in $\mathbf{R}^3$) in the neighborhood of the l trajectory, and then continues in the neighborhood of the m_1 trajectory (Figs 3.12 and 3.13a). Similarly, the 2-loop homoclinic orbit enters in the neighborhood of the

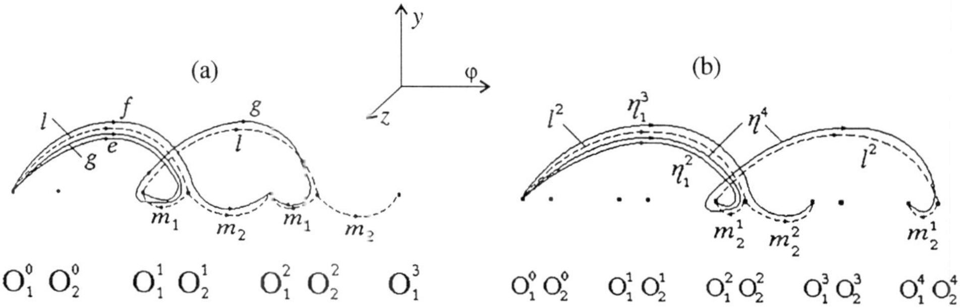

Fig. 3.13. Heteroclinic trajectories of (3.2) in the phase space G (qualitative). (**a**) Trajectories e, f, g are 1-, 2-, 2-loop homoclinic orbits for the parameters of Π^1, Π_1^2, Π^2, respectively, in the neighborhood of A^1. (**b**) Trajectories l^2, m_1^2, m_2^2, corresponding to parameters of the point A^2. $\eta_1^2, \eta_1^3, \eta^4$ are 2-, 3-, 4-loop homoclinic orbits for the parameters of Π_1^2, Π_1^3, Π^4, respectively, in the neighborhood of A^2

l trajectory, and continues in the neighborhood of the m_2 trajectory (note, that the l, m_1 and m_2 trajectories only exist for the parameter values of point A^1). Thus, the single-loop homoclinic orbit in the neighborhood of A^1 may be described schematically as $O_1^0 \to O_2^1 \to O_1^1$, and the 2-loop homoclinic orbit corresponding to the Π_1^2 line as $O_1^0 \to O_2^1 \to O_1^2$. Similarly, the 2-loop homoclinic orbit corresponding to the Π^2 line in (c, γ) plane, may be described as $O_1^0 \to O_2^1 \to O_1^1 \to O_2^2 \to O_1^2$ (the Π^2 line is only shown qualitatively in Fig. 3.8a because it lies too close to Π^1). In the neighborhood of A^2 this scheme appears as follows (Fig. 3.13b): for Π_1^2, $O_1^0 \to O_2^2 \to O_1^2$; for Π_1^3, $O_1^0 \to O_2^2 \to O_1^3$; for Π^4, $O_1^0 \to O_2^2 \to O_1^2 \to O_2^4 \to O_1^4$, etc. One may construct the phase-space structure for parameters of (3.2) on the bifurcation set in a neighborhood of the other points A^j by replacing $O_{1,2}^1$ with $O_{1,2}^{j+1}$ and adding between $O_{1,2}^1$ and $O_{1,2}^{j+1}$ the appropriate number of points $O_{1,2}^k$ $(k = 1, \dots, j)$ (Fig. 3.13).

Let us investigate the shape of the n-hump soliton solutions to (3.2) [and, consequently, the profile of the $2\pi n$-kink solutions to (3.1)] in the neighborhood of the point A^j. Let $j = 2$. The bifurcation lines in the (c, γ) plane, corresponding to 2-, 3-, 4-hump soliton solutions converge towards this point. Their profiles are depicted in Fig. 3.14. In the neighborhood of the point A^j a multi-loop homoclinic orbit "remembers" its "original" structure in A^j. Hence, in $\mathbf{R}^3$ a trajectory occurs "near" l and m_1 or m_2, which exist for the parameter values of point A^j in the (c, γ) plane. Let $\eta_1^2(\xi) \equiv \{\varphi(\xi), y(\xi), z(\xi)\}$ be a 2-loop homoclinic orbit corresponding to Π_1^2. For ξ increasing from $-\infty$, η_1^2 emerges from O_1^0 $(\varphi = \varphi_1, y_1 = 0, z_1 = 0)$, goes to the neighborhood of O_2^2 $(\varphi = 5\pi - \varphi_1, y_1 = 0, z_1 = 0)$, slows down its motion (the slower the motion, the closer the parameters of the Π_1^2 line lie to the A^2 parameters, O_i^j

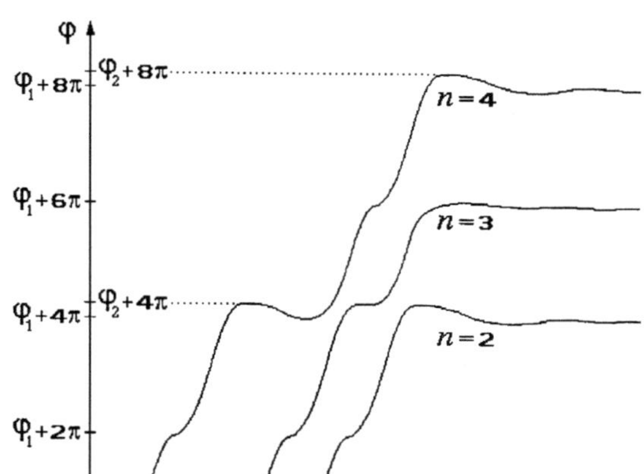

Fig. 3.14. Shape of the $2\pi n$-kink solutions to (3.1) $(n = 2, 3, 4)$ corresponding, respectively, to the η_1^2, η_1^3, η^4 solutions of (3.2) in the neighborhood of A^2

being a saddle or saddle-focus critical point), then returns to O_1^2 as $\xi \to +\infty$. A 3-loop homoclinic orbit $\eta_1^3(\xi)$ corresponding to the parameters of the Π_1^3 line in the neighborhood of A^2, also emerges from O_1^0, again goes to the neighborhood of the point O_2^2, decelerates, and then approaches O_1^3 as $\xi \to +\infty$. Thus this 3-loop homoclinic orbit looks like a $2\pi(2+1)$-kink solution to (3.1). $\eta_1^4(\xi)$ (a 4-loop homoclinic orbit corresponding to the Π^4 line) consists of two 2-loop homoclinic orbits similar to η_1^2. Its behavior in the region $O_1^0 \to O_1^2$ is similar to that of η_1^2, but it "misses" O_1^2 and has a structure close to η_1^2, which "connects" the steady states $O_1^2 \to O_1^4$ (because G is cylindrical). Thus this solution looks like a $2\pi(2+2)$ kink.

Then, an infinite number of solutions in the form of multiloop homoclinic orbits appears from A^2, such as the $2\pi\{2+(2+1)\}, 2\pi\{(2+1)+2\}, 2\pi\{(2+1)+(2+1)\}, 2\pi\{(2+2)+(2+2)\}$, etc., kink solutions. Accordingly, an infinite number of $2\pi n$-kink solutions, with $n > 1$, initiates and/or terminates at this point and other equivalent points A^j. The lines corresponding to these solutions lie in the (c, γ) plane close to Π_c^1, where the saddle value of the equilibrium state O_1 is positive. Note that the elements of the bifurcation set Π are not located according to loop number but are mixed. The bifurcation set P has a similar complex structure as Π.

The $2\pi n$-kink solutions to (3.1) have been numerically studied for the unbounded medium, corresponding to n-loop homoclinic orbits of (3.2) in the neighborhood of the points A^i. Since the bifurcation set is very complicated and the appropriate spiral-like part has a rather big decrement, it is difficult to study this problem in complete detail. We focus on its salient features.

3.5.3 The Neighborhood of A_i

First, let us detach the subset Π^* from the set Π. Let Π^* consist of the elements called Π^1 and Π_1^n ($n = 2, 3, \dots$) in Fig. 3.4. The common properties of the Π_1^n line in the (c, γ) plane and the solutions corresponding to the points of these lines are as follows: (i) in the neighborhood of $c = 1$ the branches Π_1^n ($n = 2, 3, \dots$) initiate and terminate with spiral-like parts that have foci at A^{n-1} and A^n (Figs. 3.8–10); (ii) all branches Π_1^n ($n = 2, 3, \dots$) outside the neighborhood of $c = 1$ have parabolic form (Fig. 3.4); (iii) the shape of the $2\pi n$ kinks, corresponding to Π_1^n with the parameter values from the upper part of this "parabola" looks like the $2\pi\{(n-1)+1\}$-kink solution [Figs. 3.14 ($n = 3$) and 3.15a ($n = 2$)], while those with parameter values from the lower part of this "parabola" look like the $2\pi n$-kink solution (Fig. 3.14 ($n = 2$)). The branch Π^1 does not have parabolic form. It initiates at the point ($c = 0, \gamma = 0$) and terminates with a spiral-like part with its focus at A^1. We count it in the subset Π^* because the 2π-kink solutions corresponding to the parameter values of Π^1 have similar stability properties to the $2\pi n$-kink solutions from the lower part of Π_1^n branch.

All $2\pi n$-kink solutions corresponding to the elements of the set Π not included in Π^* are unstable. In the neighborhood of $c = 1$ only those $2\pi n$-

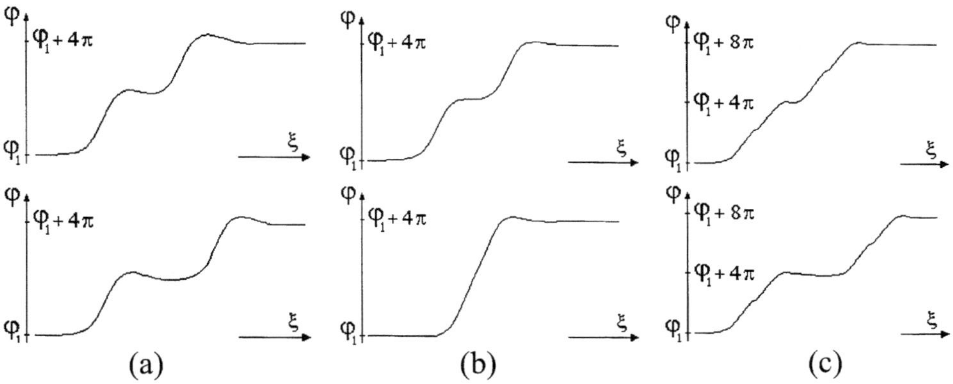

Fig. 3.15. Evolution of the initial distributions in the form of the $2\pi n$-kink solutions of (3.1). $\alpha = 0.05, \beta = 0.02$ for different values of γ. (**a**) $\gamma \approx 0.71$; (**b**) $\gamma \approx 0.677$; (**c**) $\gamma \approx 0.59$

kink solutions corresponding to parameter values from the lower part of the branches Π_1^i $(i = 2, 3, \dots)$ and the line Π^1 before the first turning point of these "spirals" are stable. The $2\pi n$-kink solutions which correspond to parameter values from the lower part of branches Π_1^i $(i = 2, 3, \dots)$ and the line Π^1 after the first turning point of these "spirals", as well as those from the upper part of these branches, are unstable (see Fig. 3.16). (Similar results exist for 2π kinks corresponding to the Π^1 branch in the neighborhood of $c = 1$.)

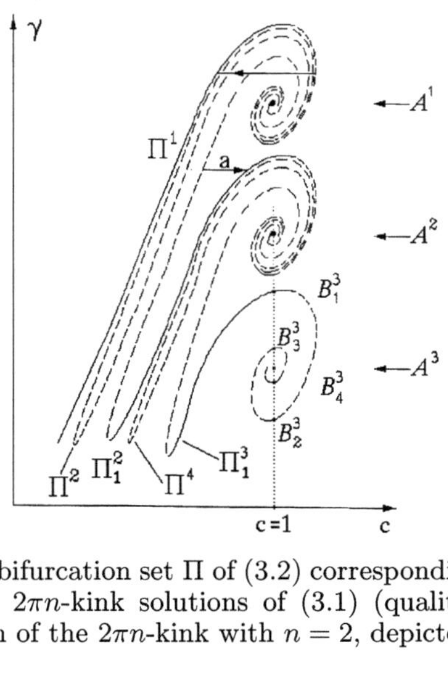

Fig. 3.16. Part of the bifurcation set Π of (3.2) corresponding to the stable (*solid*) and unstable (*dashed*) $2\pi n$-kink solutions of (3.1) (qualitatively). The *arrow a* illustrates the evolution of the $2\pi n$-kink with $n = 2$, depicted in Fig. 3.13b

The rationale for this instability of the $2\pi n$-kink solutions is the following: In the neighborhood of the point A^j, the n-loop separatrices are studied near O_2 for a rather large ξ interval. Thus, some $2\pi n$ kinks in (3.1) have a part near the steady state O_2^i (Fig. 3.14). As already mentioned, this state is unstable. Consequently, such solutions, corresponding to the n-loop homoclinic trajectories with parameters from the neighborhood of A^j, are unstable.

There are at least two different ways to destroy the $2\pi n$-kink solutions to (3.1), corresponding to the parameter values of the dashed lines in Fig. 3.16. The initial conditions for (3.1) in the form of $2\pi n$ kinks can evolve into several $2\pi n_j$ kinks ($n_1 - n_2 + \ldots + n_j = n, j \leq n$), which disintegrate as time increases (Fig. 3.15a: $n = 2$, $n_1 = n_2 = 1$; Fig. 3.15c: $n = 4$, $n_1 = n_2 = 2$). The unstable $2\pi n$-kink solution can also evolve to the $2\pi n$-kink solution, corresponding to the lower branch of Π_1^n in the region $c < 1$ (if it exists for the same parameter values of α, β, γ) (Fig. 3.15b: $n = 2$). This evolution is schematically indicated by arrows in Fig. 3.16.

All solutions of (3.1) in the form of $2\pi n$ kinks corresponding to n-loop separatrices with parameters of the bifurcation set P (or $2\pi n$ kinks of type II) are unstable, since the steady states of (3.1), which are "connected" by such $2\pi n$ kinks, are unstable.

Let us look more closely at the set of points A^j ($j = 1, 2, \ldots$), which are the foci of the spiral-like parts of the wave. Let γ^i be the γ-coordinate of the point A^i ($c = 1, \gamma = \gamma^i$) in the (c, γ) plane, and let $\tilde{\gamma}^i$ ($i = 1, 2, \ldots$) be the maximum of γ_k^i ($k = 1, 2, \ldots$), where $\{\gamma_k^i\}$ are the γ-coordinates of the points $\{B_k^i\}$. Here $\{B_k^i\}$ are the intersection points of the lower part of the branches Π_1^i and the line $c = 1$ in the (c, γ) plane (Fig. 3.16). Thus we have $\{\text{lower part of } \Pi_1^i\} \cap \{c = 1\} = \{B_k^i(c = 1, \gamma = \gamma_k^i)\}, \tilde{\gamma}_k^i = \max \gamma_k^i$, $i, k = 1, 2, \ldots$ According to Figs. 3.7–10 and 3.16, the following inequalities hold: $\tilde{\gamma}_k^i > \gamma^1 > \tilde{\gamma}^2 > \gamma^2 > \ldots > \tilde{\gamma}^n > \gamma^n$ and so on.

We now investigate the main properties of the sets $\{\gamma^i\}$ and $\{\tilde{\gamma}^i\}$. Mathematically $\{\gamma^i\}$ is important, because a bifurcation of co-dimension two takes place for the parameter values of the points A^i in the phase space G of (3.2) (Fig. 3.12). Also from the physical point of view $\{\tilde{\gamma}^i\}$ is important. On the LJJ there are no traveling wave solutions which can propagate with velocity $c > 1$. Thus stable $2\pi n$ kinks require $\gamma < \tilde{\gamma}^i$, where $\tilde{\gamma}^i$ is the maximum of the bias γ. Consequently, $\tilde{\gamma}^i$ is the height of the appropriate zero field step (ZFS) for the LJJ.

The system (3.2) for $c = 1$ is invariant under the transformation: $\xi \to \xi_n = D\xi$, $\alpha \to \alpha_n = D\alpha$, $\beta \to \beta_n = D^3\beta$, $\varphi \to \varphi_n = \varphi$, $y \to y_n = D^{-1}y$, $z \to z_n = D^{-2}z$, ($D = \text{const.} > 0$). Thus for all $\alpha^{i,j}$ and $\beta^{i,j}$ satisfying the condition $(\beta^i/\beta^j)^{1/3}(\alpha^i/\alpha^j)^{-1} = D_1 = \text{const.}$, the γ-coordinates of the intersections of all appropriate elements of the sets Π and P with the line $c = 1$ for different α and β are equal. For $2\pi n$ kinks, which can propagate with the velocity $c = 1$ along the LJJ, the quantity $M \equiv \beta^{1/3}\alpha^{-1}$ is essential. In Fig. 3.17 the dependence of $\tilde{\gamma}^i$ on M is shown.

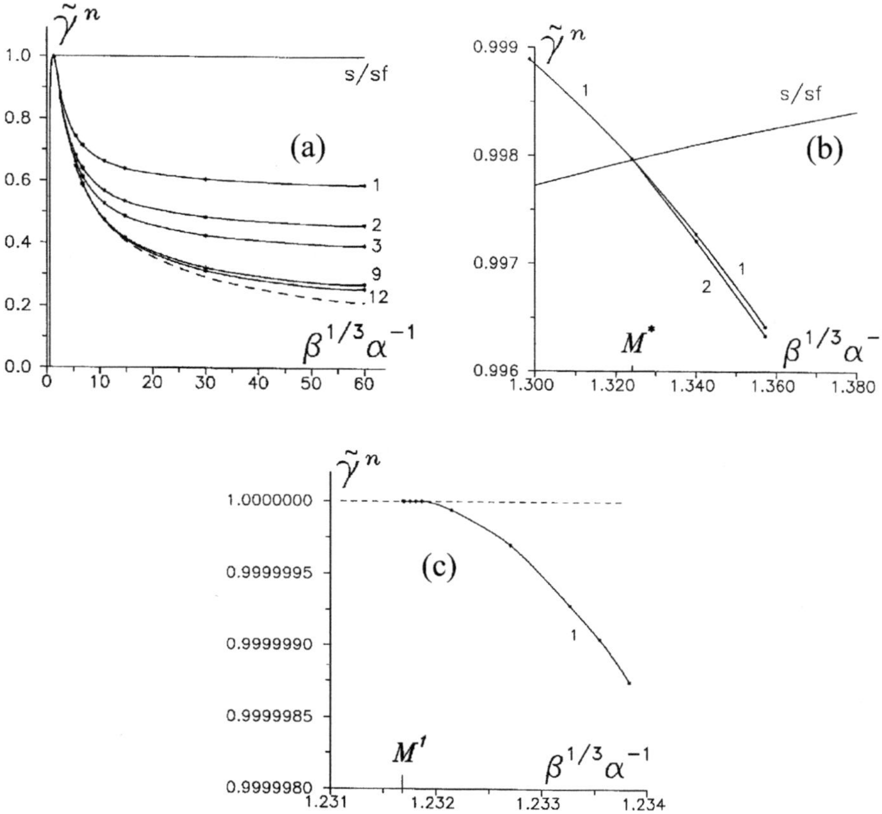

Fig. 3.17. Dependence of $\tilde{\gamma}^n$ on M for n-hump (stable) soliton solutions ($n = 1, 2, 3, 9, 12$) (the *dashed line* corresponds to the estimated $\tilde{\gamma}^\infty$). The line s/sf separates parameter regions where O_1 is a saddle and saddle-focus for $c = 1, 0.005 < \alpha < 0.1$, $0.02 < \beta < 0.3$. (**a**) Overview; (**b**) neighborhood of the value M^*, (**c**) neighborhood of the value M^1

From the analysis of the parameter region where the steady state O_1 is a saddle-focus with a positive saddle value, and from the analysis of the results shown in Fig. 3.17, we see that, if $M > M^* \approx 1.324$, $2\pi n$-kink solutions to (3.1) exist. In the latter case, the bifurcation sets Π and P consist of only one element, Π^1 and P^1, respectively. For $M < M^*$ the Π^1 lines in the (c, γ) plane are depicted in Fig. 3.18. These lines initiate at the point $(c = 0, \gamma = 0)$ and terminate on the straight line $\gamma = 1$. If $M^1 \leq M < M^*$ (where $M^1 \approx 1.232$) a part of Π^1 extends into the region $c > 1$ (Fig. 3.18, lines 3–5). If $0 < M < M^1$, the Π^1 line does not intersect the straight line $c = 1$ (Fig. 3.18, line 1). For such M the maximum value of c for Π^1, c_{max}, decreases with increasing α or with decreasing β, and satisfies the inequality $c^* \leq c_{max} < 1$ [where $c^* = r(r^2 + \alpha^2)^{-1/2}, r \approx 1.193$]. The value of c^* is obtained from the properties of Tricomi's curve. This curve comes from

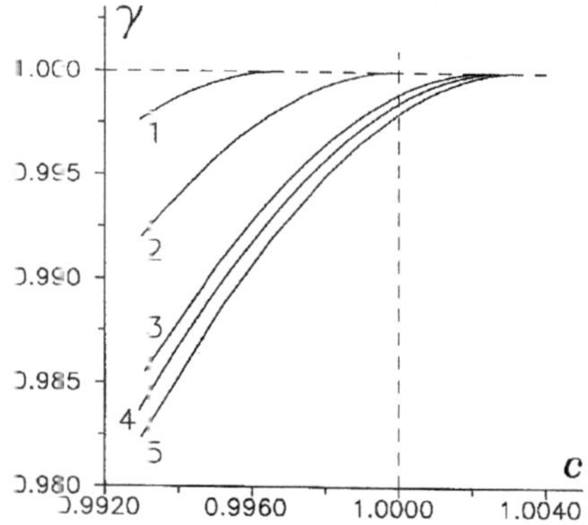

Fig. 3.18. Branches Π^1 for $\beta = 0.02$: $\alpha_1 = 0.235, \alpha_2 = 0.220382, \alpha_3 = 0.209$, $\alpha_4 = 0.207, \alpha_5 = 0.20502$ ($M_1 < M_2 \approx M^1 < M_3 < M_4 < M_5 < M^*$)

the existence of the connection of separatrices in the phase plane of the dynamical system, which describes, in particular, a physical pendulum under constant torque in a viscous medium. Thus, for $0 < M < M^1$ the maximum propagation velocity of 2π kinks in LJJ is smaller than unity.

3.5.4 The Sets $\{\gamma^i\}$ and $\{\tilde{\gamma}^i\}$

To study the properties of the sets $\{\gamma^i\}$ and $\{\tilde{\gamma}^i\}$ $(i = 1, 2, \ldots)$ we introduce the following ratios:

$$\delta(i) = \frac{\gamma^{i+1} - \gamma^i}{\gamma^{i+2} - \gamma^{i+1}}, \quad \tilde{\delta}(i) = \frac{\tilde{\gamma}^{i+1} - \tilde{\gamma}^i}{\tilde{\gamma}^{i+2} - \tilde{\gamma}^{i+1}},$$

$$i = 1, 2, \ldots$$

In Fig. 3.19 the dependence of δ on i, for the first 12 points, A^i, is depicted.

The value of γ^i $(i = 1, 2, \ldots)$ can be obtained not as the focus of the spiral-like parts of Π, but in a simpler way. As follows from Fig. 3.12 we have bifurcation of co-dimension two in the points A^i. Generally, it is rather difficult to obtain a bifurcation set of this type because in a N-dimensional parameter space the dimension of this bifurcation is $N - 2$. But here this task simplifies since the c-coordinate of A^i is known. Thus, we only need to vary a single parameter. Therefore, the algorithm for finding γ^i may be the following: construct in G the plane $D : \{\varphi = \varphi^0, y, z \in R\}$ between O_1^0 and O_2^i. Then, by numerical integration of (3.2) determine the points $L_{1,2}$: $W_1^u \cap D = L_1, W_2^s \cap D = L_2$. Introducing the function $d(\gamma) = |L_1 L_2|$, we obtain $d(\gamma^i) = 0$.

The δ-dependence has the asymptote $\delta = \delta^*$. We also observe that the ratios, $\delta(i)$ and $\tilde{\delta}(i)$, for the values of M investigated, are close to each other.

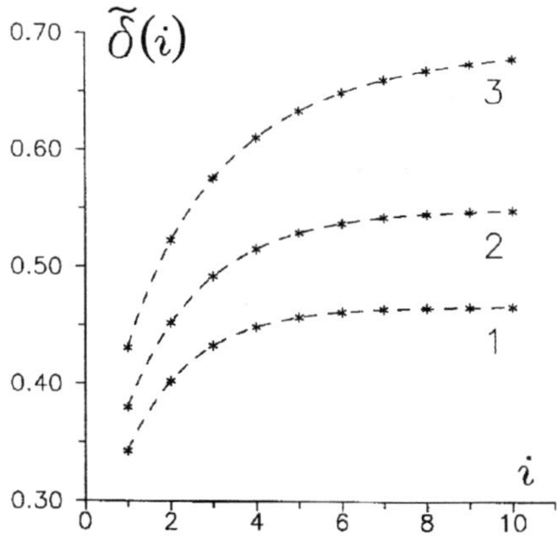

Fig. 3.19. Dependence of $\tilde{\delta}(i)$ on i. $1-\alpha = 0.05, \beta = 0.02\,(M = 5.43)$; $2-\alpha = 0.05$, $\beta = 0.04\,(M = 6.84)$; $3 - \alpha = 0.025, \beta = 0.02\,(M = 10.86)$

Their difference is less than 0,1% and does not increase with increasing i or M. Thus, the limit, $\lim_{i\to+\infty} \tilde{\delta}(i) = \tilde{\delta}^*$, also exists and $\tilde{\delta}^* \approx \delta^*$. Accordingly, we can use $\tilde{\delta}$ instead of δ. This is important, because $\{\tilde{\gamma}^i\}$ is more easily obtained than $\{\gamma^i\}$. As follows from the asymptotic behavior of $\delta(i)$ and $\tilde{\delta}(i)$, $\{\gamma^i\}$ and $\{\tilde{\gamma}^i\}$ approximate a scaling law.

In Fig. 3.20 the dependence of δ^* (and, consequently, $\tilde{\delta}^*$) on M is depicted. With the method used in this chapter it is impossible to obtain the exact value of δ^*. Thus, we approximate it by $\tilde{\delta}(j)$ ($j \sim 10 - 30$, depending on the value of M). Then δ^* monotonically increases with increasing M. However, $\delta^*(M) < 1$ even when M tends to infinity. Using the results shown in Figs. 3.17 and 3.20, we can estimate the height of the nth ZFS of the LJJ for $n \to \infty$ (Fig. 3.17, dashed line).

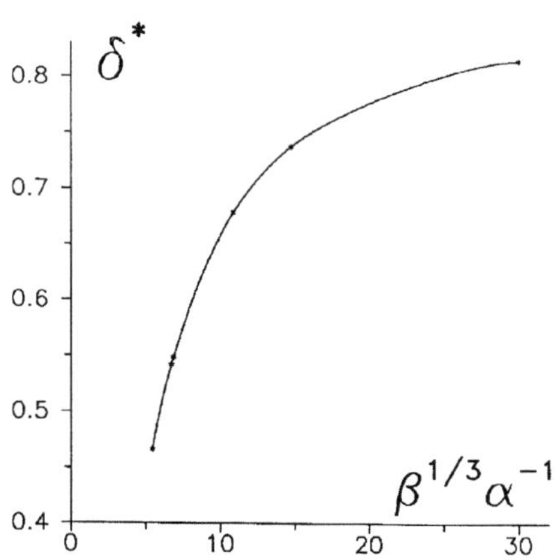

Fig. 3.20. Dependence of δ^* on M. $0.005 < \alpha < 0.1, 0.02 < \beta < 0.3$

3.6 Existence of Homoclinic Orbits

For simplicity we change the variables in (3.2) and take $\Theta = \pi - \varphi$, $y_H = -y$, and $z_H = -z$. Then, (3.2) (the subscript H is omitted) becomes

$$\begin{gathered}\dot{\Theta} = y\,, \quad \dot{y} = z\,, \\ \beta c\dot{z} = -(1-c^2)z + c\alpha y - \sin\Theta + \gamma\,.\end{gathered} \tag{3.4}$$

Consider (3.4) in a cylindrical phase space $G = S^1 \times R^2$. For $\gamma < 1$ in G, (3.4) has two saddle steady states: O_1 ($\Theta = \Theta_1 = \arcsin\gamma$, $y = z = 0$) and O_2 ($\Theta = \Theta_2 = \pi - \arcsin\gamma$, $y = z = 0$). The O_1 state has a stable 1D and an unstable 2D manifold, while O_2 has a stable 2D and an unstable 1D manifold. Apparently, if any of the unstable trajectories of the steady state O_j^0 ($\Theta = \Theta_j$, $y = z = 0$) belongs to a 2D stable manifold of the point O_j^n ($\Theta = \Theta_j - 2\pi n$, $y = z = 0$), where $j = 1, 2$, $n = \pm 1, \pm 2, \ldots$, then there exists in G a rotational homoclinic loop (see Fig. 3.21 where $j = 2$ and

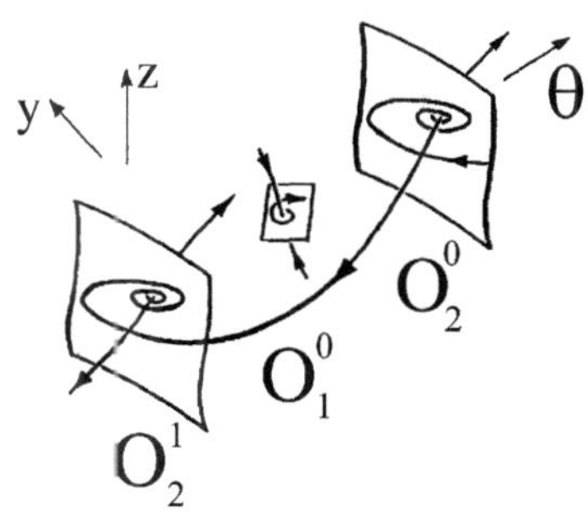

Fig. 3.21. Rotational homoclinic loop in the phase space G of (3.4)

$n = 1$). A homoclinic trajectory of (3.2) that corresponds to this loop is shown in Fig. 3.2a. The points O_2^0 and O_2^1 are taken to be identical, and instead of Θ the coordinate φ is introduced again. We draw all other rotational homoclinic loops as shown in Fig. 3.2b,c because it is very difficult to represent them in G as in Fig. 3.21.

When $\beta = 0$ ($\alpha \neq 0$, $\gamma \neq 0$), (3.4) reduces to a second-order system. It describes, in particular, the dynamics of a physical pendulum in a viscous medium. In this case, for $\gamma < 1$, the bifurcation diagram contains two curves, $\gamma = \gamma^*(\alpha c/\sqrt{1-c^2})$ and $\gamma = \gamma^*(\alpha c/\sqrt{c^2-1})$, which correspond to different rotational homoclinic trajectories, with γ^* assumed to be known (Fig. 3.22).

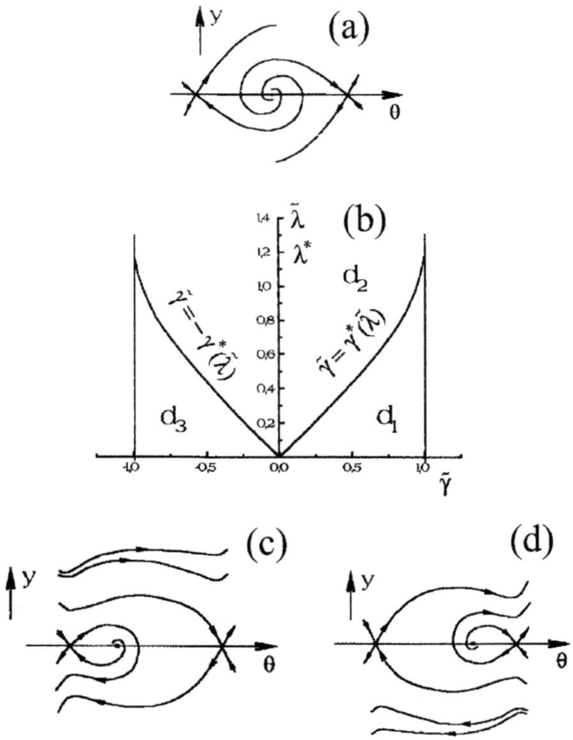

Fig. 3.22. Bifurcation diagram of (3.12)

3.6.1 Lyapunov Function

Let α, β and γ be all different from zero. The system (3.4) has no oscillatory trajectories different from O_1 or O_2. Consider the function

$$\begin{aligned} V(\Theta, y, z) =& \beta c \frac{z^2}{2} + (1 + \alpha\beta + \beta/\alpha) zy + (1 - c^2)(1 + \alpha\beta + \beta/\alpha) \frac{y^2}{2\beta c} \\ &+ (\sin\Theta - \gamma) y + (1 + \alpha\beta + \beta/\alpha)/\beta c \int_{\Theta_1}^{\Theta} (\sin\Theta - \gamma)\, \mathrm{d}\Theta\, . \end{aligned} \tag{3.5}$$

Using (3.4) we obtain

$$\dot{V}(\Theta, y, z) = (c^2 + \alpha\beta + \beta/\alpha) z^2 + \alpha c y z + y^2(\alpha^2 + \alpha/\beta) + (1 + \cos\Theta) y^2\, .$$

Then $\dot{V}(\Theta, y, z) \geq 0$, with $\dot{V} = 0$ only on the straight line $\{y = z = 0\}$. Consequently, (3.4), indeed, has no oscillatory trajectories different from O_1 and O_2.

3.6.2 The Vector Field of (3.4) on Two Auxiliary Surfaces

Consider the following two surfaces:

$$\begin{aligned} W_1 &\equiv z - qy + q/\alpha c(\sin\Theta - \gamma) = k\,, \\ W_2 &\equiv z - qy + q/\alpha c(\sin\Theta - \gamma) = -k\,, \end{aligned} \tag{3.6}$$

with

$$k = \text{const.} > 0\,,$$
$$q = \{[(1-c^2)^2 + 4\alpha\beta c^2]^{1/2} - (1-c^2)\}/2\beta c\,.$$

The derivatives of the functions $W_1 = k$ and $W_2 = -k$, using (3.4), are

$$\dot{W}_1 \mid_{W_1=k} = -\alpha k/\beta q + q/\alpha c \cos\Theta y\,,$$
$$\dot{W}_2 \mid_{W_2=-k} = \alpha k/\beta q + q/\alpha c \cos\Theta y\,.$$

If

$$|y| < \frac{\alpha^2 ck}{\beta q^2} \tag{3.7}$$

holds, it follows that

$$\dot{W}_1 \mid_{W_1=k} < 0 \quad \text{and} \quad \dot{W}_2 \mid_{W_2=-k} > 0\,. \tag{3.8}$$

Let us denote by G^+ the region of the phase space, G, limited by the surfaces $W_1 = k$ and $W_2 = -k$ and the planes

$$Q_1 : \left\{ y = \frac{\alpha^2 ck}{\beta q^2} \right\} \quad \text{and} \quad Q_2 : \left\{ y = -\frac{\alpha^2 ck}{\beta q^2} \right\}\,.$$

Using (3.8) we see that the vector field of (3.4) on the boundaries of G^+ formed by $W_1 = k$ and $W_2 = -k$ is oriented inwards. Consequently,

$$qy - \frac{q}{\alpha c}(\sin\Theta - \gamma) - k < z < qy - \frac{q}{\alpha c}(\sin\Theta - \gamma) + k \tag{3.9}$$

holds for all solutions $\{\Theta(\xi),\ y(\xi),\ z(\xi)\}$ of (3.4) in G^+.

3.6.3 Auxiliary Systems

We introduce auxiliary systems of the form

$$\dot{\Theta} = y\,, \quad \dot{y} = qy - \frac{q}{\alpha c}(\sin\Theta - \gamma) + k\,, \tag{3.10}$$

$$\dot{\Theta} = y\,, \quad \dot{y} = qy - \frac{q}{\alpha c}(\sin\Theta - \gamma) - k\,, \tag{3.11}$$

which appear like the equations for an oscillator:

$$\dot{\Theta} = y\,, \quad \dot{y} = \tilde{\lambda} y - \sin\Theta + \tilde{\gamma}\,. \tag{3.12}$$

The part of the $(\tilde{\gamma}, \tilde{\lambda})$-parameter plane on the regions corresponding to the different qualitative behavior of the trajectories of (3.12) is presented in Fig. 3.22. This partition is conditioned by the lines $\tilde{\gamma} = \pm 1$, corresponding to the saddle-node bifurcations, and by the lines $\tilde{\gamma} = \pm\gamma^*(\tilde{\lambda})$, associated with the rotational homoclinic bifurcations. Let us require (3.10) and (3.11) to have, simultaneously, steady states. This requirement holds if

$$k < \frac{q(1-\gamma)}{\alpha c}\,. \tag{3.13}$$

It follows from (3.10–11) that the homoclinic orbits and the limit cycles of these systems are located within the region

$$g^+ : \left\{ \Theta \in S^1, -\frac{1+\gamma}{\alpha c} + \frac{k}{q} < y < \frac{1-\gamma}{\alpha c} - \frac{k}{q} \right\}. \tag{3.14}$$

Let us require that $g^+ \subset G^+$. Using (3.7) and (3.14) we find that this condition holds if

$$k \geq k_0\,, \quad k_0 \equiv \frac{(1+\gamma)\beta q^2}{\alpha c(\alpha^2 c + \beta q)}\,. \tag{3.15}$$

Let $k = k_0$. Then it follows from (3.13) that (3.10–11), simultaneously, have steady states for the parameter values in the region d determined by

$$\gamma < \frac{\alpha^2 c}{\alpha^2 c + 2\beta q}\,. \tag{3.16}$$

Thus, for points in the region d, (3.10–11), simultaneously, have steady states, and the vector field of (3.4), on the surfaces $W_1 = k_0$ and $W_2 = -k_0$, is oriented inwards to G^+ under the "stripe" g^+.

3.6.4 "Tunnels" for Manifolds of the Saddle Steady State O_2

Since (3.10–11) are z-independent, their trajectories in G^+ form cylindrical surfaces connecting $W_1 = k_0$ and $W_2 = -k_0$. These cylindrical surfaces inside G^+ are intersected by the trajectories of (3.4), transversally in one direction, due to (3.9–11). Let us compare the vector fields of (3.4) and (3.10–11):

$$R^+ = \left(\frac{\mathrm{d}y}{\mathrm{d}\Theta}\right)_{(3.4)} - \left(\frac{\mathrm{d}y}{\mathrm{d}\Theta}\right)_{(3.10)} = \frac{z - qy + q/\alpha c(\sin\Theta - \gamma) - k_0}{y}\,,$$

$$R^+ = \left(\frac{\mathrm{d}y}{\mathrm{d}\Theta}\right)_{(3.4)} - \left(\frac{\mathrm{d}y}{\mathrm{d}\Theta}\right)_{(3.11)} = \frac{z - qy + q/\alpha c(\sin\Theta - \gamma) + k_0}{y}\,.$$

Taking into account (3.9) we find that

$$R^+ y < 0 \quad \text{and} \quad R^- y > 0 . \tag{3.17}$$

Let us consider in G^+ the cylindrical surfaces L^+, L^- formed by the saddles of (3.10–11) and their separatrices. Note, that the steady state O_2 of (3.4) on the circle S^1 is located between the saddles of (3.10) and (3.11). Consequently, using (3.8) and (3.17) the regions of the phase space restricted by the cylindrical surfaces L^+ and L^- and by the surfaces $W_1 = k_0$ and $W_2 = -k_0$ form tunnels. The trajectories of the manifolds of the point O_2 extend "along" these tunnels. The relative arrangement of the tunnels for $y < 0$ depends on which of the regions $\{d_2 \cup d_3\}$ or d_1 (Fig. 3.22) the parameters of (3.10–11) are, simultaneously, contained. Let us require these parameters to be in the region $\{d_2 \cup d_3\}$. This condition will be fulfilled for the values of the parameters of (3.4) contained in the region d_2^+ given by the inequalities

$$0 \le \gamma < \begin{cases} \dfrac{\alpha^2 c}{\alpha^2 c + 2\beta q} & \text{if} \quad \sqrt{q\alpha c} \ge \lambda^* , \\[2ex] \dfrac{(\alpha^2 c + \beta q)\gamma^*(\sqrt{q\alpha c}) - \beta q}{\alpha^2 c + 2\beta q} & \text{if} \quad \sqrt{q\alpha c} < \lambda^* , \end{cases} \tag{3.18}$$

where $\lambda^* = 1.193\ldots$ and $\gamma^*(\lambda^*) = 1$. The cross-sections ($z = \text{const.}$) of the tunnels, and the corresponding orientation of the vector field of (3.4), are shown in Fig. 3.23a. Let us now require the parameter values of (3.10–11) to be contained in the region d_1 (Fig. 3.22). This condition will be fulfilled if those parameters of (3.4) are contained in the region d_2^- given by

$$\begin{cases} \dfrac{\beta q + (\alpha^2 c + \beta q)\gamma^*(\sqrt{q\alpha c})}{\alpha^2 c} < \gamma < \dfrac{\alpha^2 c}{\alpha^2 c + 2\beta q} , \\[2ex] \sqrt{q\alpha c} < \lambda^* . \end{cases} \tag{3.19}$$

When the parameters lie within the region d_2^-, the tunnels are arranged as in Fig. 3.23b. Thus, there exists in the parameter space of the system (3.4) the region d (3.16), such that in the phase space G there are corresponding tunnels controlling the arrangement of the manifolds of the point O_2. In addition, there exist in d the subregions d_2^+ and d_2^- corresponding to the different relative arrangement of the tunnels for $y < 0$.

3.6.5 Homoclinic Orbits

Since the 1D and 2D manifolds of the steady state O_2 lie inside the tunnels and are located inversely at the points of the regions d_2^+ and d_2^- when $y < 0$ (Fig. 3.23), there exists between d_2^+ and d_2^-, in the parameter space of (3.4), a bifurcation set Π^1 of co-dimension 1 that corresponds to the existence of a rotational homoclinic loop connecting the points $O_2^0\,(\Theta = \Theta_2,\ y = z = 0)$ and $O_2^1\,(\Theta = \Theta_2 - 2\pi,\ y = z = 0)$ (i.e. $n = 1$ in this case) (Fig. 3.22). Returning to the initial variable φ, we find that the points of the bifurcation set Π^1 correspond to the rotational homoclinic loop (Fig. 3.2a) connecting saddle steady states with the coordinates $(\varphi = \varphi_1,\ \dot{\varphi} = \ddot{\varphi} = 0)$ and $(\varphi = \varphi_i + 2\pi,\ \dot{\varphi} = \ddot{\varphi} = 0)$ and, consequently, to a solitary front or a single-hump soliton in the model (3.1) (Fig. 3.2a).

The existence, for $c > 1$, of a bifurcation set, P^1, of co-dimension 1 that corresponds to the existence of a rotational loop ($n = 1$) connecting the points $(\varphi = \varphi_2,\ \dot{\varphi} = \ddot{\varphi} = 0)$ and $(\varphi = \varphi_2 + 2\pi,\ \dot{\varphi} = \ddot{\varphi} = 0)$ can be shown in a similar way. Also in this case the parameter regions d_1^+ and d_1^-, analogous to d_2^+ and d_2^-, exist in the space of parameters of (3.4). The cross-sections of the regions $d_{1,2}^{+;-}$ are depicted in the (c, γ) plane ($\alpha = 0.5,\ \beta = 0.02$) in Fig. 3.24, which also shows the Π^1 and P^1 curves numerically constructed.

The bifurcation diagram in the (c, γ) plane, for small β, contains only the Π^1 and P^1 components and looks like the diagram in Fig. 3.24. As the parameter β is further increased, the curves Π^1 and P^1 evolve as follows: Starting from some value of β, these curves no longer intersect the line $\gamma = 1$ and interlock, when $c = 1$ and $\gamma < 1$. The interlocking of curves at $\gamma < 1$ occurs because (3.4) is reversible when $c = 1$ (it is invariant under the transformation $\Theta \to \pi - \Theta,\ \xi_H \to -\xi_H,\ z \to -z$). Consequently, if

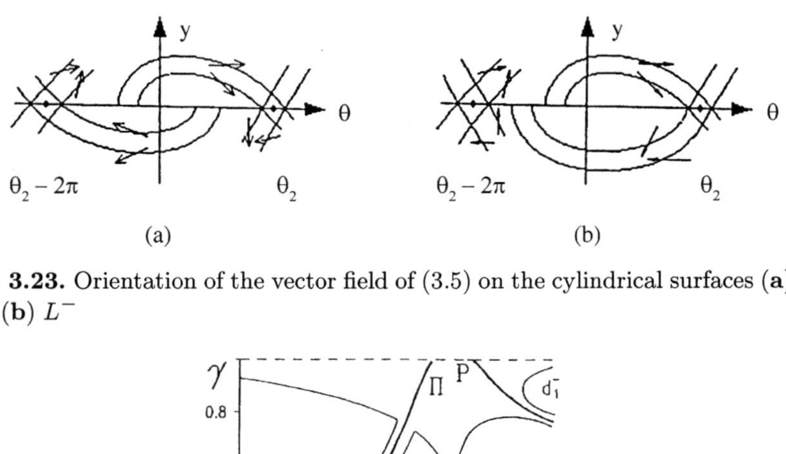

Fig. 3.23. Orientation of the vector field of (3.5) on the cylindrical surfaces (**a**) L^+ and (**b**) L^-

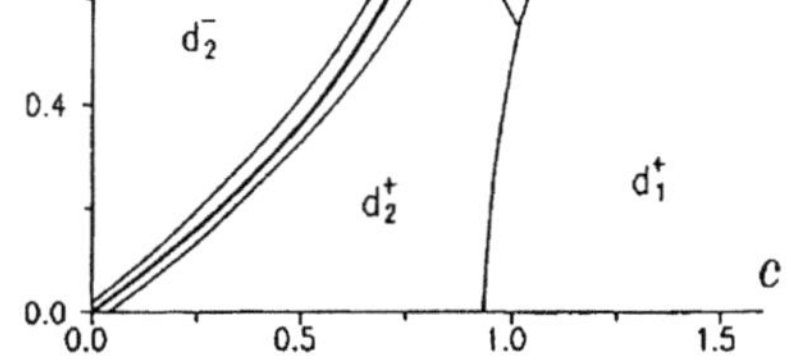

Fig. 3.24. Cross-sections of the $d_{1,2}^{+;-}$ parameter domain ($\alpha = 0.5, \beta = 0.02$)

there exists a trajectory connecting O_2^0 and O_2^1, then, for the same values of the parameters, there also exists a trajectory connecting O_1^0 and O_1^1. As β is increased still further, the components Π_c^1 and P_c^1 appear on the curves Π^1 and P^1 (Fig. 3.4b). If the parameters of (3.4) belong to the components Π_c^1 and P_c^1, the corresponding steady state is a saddle-focus with positive saddle value. Then, the bifurcation set of (3.4) in the neighborhood of Π_c^1 and P_c^1 contains, besides Π_c^1 and P_c^1, an infinite number of other elements, and there is a nontrivial hyperbolic set in the phase space. In particular, there exists a countable set of bifurcation surfaces that correspond to the separatrix loops of increasing envelope ability with $n > 1$ to which $2\pi n$ fronts (multihump solitons) correspond in (3.1). For example, when $c < 1$, there exists a countable set of curves, $\{\Pi_j^2\}_{j=1}^{\infty}$, which correspond to a 2-envelope rotational separatrix loop connecting O_1^0 and O_1^2 and performing j oscillations in the neighborhood of O_1^1. Thus, a common property of all the Π_j^2 curves is that a 2-hump soliton corresponds to any of these curves, and, consequently, a train with two bound solitons may propagate for values of the parameters in $\{\Pi_j^2\}_{j=1}^{\infty}$ in (3.1) (Fig. 3.2b,c). The solitons corresponding to the Π_j^2 curves, with different j, differ according to the number of oscillations at the level $\varphi = \varphi_1 + 2\pi$. Each of the Π_j^2 curves on the (c, γ) plane is parabolic. The tops of these "parabolas", as $j \to \infty$, tend to the point A, i.e. to the point on Π where the saddle value vanishes.

The Π_j^2 curves, with $j > 1$, are located inside a rather small region between Π^1 and Π_1^2 on the (c, γ) plane (Fig. 3.4a). The Π_1^2 curve is shown in Fig. 3.4b in an enlarged scale. Note that, although to all points of the Π_1^2 curve correspond solitons with $n = 2,\ j = 1$, their profiles are qualitatively different, depending on the choice of the parameters in Π_1^2. If the parameter values are taken in the portion of Π_1^2 close to the curve Π^1 (Fig. 3.4b), the soliton profile looks like a bound state of two solitons with $n = 1$ (Fig. 3.2b), while for the values of the parameters belonging to the other part of the parabola Π_1^2, the soliton shape is as shown in Fig. 3.2c. Although Π and Π_1^2 are rather close to each other, they do not interlock (there are other cases in which they *do* interlock). In addition, between each pair of curves Π_j^2 and Π_{j+1}^2 there exists a bifurcation set of curves Π_{jl}^3 (where l depends on j) which correspond to 3-envelope homoclinic loops connecting the points O_1^0 and O_1^3. The subscripts j and l indicate that the corresponding homoclinic trajectory performs j oscillations near O_1^1 and l oscillations near O_1^2. The number of Π_{jl}^3 curves is finite for each fixed j and increases without bound as $j \to \infty$. Each Π_{jl}^3 curve consists of two connectivity components. Solitons with $n = 3$ correspond to the points of the Π_{jl}^3 curves. Consequently, trains of three bound single-hump solitons may propagate within (3.1) for these values of the parameters.

The bifurcation set of the Π_{jl}^3 curves does not cover the entire set of 3-envelope homoclinic loops. Numerical integration of (3.4) reveals that below the Π_1^2 curve a parabolic Π_1^3 curve exists to which, again, solitons with $n = 3$ correspond. The Π_1^3 curve is enclosed between two parabolic curves (Π_1^4 and

Π^4 in Fig. 3.4b) to which solitons with $n = 4$ correspond (Fig. 3.2f–i). The bifurcation diagram also contains the curves that correspond to solitons with $n > 4$, but they are not shown in Fig. 3.2 because the corresponding parabolas are very narrow and too close to the curves of lower envelope ability.

A similar situation occurs for $c > 1$. In this case, there exist, besides P^1, the bifurcation sets P_j^2, P_{jl}^3, etc., which correspond to solitons whose tails tend to the values $\varphi = \varphi_2$ and $\varphi = \varphi_2 - 2\pi$. However, numerical experiments show that all these solitons are unstable.

Thus, the bifurcation diagram of (3.4) contains an infinite number of elements and, therefore, cannot be represented in its complete form. But its basic features are clear. The (c, γ) plane contains two families of Π and P curves. Each bifurcation curve of the family Π interlocks, when $c = 1$, with an analogous curve of the family P [(3.4) is reversible when $c = 1$]. We have already mentioned that the envelope ability of the bifurcation curves, to which the bifurcation curve of interest is close, influences the soliton shape. Take, for instance, the Π^4 and Π_1^4 curves. The points on the Π^4 branch located near the Π_1^2 curve correspond to a soliton which is a bound state of a pair of 2-hump solitons (Fig. 3.2f) with $n = 2$. The points on the Π_1^4 branch located near the Π_1^3 curve correspond to the fronts that consist of a "train" of two solitons with $n = 3$ and $n = 1$ (Fig. 3.2h).

3.7 Salient Features of the Perturbed Sine–Gordon Equation

We have obtained analytically and numerically the $2\pi n$-kink solutions ($n \geq 1$) to the PSG equation (3.1) with propagation velocity c. Equation (3.1) models, in particular, the LJJ, and hence its solutions correspond to fluxons in LJJ. By introducing a moving frame coordinate, ξ, the PSG equation reduces to its underlying nonlinear dynamical system given by the ordinary differential equations (3.2). Its n-loop homoclinic solutions correspond to $2\pi n$-kink solutions of the PSG equation and, consequently, to the fluxons in LJJ. Thus, the analytical and numerical exploration of (3.2) provided the shape and the evolution of kink solutions for different parameter values in the PSG equation. Numerical computations with the PSG equation also permitted us to assess the stability of these solutions. The $I - V$ characteristics thus constructed agree well with the available experimental data (Fig. 3.7).

The quantity $M \equiv \beta^{1/3}\alpha^{-1}$ is essential for the PSG equation and, consequently, for the LJJ. If $M < M^* \approx 1.324$, only $2\pi n$ kinks, with $n = 1$, can propagate along the LJJ. The dependence of the propagation velocity, c, of these solutions on the bias current, γ, has a simple monotonic form. If $M > M^*$, $2\pi n$ kinks with $n \geq 1$ can propagate along the LJJ. Then the dependence of c on γ has a very complex form. It has an infinite number of branches. The branches corresponding to $2\pi n$ kinks, with different n, are located on the (c, γ) plane not sequentially ordered by n but mixed. All these branches have a spiral-like part in the neighborhood of the straight line $c = 1$.

The dependence of the heights of the ZFS on the LJJ parameters has also been investigated. We have found that, if $M \leq M^1 \approx 1.232$, the height is constant and equals unity (in the normalized variables of the LJJ). If $M > M^1$, the height depends only on the number of the ZFS and on the parameters α and β, but only through the quantity M. It was also shown that the dependence of the ZFS heights corresponding to the propagation of $2\pi n$ kinks on n approximately obeys a scaling law. The value of the scaling parameter δ^* depends only on the quantity M. This permitted us to estimate the value of the height of the nth ZFS in the limit $n \to \infty$.

Let us conclude by saying that various aspects of the dynamics of Josephson junctions, with broad coverage of the phenomena and mathematics underlying them, have been studied by numerous authors. In relation to the content of this chapter, we should mention the works of Barone, Paterno, Likharev, Lonngren, Ustinov and Scott [3.1–3,3.7].

4. Spatial Structures, Wave Fronts, Periodic Waves, Pulses and Solitary Waves in a One-Dimensional Array of Chua's Circuits

4.1 Introduction and Motivation

Starting with the discovery of deterministic chaos in a 3D dynamical system made by Lorenz in 1963, a great deal of effort was made to build electronic circuits exhibiting chaotic oscillations. At present a large number of such items have been proposed. In particular, special interest has been devoted to the so-called Chua's circuit (oscillator). Although earlier related proposals existed, this circuit was introduced – in the proper context – by L.O. Chua at the opening lecture given in the Workshop on Nonlinear Theory and its Applications (NOLTA'92), held at Waseda University, Tokyo, in January 1992. Chua's circuit possesses a large variety of possible dynamical behaviors. By changing the values of control parameters, one can obtain regular behavior or chaotic oscillations. [4.2, 4.6, 4.7]

Chua's circuit is shown in Fig. 4.1a. It consists of a linear inductor L, linear resistors $R_0, R = 1/G$, two linear capacitors C_1 and C_2, and a nonlinear resistor N_R (Chua's diode). The behavior of such a circuit can be described by the following evolution equations:

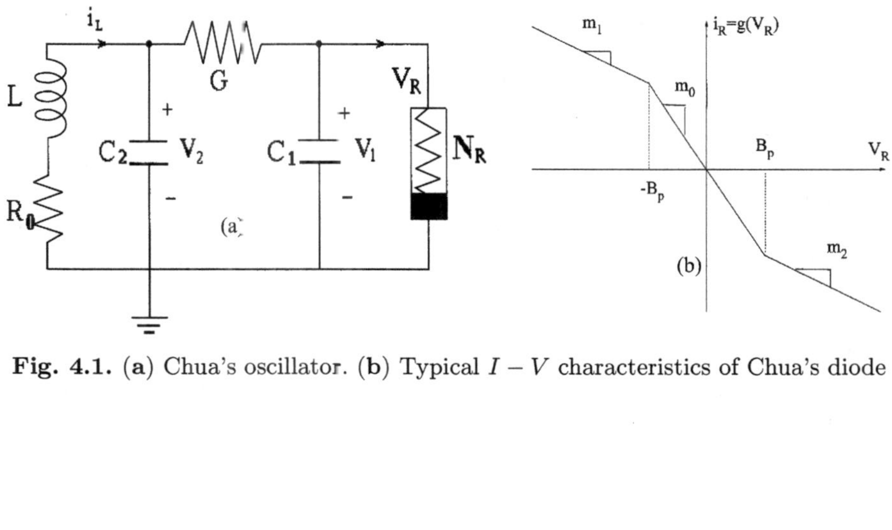

Fig. 4.1. (**a**) Chua's oscillator. (**b**) Typical $I - V$ characteristics of Chua's diode

$$\begin{cases} C_1 \dfrac{dv_1}{dt} = G(v_2 - v_1) - g(v_1)\,, \\ C_2 \dfrac{dv_2}{dt} = G(v_1 - v_2) + i_L\,, \\ L \dfrac{di_L}{dt} = -(v_j + R_0 i_L)\,, \end{cases} \tag{4.1}$$

where $g(v)$ is the $I-V$ characteristics of the nonlinear resistor, N_R (Fig. 4.1b); v_{C_1}, v_{C_2} and i_L denote the voltage across C_1, the voltage across C_2 and the current through L, respectively. In most practical cases the nonlinearity in a real circuit can be described well by a smooth function, $g(V)$.

For the universality of our arguments, let us change scales and define dimensionless quantities, using new units. We set

$$x \equiv \frac{v_1}{B_p}\,, \quad y \equiv \frac{v_2}{B_p}\,, \quad z \equiv \frac{i_L}{B_p G}\,,$$

$$\tau \equiv \frac{G}{C_2} t\,, \quad \alpha \equiv \frac{C_2}{C_1}\,, \quad \beta \equiv \frac{C_2}{LG^2}\,,$$

$$\gamma \equiv \frac{R_0 C_2}{LG}\,, \quad a \equiv \frac{m_0}{G}\,, \quad b_i \equiv \frac{m_i}{G}\,, \quad i = 1, 2\,.$$

Then (4.1) become in dimensionless form

$$\begin{cases} \dot{x} = \alpha[y - x - f(x)]\,, \\ \dot{y} = x - y + z\,, \\ \dot{z} = -\beta y - \gamma z\,, \end{cases} \tag{4.2}$$

where the dot denotes differentiation with respect to τ and $f(x)$ is the piecewise-linear function

$$f(x) = \begin{cases} b_1 x + a - b_1 & \text{if} \quad x \geq 1\,, \\ a\ x & \text{if} \quad -1 \leq x \leq 1\,, \\ b_2 x - a + b_2 & \text{if} \quad x \leq -1\,, \end{cases} \tag{4.3}$$

characterizing Chua's circuit.

Figure 4.2a illustrates an implementation of Chua's circuit (approximately 5 cm $\times$ 8 cm in size) with a microelectronic chip (Fig. 4.2b). The chip has been designed and built by using 2-mm CMOS technology, with the circuit itself occupying a silicon area of 2,5 mm $\times$ 2,8 mm.

Building upon the Chua's circuit, cellular neural networks (CNN) have been constructed and used to mimic visual information transmission, image recognition, etc. A CNN consists of a large array of identical or almost identical active elements (e.g. Chua's cells) joined together with suitable couplings. The type of coupling varies from a simple linear local resistor to complex nonlinear connections. Accordingly, a CNN can be considered as a spatially

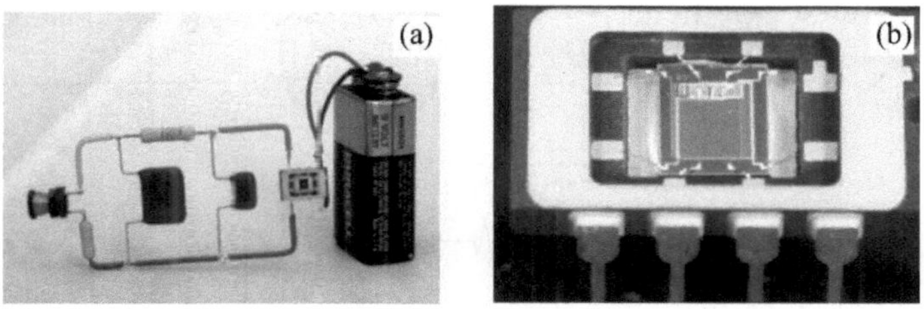

Fig. 4.2. (**a**) Demo Chua's circuit. (**b**) Photograph of the microelectronic chip implementation of Chua's circuit (designed by N. Delgado-Restituto and A. Rodriguez-Vazquez, Seville, Spain)

distributed system with discrete spatial coordinates, and hence a CNN composed of a large number of coupled Chua's circuits represents a very complex dynamical system. Then, for its effective application it is necessary to study the basic features of possible self-organization phenomena. We shall take as bonds or connectors either resistors or inductances. The arrays can have different structure, circular rings or unbounded lattices (see Fig. 1.4). We shall show in this chapter that even the 1D array of Chua's circuits has all the features of nonlinear media, such as supporting spatial structures of both regular and chaotic form, wave fronts, periodic and solitary waves, pulses and waves of chaotic profile. Hence, by means of self-consistent, cooperative action of active elements compensating the dissipation, this medium may self-organize. Various coherent space–time structures observed are the effects of this self-organization.

4.2 Spatio-Temporal Dynamics of an Array of Resistively Coupled Units

In this section we consider an array of resistively coupled identical units (4.2) [4.8]. A schematic of such a system is presented in Fig. 4.3. The dynamics of the array is described by the following system of equations:

$$\begin{cases} \dot{x}_j = \alpha[y_j - x_j - f(x_j)] + d(x_{j-1} - 2x_j + x_{j+1}) \,, \\ \dot{y}_j = x_j - y_j + z_j \,, \\ \dot{z}_j = -\beta y_j - \gamma z_j \,, \end{cases} \tag{4.4}$$

where d is the (intralattice) coupling coefficient or bond which is a resistor, hence diffusive and j denotes the site where the element is placed. It is treated as a spatial coordinate. N accounts for the number of units in the array, hence its length. For simplicity, we restrict our consideration to the dynamics of (4.4) with (4.3) for the parameters satisfying the conditions $b_1 > 0$, $b_2 > 0$, $a < 0$. This choice of parameter range leads to interesting behavior by the unit. We consider, first, the case of an unbounded array, i.e. a chain where j

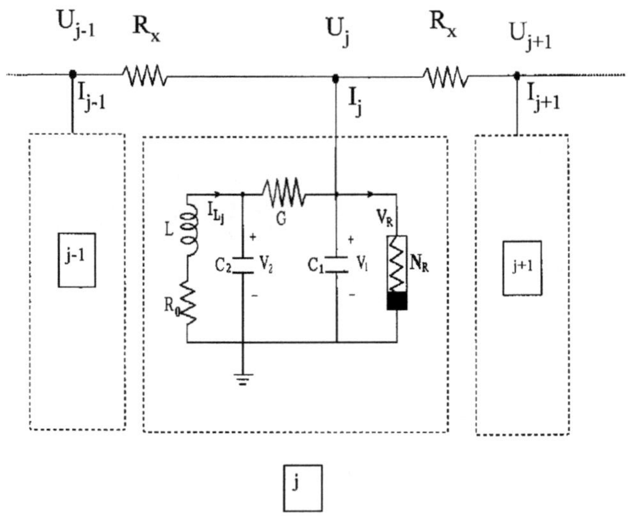

Fig. 4.3. Block diagram, with unit j, of the 1D array of Chua's circuits coupled by the resistor R_x

goes through all positive, zero, and negative integers ($j \in \mathbf{Z}$). Subsequently, we investigate the influence of boundary conditions on the dynamics of a finite chain.

4.2.1 Steady States and Spatial Structures

Existence of Steady States. Smale's Horseshoe Map. The steady states of (4.4) are determined by

$$\begin{cases} \alpha[y_j - x_j - f(x_j)] + d(x_{j-1} - 2x_j + x_{j+1}) = 0\,, \\ x_j - y_j - \beta/\gamma y_j = 0\,, \\ z_j - \beta/\gamma y_j = 0\,. \end{cases} \tag{4.5}$$

Hence

$$\frac{\alpha\beta}{\gamma+\beta}x_j - \alpha f(x_j) + d(x_{j-1} - 2x_j + x_{j+1}) = 0\,,$$

which can be recast in the following 2D system:

$$\begin{cases} u_{j+1} = x_j\,, \\ x_{j+1} = -u_j + 2x_j + \dfrac{\alpha}{d}F(x_j)\,, \end{cases} \tag{4.6}$$

with $x_{j-1} = u_j$,

$$f(x) = \begin{cases} k_1 x + k_1 + k_0 \,, & \text{if} \quad x \leq 1 \,, \\ -k_0 x \,, & \text{if} \quad |x| \leq 1 \,, \\ k_2 x - k_2 - k_0 \,, & \text{if} \quad x \geq -1 \,, \end{cases}$$

and

$$k_i \equiv \frac{b_i \gamma + (b_i + 1)\beta}{\gamma + \beta} \,, \qquad i = 1, 2 \,,$$

$$k_0 \equiv \frac{\gamma}{\gamma + \beta} - (a + 1) \,.$$

The system (4.6) can be considered as a dynamical system on the plane, defined by the point map S

$$(u, x) \rightarrow \left(x, -u + 2x + \frac{\alpha}{d} F(x) \right) \,.$$

Any bounded trajectory of S corresponds to some steady state of (4.4). S has three fixed points: $O_1\,(u = x = x_1^*)$, $O_2\,(u = x = x_2^*)$, and $O_0\,(u = x = 0)$, where

$$x_1^* = -\frac{k_0 + k_1}{k_1} \quad \text{and} \quad x_2^* = \frac{k_0 + k_2}{k_2} \,.$$

Points O_1 and O_2 are saddles with positive eigenvalues, and O_0 either is a saddle with negative eigenvalue or has complex eigenvalues of unit modulus. Accordingly, (4.4) has three spatially homogeneous steady states which correspond to these fixed points. Moreover, for some values of parameters there is a region Π in the (u, x) plane, where the map S acts like a "Smale's horseshoe" map. This can be shown using the technique for 2D maps given in Chap. 7.

Let us consider the following rectangle in the phase plane of S:

$$\Pi = \{(u, x) \,|\, |u| \leq p, |x| \leq q\} \,;$$

p and q are parameters satisfying the following conditions:

$$\begin{aligned} & p > \max\left\{ \frac{k_1 + k_0}{k_1}, \frac{k_2 + k_0}{k_2} \right\}, \\ & q < \max\left\{ \frac{k_1 + k_0}{k_1}, \frac{k_2 + k_0}{k_2} \right\}, \\ & p > q \,, \quad p < -q - 2 + \frac{\alpha}{d} k_0 \,, \\ & p < q + \frac{\alpha k_1}{d} q - \frac{\alpha}{d}(k_1 + k_0) \,, \\ & p < q + \frac{\alpha k_2}{d} q - \frac{\alpha}{d}(k_2 + k_0) \,. \end{aligned} \tag{4.7}$$

If (4.7) has a solution, then after one iteration of S the rectangle Π maps into a region $\tilde{\Pi}$ which intersects Π in three regions, Ω_1, Ω_0 and Ω_2 (Fig. 4.4).

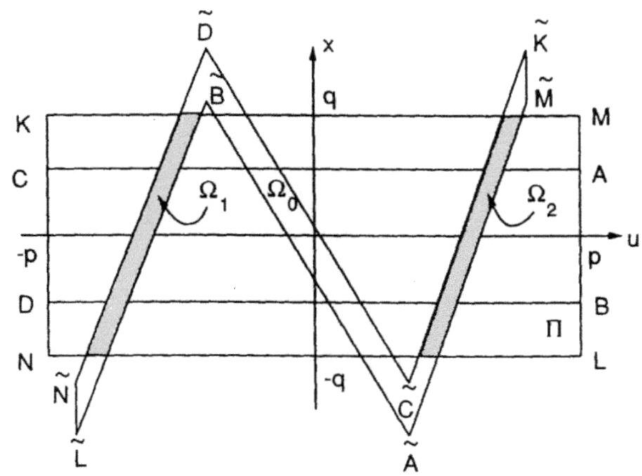

Fig. 4.4. The mapping of the region Π is due to the iteration of the map S. S defines the coordinates of the steady states of (4.4). $\tilde{A}, \tilde{B}, \tilde{C}, \ldots$ are the images of the corresponding points $A, B, C, \ldots$

Then (4.7) are satisfied if the parameters of (4.4) belong to the region defined by

$$\frac{d}{\alpha} < \min\left\{\frac{k_0 k_1}{2(k_0+2k_1)}, \frac{k_0 k_2}{2(k_0+2k_2)}\right\} . \tag{4.8}$$

Let us denote by ω_i the image of Ω_i under the action of the inverse map S^{-1}, i.e. $\omega_i = S^{-1}(\Omega_i)$, $i = 1, 2$. Thus we consider only those trajectories of S which do not enter Ω_0, since, as it will be shown later, any trajectory of S which intersects Ω_0 is unstable. The sufficient conditions for hyperbolicity of S on ω_i yield that any trajectory of S which lies completely in $\Omega_1 \cup \Omega_2$ for all values of j is hyperbolic (saddle). In other words, when (4.8) is satisfied and there is a region Π, in the (u, x) plane, where S acts like the "Smale's horseshoe" map. Let Δ be a set of hyperbolic trajectories lying in $(\Omega_1 \cap \omega_1) \cup (\Omega_2 \cap \omega_2)$. Using symbolic dynamics we can find that $S_{|\Delta}$ is topologically conjugate to the map of the Bernoulli's shift of two symbols. Thus S has some set of bounded trajectories each in one-to-one correspondence with any doubly infinite sequence of two symbols (i.e. infinite in both directions). Since all trajectories of this set correspond to steady states of (4.4), there is a continuum of such states. It follows that the profile of such states "along" the spatial coordinate j can be highly diverse: it includes simple homogeneous distributions (corresponding to fixed points of S), periodic distributions (corresponding to periodic trajectories of S), "soliton" distributions (corresponding to homoclinic trajectories of S), etc. Moreover, the profiles of steady states can be extremely complex, since there is a trajectory of S that corresponds to any doubly infinite sequence of two symbols, generated by Bernoulli's shift.

Stability of Steady States. Let us now investigate the stability of the steady states. If the chain is unbounded ($j \in \mathbf{Z}$), we can apply the method of the finite-dimensional approximation spectrum for arbitrary linear operators L. With this method an infinite matrix representing an operator L is approximated by a sequence of finite-dimensional matrices of increasing dimension. As we shall prove in Chap. 7, spec L belongs to the closure of the union of the set of the eigenvalues of all approximating matrices. These finite-dimensional matrices are constructed using the boundary conditions for the corresponding finite chain. Consequently, it suffices to study the stability of the steady states of (4.4) for a finite chain $j = 1, 2, \ldots, N$. Let us add to (4.4) the boundary conditions

$$x_0 \equiv x_1 \quad \text{and} \quad x_{N+1} \equiv x_N \,. \tag{4.9}$$

Condition (4.9) implies that there will be only a discrete set of steady states for (4.4) with (4.9) and not a continuum, as for an unbounded chain. However, in the case of a "long" chain ($N \gg 1$), the systems (4.4) and (4.9) have indeed a large number of steady states and multistability. Let

$$S_j = \left\{ x_j = \psi_j,\ y_j = \frac{\gamma}{\gamma+\beta}\psi_j,\ z_j = -\frac{\beta}{\gamma+\beta}\psi_j \right\}$$

be one of these states. Its corresponding characteristic determinant is $Q_N = \det(P_N - \lambda I_N)$, where I_N is the $N \times N$ identity matrix and P_N is the block matrix:

$$P_N = \begin{pmatrix} \Sigma_1 & D & 0 & & \cdots & 0 \\ D & \Sigma_2 & D & & \cdots & 0 \\ 0 & D & \Sigma_3 & & \cdots & 0 \\ \vdots & & & \ddots & & \\ \vdots & & & & \Sigma_{N-1} & D \\ 0 & \cdots & \cdots & & D & \Sigma_N \end{pmatrix},$$

with

$$D = \begin{pmatrix} d\,0\,0 \\ 0\,0\,0 \\ 0\,0\,0 \end{pmatrix}, \quad \Sigma_j = \begin{pmatrix} -\sigma_j & \alpha & 0 \\ 1 & -1 & 1 \\ 0 & -\beta & -\gamma \end{pmatrix} \quad j = 1, 2, \ldots, N,$$

$$\begin{aligned} \sigma_l &= \alpha + d + \alpha f'(\psi_l)\,, & l &= 1, N\,, \\ \sigma_k &= \alpha + 2d + \alpha f'(\psi_k)\,, & k &= 2, \ldots, N-1\,. \end{aligned}$$

Let us consider, first, the stability of the spatially *homogeneous* states, i.e. $\psi_j = x^0$, where $x^0 = x^*_1, x^*_2, 0$. In this case $\sigma_1 = \sigma_N \equiv \sigma, \sigma_k \equiv \sigma$. Expanding Q_N along the first three rows, we obtain

$$Q_N = \epsilon(2z+1)A_{N-1} - \epsilon B_{N-1}\,, \tag{4.10}$$

where A_{N-1} and B_{N-1} are determinants, and

$$\begin{aligned} \epsilon &\equiv d[\lambda^2 + (1+\gamma)\lambda + \gamma + \beta]\,, \\ 2z\epsilon &\equiv \alpha(\lambda+\gamma) - (\lambda+\sigma_0)[\lambda^2 + (1+\gamma)\lambda + \gamma + \beta]\,. \end{aligned} \tag{4.11}$$

Next, let us expand the determinants A_{N-1} and B_{N-1} along the corresponding rows, such that

$$\begin{aligned} A_{N-1} &= \epsilon(2z+1)D_{N-2} - \epsilon^2 D_{N-3}\,, \\ B_{N-1} &= \epsilon^2(2z+1)D_{N-3} - \epsilon^3 D_{N-4}\,. \end{aligned} \tag{4.12}$$

where $D_m = \det(C_m - \lambda I_m)$, and the block matrix of size $m \times m$ ($m \leq N-2$), C_m, is

$$C_m = \begin{pmatrix} \Sigma_0 & D & 0 & & \cdots & 0 \\ D & \Sigma_0 & D & & \cdots & 0 \\ 0 & D & \Sigma_0 & & \cdots & 0 \\ \vdots & & & \ddots & & \\ \vdots & & & & \Sigma_0 & D \\ 0 & \cdots & \cdots & & D & \Sigma_0 \end{pmatrix},$$

$$\Sigma_0 = \begin{pmatrix} -\sigma_0 & \alpha & 0 \\ 1 & -1 & 1 \\ 0 & -\beta & -\gamma \end{pmatrix}.$$

Expanding D_m along the first row we obtain the following recurrent relation:

$$D_m = 2z\epsilon D_{m-1} - \epsilon^2 D_{m-2}\,. \tag{4.13}$$

We shall treat (4.13) as a difference equation with initial conditions $D_1 = 2z\epsilon$ and $D_2 = \epsilon^2(4z^2-1)$. Solving this equation we find

$$D_m = \epsilon^m U_m(z), \tag{4.14}$$

where $U_m(z)$ is the Chebyshev polynomial of the second kind. Equations (4.10–14) yield

$$Q_N = \epsilon^N[(2z+1)^2 U_{N-2}(z) - 2(2z+1)U_{N-3}(z) + U_{N-4}(z)]\,.$$

Hence, using the property of Chebyshev polynomials

$$U_{r+1}(z) = 2zU_r(z) - U_{r-1}(z)\,,$$

we have

$$Q_N = 2\epsilon^N(z+1)U_{N-1}(z)\,. \tag{4.15}$$

Then $Q_N = 0$, if

$$z = -1\,, \quad U_{N-1}(z) = 0\,. \tag{4.16}$$

Since the roots of the Chebyshev polynomials are $z_s = \cos s\pi/N$, (4.11) and (4.16) easily yield the equation for the Lyapunov characteristic eigenvalues:

$$\begin{gathered}\lambda^3 + [1+\gamma+\alpha(1+f'(x^0)) + 2d(1+z_s)]\lambda^2 \\ +[\beta+\gamma(1+\alpha)\alpha(1+\gamma)f'(x^0) + 2d(1+\gamma)(1+z_s)]\lambda \\ +[\alpha\beta+\alpha(\beta+\gamma)f'(x^0) + 2d(\beta+\gamma)(1+z_s)] = 0\,, \\ s = 1, 2, \ldots, N\,.\end{gathered} \tag{4.17}$$

A remarkable fact is that when $s = N$, (4.17) coincides with the characteristic equation for the steady states of the system composed of a single Chua circuit.

Analyzing for each s the position of the roots of (4.15) in the complex plane, it appears that for each set of parameter values the "outer" steady states are asymptotically stable, while the trivial steady state is unstable.

Let us now consider the stability of spatially inhomogeneous steady states. If $b_1 = b_2$, the conditions for the stability of such states coincide with the conditions for stability of the nontrivial spatially steady inhomogeneous states. When $b_1 = b_2$ the value of $f'(\psi_j)$ does not depend on the homogeneous steady states. Unfortunately, if $b_1 \neq b_2$, the principal diagonal of the determinant Q_N contains different elements, and one cannot obtain an equation of the same type as (4.17). However, in this case we can use the Gershgorin disks (Appendix G) [4.4] to find sufficient conditions for stability. Namely, when the inequalities

$$\gamma > 1\,, \quad \beta < 1-\alpha\,, \quad \alpha > \max\left\{\frac{1}{b_1+1}, \frac{1}{b_2+1}\right\}\,, \tag{4.18}$$

are fulfilled, all spatially inhomogeneous states corresponding to the regions Ω_i are stable.

Note that using the method of finite-dimensional approximations we can show that these stability properties are also valid for an unbounded chain ($N \to \infty$). Thus, we single out the parameter values such that (4.4) contains a set (continuum as $N \to \infty$) of stable steady states which are in one-to-one correspondence with the topological Bernoulli's shift of two symbols. This means that, depending on the initial conditions, different spatial structures are established whose profiles are determined by a given arbitrary doubly infinite sequence of two symbols. It follows that for large but finite N, (4.4) has a very large number of steady states, and as $N \to \infty$ we have spatial chaos. We shall return to this matter in Chap. 7.

Spatial Structures. Any steady state of (4.4) is defined in $\{\mathbf{Z}, \mathbf{R}\}$ as a 1D stationary spatial structure. It represents the distribution of steady-state coordinates, for example x_j, "along" index j. Since this distribution is determined by the trajectories of S having chaotic dynamics, then there exists in (4.4) a very large number of highly diverse spatial structures (from spatially homogeneous to spatially chaotic). When the array has a finite length, with N units, the number of spatial structures equals 2^N, and hence (4.4) represents a discrete medium capable of exhibiting high multistability. For illustration, we consider (4.4) with $N = 50$ elements. For the parameter values satisfying the conditions (4.8) and (4.18), different initial distributions bring the system to one of the spatial structures. Figure 4.5a,b illustrate the spatial structures obtained in the evolution of the array from two different initial distributions (Fig. 4.5a,b).

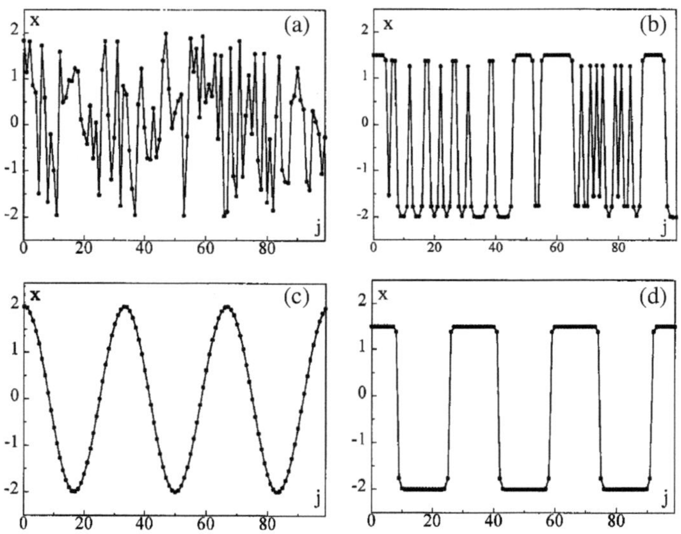

Fig. 4.5. Examples of the spatial structures (**c,d**) formed from the initial distributions (**a,b**). The parameters are $m_1 = 5, m_2 = 2, m_0 = -0.5, \alpha = 0.8, \beta = 0.1, \gamma = 1.1, d = 0.1$

4.2.2 Wave Fronts in a Gradient Approximation

The Case When $\gamma \gg 1$: a Gradient System. Consider (4.4) and (4.9) when $\gamma \gg 1$. We can take advantage of the smallness parameter, $\mu = 1/\gamma$, in the third equation of (4.4). There exists a stable manifold of slow motions (dimension N) which attracts all trajectories. With accuracy up to terms of higher order, this manifold is defined by $z_j = 0$, and hence all motions restricted to this manifold are governed by

$$\begin{cases} \dot{x}_j = \alpha[y_j - x_j - f(x_j)] + d(x_{j+1} - 2x_j + x_{j-1}), \\ \dot{y}_j = x_j - y_j. \end{cases} \tag{4.19}$$

By defining a scalar "potential" function

$$U = \sum_{j=1}^{N} \left(\frac{y_j^2}{2} - x_j y_j + \frac{x_j^2}{2} + \int_0^{x_j} f(x)\,\mathrm{d}x + \frac{d}{2\alpha}(x_{j+1} - x_j)^2 \right),$$

(4.19) can be recast in terms of U as follows:

$$\dot{x}_j = -\alpha \frac{\partial U}{\partial x_j} \quad \text{and} \quad \dot{y}_j = -\frac{\partial U}{\partial y_j}. \tag{4.20}$$

Thus (4.19) is a gradient system. Note, that the extrema of the function U coincide with the steady states of the system, which in turn are determined by the trajectories of the map S. It follows that the minima of U correspond to the stable steady states. Since every trajectory in a gradient system must tend to a steady state as $t \to \infty$, it follows that each initial condition must tend to one of the steady states, which corresponds to some spatial structure. Let us now consider the dynamical process of establishing spatially homogeneous steady states in the chain (4.19).

Continuous Approximation. Traveling Waves. We look for traveling wave solutions of (4.19) of the form

$$x_j = x(\xi), \quad y_j = y(\xi), \quad \xi = t - jh, \tag{4.21}$$

where $h > 0$ is a parameter. Substituting (4.21) into (4.19), we obtain the following system of differential-difference equations:

$$\begin{cases} \dot{x} = \alpha(y - x - f(x)) + d[x(\xi + h) - 2x(\xi) + x(\xi - h)], \\ \dot{y} = x - y. \end{cases} \tag{4.22}$$

where the dot denotes differentiation with respect to the "moving coordinate" ξ. For small enough h the difference term in (4.22) can be approximately replaced by the second derivative $\ddot{x}$ (with respect to ξ), thereby obtaining a system of third-order *ordinary* differential equations (ODE). Introducing two new variables

$$\begin{cases} w = \dot{x}, \\ v = \dfrac{\alpha c^2}{d}(x - y), \end{cases}$$

with $c \equiv 1/h$, we obtain the following equivalent third-order *autonomous* system of ODE

$$\begin{cases} \dot{x} = w\,, \\ \dot{w} = v + \dfrac{\alpha c^2}{d} f(x) + \dfrac{c^2}{d} w\,, \\ \dot{v} = -v + \dfrac{\alpha c^2}{d} w\,. \end{cases} \tag{4.23}$$

Equations (4.23) describe traveling wave motions in the chain with velocity c. Since the approximation of (4.22) by (4.23) is not acceptable when $c \to 0$ (or $h \to \infty$), we assume a large enough c. System (4.23) has three steady states: $A_1(x_1^0, 0, 0), A_0(0,0,0)$ and $A_2(x_2^0, 0, 0)$, with

$$x_1^0 = -\frac{b_1 - a}{b_1} \quad \text{and} \quad x_2^0 = \frac{b_2 - a}{b_2}\,.$$

The state A_1 (or A_2) is either a saddle with one positive and two negative real eigenvalues or a saddle-focus with one real positive eigenvalue and a pair of complex conjugate eigenvalues with negative real part. Therefore the latter has a 1D unstable manifold W^u (S^u) and a 2D stable manifold W^s (S^s). The manifold W^u (S^u) consists of the point A_1 (A_2) and two outgoing trajectories W_1^u and W_2^u (S_1^u and S_2^u). In the (x, w, v) phase space one of the trajectories W_1^u (S_1^u) extends to the region $x > x_1^0$ and the other, W_2^u (S_2^u), to the region $x < x_1^0$ ($x < x_2^0$). The steady state A_0 is also a saddle, but its 1D manifold is stable, while its 2D manifolds are unstable.

The Surfaces of Localization of One-Dimensional Unstable Manifolds. For (4.23), let us define the scalar function V:

$$V(x, w, v) = \frac{v^2}{2} + \frac{\alpha c^2}{d}\frac{w^2}{2} + \frac{\alpha^2 c^4}{d^2} \int_{x_1^0}^{x} f(x)\,\mathrm{d}x,$$

whose derivative along the flow is given by

$$\dot{V} = -v^2 - \frac{\alpha c^4}{d^2} w^2 \leq 0\,. \tag{4.24}$$

Note that $\dot{V} = 0$ only along the line $\{w = v = 0\}$, i.e. along the x-axis. Consequently, on the 2D surface in $\mathbf{R}^3$ defined by $V = C =$ const. (excluding the points A_1, A_2 and A_0), shown in Fig. 4.6 as a 2D constant-level surface in $\mathbf{R}^3$, parametrized by C, the vector field (4.23) is directed into the region bounded by this surface. It follows that the trajectories of (4.23) originating from all points $(x, y, w) \in \mathbf{R}^3$, excluding those points where $\{V = C\} \cap \{w = v = 0\} \neq 0$, must transversely cross this constant-level surface. Taking into consideration these properties of the trajectories of (4.23) and analyzing the configuration of the surfaces $V =$ const. in the phase space, it follows that (4.23) cannot have periodic and homoclinic trajectories. Since our goal in this section is to prove the existence of wave fronts which correspond to heteroclinic orbits "connecting" points A_1 and A_2 or vice versa, let us

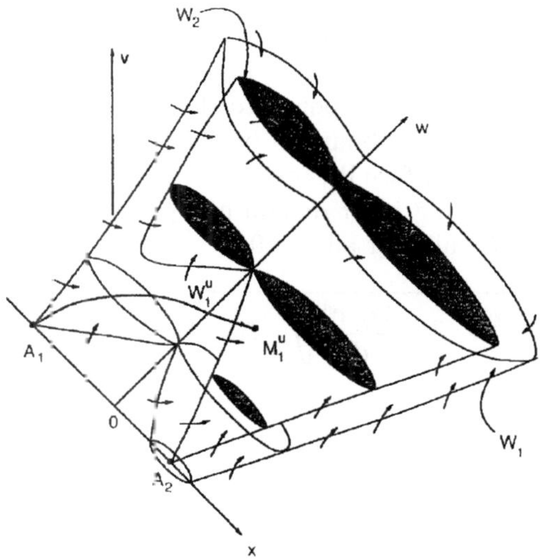

Fig. 4.6. The orientation of the vector field of (4.22) through the surfaces of localization. Tracking is shown by *arrows*

therefore assume that $b_1 > b_2$. In this case there exist two heteroclinic orbits "joining" points A_1 and A_2. Indeed, consider first the case $b_1 > b_2$ and denote by W_2 the constant-level surface $V = C_2$, passing through the steady state A_2, i.e.

$$W_2 : \{V(x, w, v) = C_2\},$$

with

$$C_2 \equiv \frac{\alpha^2 c^4}{d^2} \int_{x_1^0}^{x_2^0} f(x)\,\mathrm{d}x.$$

Qualitatively, W_2 consists of two double-well "conic" surfaces with the apex located at A_2. One of them, shown in Fig. 4.6, and denoted by W_2^+, is completely confined to the half–space $w > 0$, while the other (not shown), W_2^-, is confined to the half-space $w < 0$. Accordingly, for all points of the surface W_2 (except the point A_2), as well as all points on the surfaces $V = C$, for all $C < C_2$ the inequality $\dot{V} < 0$ is satisfied. Consequently, the trajectories of (4.23) cross each of these surfaces transversely (not tangentially) into the region bounded by each surface. Therefore, all trajectories of (4.23) crossing the surface W_2^+ (W_2^-), as well as the trajectories S_1^u (S_2^u) of the saddle point A_2, must remain in the half-space $w > 0$ ($w < 0$) and tend to ∞ as $\xi \to +\infty$. Thus for $b_1 > b_2$, the two trajectories of the point A_2 cannot form heteroclinic orbits with A_1. In a similar way, considering the constant-level surface $V = C$, $C \le 0$, one can show that the trajectory W_2^u also tends to infinity at $\xi \to \infty$.

Let us now investigate the intersections between the trajectory W_1^u of the saddle point A_1 and the surface W_2^+. Let $\eta_1^u(\xi) : \{x_1^u(\xi), w_1^u(\xi), v_1^u(\xi)\}$ be the solution of (4.23) that corresponds to the trajectory W_1^u, and let B_1 be the region of the phase space bounded by the surface $W_1 : \{V(x,w,v) = 0\}$. Since everywhere on the surface W_1 (excluding the point A_1) (4.23) is oriented inwards, the trajectory W_1^u must remain in B_1 for any $\xi > 0$. Therefore, the following inequality is valid for all $\xi > 0$:

$$|v_1^u(\xi)| < \sqrt{\frac{\alpha c^2}{d} w^2 - \frac{2\alpha^2 c^4}{d^2} \int_{x_1^0}^{x} f(x)\,\mathrm{d}x} \leq \sqrt{\frac{\alpha c^2}{d} y^2 + \frac{2\alpha^2 c^4}{d^2} r}\,, \quad (4.25)$$

with

$$r = -\int_{x_1^0}^{x_2^0} f(x)\,\mathrm{d}x = \frac{a^2(b_1 - b_2)}{2b_1 b_2}\,.$$

Heteroclinic Orbits.

Auxiliary Systems. Let us now show that there is a region where the trajectory W_1^u crosses the "conic" surface W_2^+ at some point M_1^u (Fig. 4.6). To do this let us introduce the 2D system

$$\begin{cases} \dot{x} = w\,, \\ \dot{w} = \dfrac{\alpha c^2}{d} f(x) + \dfrac{c^2}{d} w - \sqrt{\dfrac{\alpha c^2}{d} w^2 - \dfrac{2\alpha^2 c^4}{d^2} r}\,, \end{cases} \quad (4.26)$$

which will be used as the "comparison" system. Since at variance with (4.23) here (4.26) do not depend on v, its trajectories lie on cylindrical surfaces in the (x, v, w) phase space. Let us find the orientation of the vector field (4.23) on these surfaces. First, let us calculate

$$R_1 \equiv \left(\frac{\mathrm{d}w}{\mathrm{d}x}\right)_{(4.23)} - \left(\frac{\mathrm{d}w}{\mathrm{d}x}\right)_{(4.26)} = \frac{v + \sqrt{\dfrac{\alpha c^2}{d} w^2 - \dfrac{2\alpha^2 c^4}{d^2} r}}{w}\,. \quad (4.27)$$

It follows from (4.25) and (4.27) that for points of the region B_1 with $w > 0$ the trajectories of (4.23) must cross the cylindrical sheets transversely in the direction in which w increases. For $b_2 < 1$ and $(b_1 + 1)(1 - b_2) < 1$, (4.26) has three steady states. The state farthest away from the origin is a saddle. Then let $O_1^-(x_1^0, 0)$ be a saddle point of (4.26) for which $x_1^- > x_1^0$. We denote by w^- the unstable separatrix of O_1^- that lies in the half-plane $w > 0$. In a similar way we can show that there exists a function $\varphi^*(\alpha)$ such that, if the parameters α, c and d satisfy the condition $c^2/d = \varphi^*(\alpha)$, then the separatrix w^- lies completely in the half-plane $w > 0$ and tends to infinity after crossing the straight line $x = x_2^0$. We denote by W^- the cylindrical surface formed in the phase space of (4.23) by the saddle point O_1^- and its

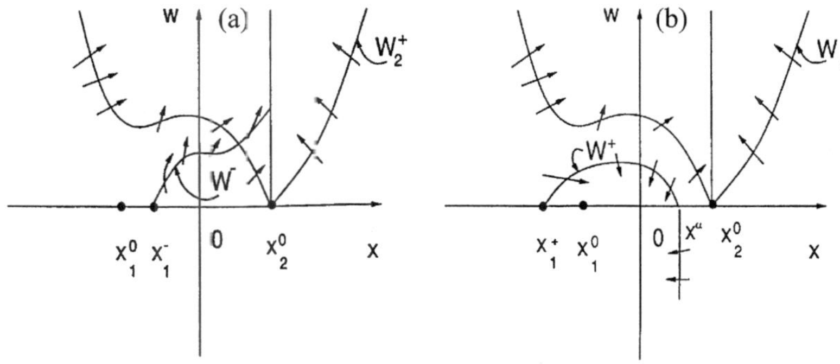

Fig. 4.7. The intersections of the "conic" surface W_2^+ and the cylindrical surfaces (**a**) W^+ and (**b**) W^- by the plane $v = 0$. W^- and W^+ are the separatrices of the saddle points from the 2D systems (4.25) and (4.27). The *small arrows* show the vector field orientation of (4.22) through these surfaces

separatrix w^-. Then when $c^2/d > \varphi^*(\alpha)$ the surface W^- crosses the surfaces $\{x = x_2^0\}$ and $\{w = 0\}$ (Fig. 4.7a) and, being completely confined within the region $w \geq 0$, must tend to infinity. Moreover, W^- must cross W_1 in such a way that the section is a curve confined completely within the region $w > 0$. Therefore, the separatrix W_1^u leaving to region B_1 cannot cross the plane $w = 0$ and must lie completely within the region $w > 0$. It follows from the inequality $\dot{V} < 0$, satisfied for $w > 0$, that the separatrix W_1^u must cross the "conic" surface W_2^+ at some point M_1^u and must tend to infinity while staying within the region of the phase space bounded by W_2^+. Thus, for points of the parameter region $d_1^+ \equiv \{c^2 > d\varphi^*(\alpha)\}$ the trajectory W_1^u of the saddle point A_1 must cross the conic surface W_2^+ at a single point M_1^u (Fig. 4.6).

Let us derive, next, the parameter values such that the trajectory W_1^u does not cross the conic surface W_2^+. To do this, let us introduce the auxiliary system

$$\begin{cases} \dot{x} = w\,, \\ \dot{w} = \dfrac{\alpha c^2}{d} f(x) + \dfrac{c^2}{d} w + \sqrt{\dfrac{\alpha c^2}{d} w^2 - \dfrac{2\alpha^2 c^4}{d^2} r}\,. \end{cases} \tag{4.28}$$

For $b_2 < 1$ and $(b_1+1)(1-b_2) < 1$, (4.28) has the same steady states as (4.26). Let $O_1^+(x_1^+, 0)$ be a saddle point with $x_1^+ < x_1^0$. Following the procedure used for the system (4.26) we can show that in this case there exists a function $\varphi^0(\alpha) < \varphi^*(\alpha)$ such that, if the parameters α, d and c satisfy the condition $c^2/d = \varphi^0(\alpha)$, then the unstable separatrix w^+ of the saddle O_1^+ must cross the straight line $x = x_2$ at some point $w = 0$. When $c^2/d < \varphi^0(\alpha)$, the unstable separatrix w^+ of O_1^+, which traverses initially the region $w > 0$, must eventually cross $w = 0$ at $x = x^\alpha < x_2^0$ (Fig. 4.7b) and, moreover, must diverge to minus infinity in the half-plane $w < 0$. Let us denote by

W^+ the cylindrical surface in the phase space of (4.23) when $c^2 < d\varphi^0(\alpha)$, formed by a piece of the separatrix w^+ confined between the points x_1^+ and x^α (Fig. 4.7b). For points in the region $d_2^+ \equiv \{c^2 < d\varphi^0(\alpha)\}$ of the (x, w, v) phase space, W_1 and W^+ exist simultaneously and cross each other. On the part of W^+ confined within the region B_1, (4.23) for $w > 0$ is oriented toward decreasing w-coordinate values. This implies the following inequality (4.25):

$$R_2 \equiv \left(\frac{dw}{dx}\right)_{(4.23)} - \left(\frac{dw}{dx}\right)_{(4.28)} = \frac{v - \sqrt{\dfrac{\alpha c^2}{d} w^2 - \dfrac{2\alpha^2 c^4}{d^2} r}}{w} < 0 .$$

Now consider in the (x, w, v) phase space the cylindrical surface P consisting of W^+ and the half-plane $P^+ : \{x = x^\alpha, w < 0, v \in R^1\}$. At all points of P^+ (4.23) is directed toward decreasing x-coordinate values, because P^+ is in the region $w < 0$, where, in view of the first equation in (4.23), we have $\dot{x} < 0$. The surface P crosses W_1 and separates some region $B_0 \in B_1$ in the (x, w, v) phase space. Taking into account the orientation of (4.23) at the boundary of B_0, and the fact that point A_1 belongs to this region, we find that in the region d_2^+ the following relation holds:

$$\eta_1^u(\xi) \in B_0 \, , \; \forall \xi \in R^1 \, . \tag{4.29}$$

Since region B_0 does not have common points with the "conic" surface W_2^+, (4.29) implies that, in the regions of the parameters d_2^+, the trajectory W_1^+ does not cross the surface W_2^+.

Existence of Heteroclinic Orbits. In the (x, w, v) phase space there is a "conic" surface W_2^+ (Fig. 4.6), relative to which the behavior of the trajectory W_1^u is different. If the parameters of (4.23) belong to the region d_1^+, the trajectory W_1^u of the steady state A_1 crosses W_2^+ at a single point, M_1^u. If, however, they belong to the region d_2^+, the surface W_1^u does not cross the surface W_2^u, and the intersection point M_1^u does not exist. Hence, in any continuous transition in the space of parameters from d_1^+ to d_2^+, the point M_1^u must disappear from the "conic" surface W_2^+. Since the trajectories of (4.26) cross the surface W_2^+ at all points, except A_2, and since within the (x, w, v) phase space there is a family of surfaces $V = C =$ const., where (4.23) is directed toward the region $V < C$, the point M_1^u can disappear only when the trajectory W_1^u passes through A_2. Thus there is at least some range of parameter values in which the trajectory W_1^u forms a heteroclinic orbit Γ^+ "joining" the points A_1 and A_2. In the function space of dynamical systems, those having such a trajectory linkage form a smooth surface of co-dimension one. A cross-section of this surface with the family of systems (4.26) [we assume that (4.26) is generic] is also some smooth surface, each of whose points corresponds to a heteroclinic orbit Γ^+. Therefore, the range of parameters of (4.23), where a heteroclinic orbit Γ^+ exists, forms a smooth surface Π^+. A view of the cross-section of this surface is shown in the (d, c) plane in Fig. 4.8. The regions d_1^+,

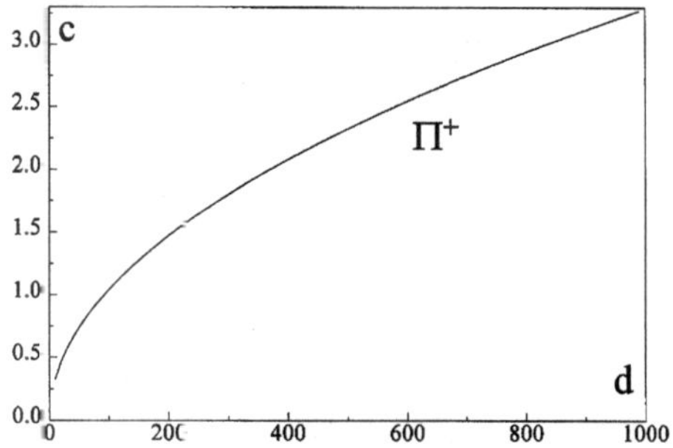

Fig. 4.8. The bifurcation curve Π^+ corresponding to the heteroclinic orbit Γ^+. The curve Π^+ defines the velocity of the wave front from a chain of coupled Chua's circuits

d_2^+ and Π^+ are identified for $c > 0$, since, as earlier mentioned, the continuous model (4.23) is not valid as $c \to 0$. Similarly, we can prove the existence, for $m_1 < m_2$, of some curve, Π^-, corresponding to the existence of a heteroclinic orbit, Γ^-, which "joins" A_2 and A_1.

Wave-Front Profiles. The existence of the heteroclinic orbits Γ^+ and Γ^- of (4.23) ensures the existence of traveling wave fronts of (4.19), and hence along the original chain with $\gamma \gg 1$. For $b_1 > b_2$ the propagation of a wave front leads to the establishment in the entire chain of a spatially homogeneous state $\{x_j = x_1^0, y_j = y_1^0\}$. Similarly, for $b_1 < b_2$, the traveling wave settles to the spatially homogeneous state $\{x_j = x_2^0, y_j = y_2^0\}$. The dependence of the wave-front velocity, c, on the diffusion coefficient, d, is determined by the

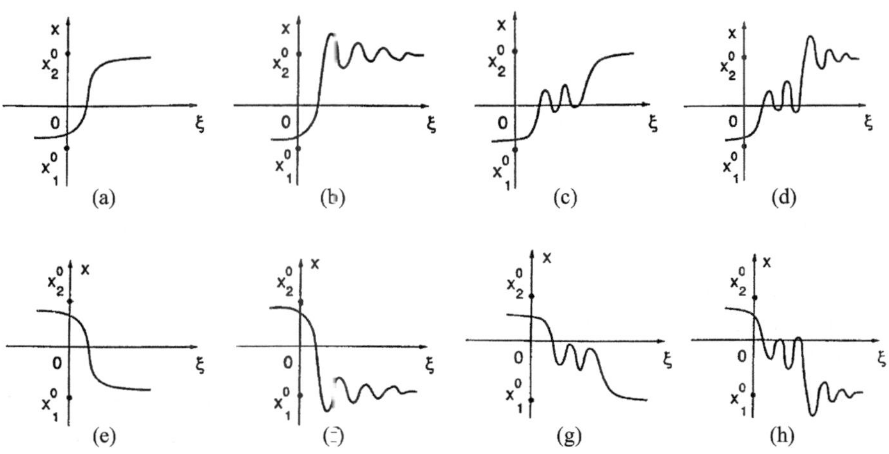

Fig. 4.9. The approximate shape of the wave fronts in a chain of coupled Chua's circuits. **(a–d)** The propagating wave fronts which give rise to the homogeneous steady state $\{x_j = x_1^0, y_j = x_2^0\}$. **(e–h)** The propagating wave fronts which give rise to the homogeneous steady state $\{x_j = x_2^0, y_j = x_2^0\}$

curves Π^+ and Π^-. Figure 4.9a–d show a qualitative view of the wave fronts corresponding to the curve Π^+, while those of Π^- are shown in Fig. 4.9e–h. The shape of the wave fronts is determined by the location of the separatrix forming the heteroclinic orbit in the (x, w, v) phase space. If, as $\xi \to +\infty$, the separatrix tends to a saddle-focus, the wave front contains an *oscillatory tail* (Fig. 4.9b,d,f,h). If it tends to a saddle, the corresponding tail will decay *monotonically* (Fig. 4.9a,c,e,g). The wave front itself can be either monotonic (Fig. 4.9a,e) or exhibit a finite number of oscillations (Fig. 4.9c,g). This is determined by the behavior of the trajectories W_1^u and S_2^u in the vicinity of A_0.

4.2.3 Pulses, Fronts and Chaotic Wave Trains

Let us now consider the spatio-temporal dynamics of (4.4) for arbitrary values of γ and two types of boundary conditions: zero-flux conditions (4.9) and periodic conditions (circular array)

$$x_{j+N}(t) = x_j(t)\,. \tag{4.30}$$

We assume that the function $f(x)$ satisfies (4.3) with $b_1 = b_2 = b > 0$ and $a < 0$. As shown in Sect. 4.2.1, (4.4) has three fixed points corresponding to the homogeneous steady states of the array. To simplify notations we introduce

$$\begin{aligned} O\ &: \{x_j = y_j = z_j = 0\}\,, \\ P^+ &: \{x_j = x_0, y_j = y_0, z_j = z_0\}\,, \\ P^- &: \{x_j = -x_0, y_j = -y_0, z_j = -z_0\}\,, \end{aligned}$$

with

$$x_0 = \frac{(b-a)(\gamma+\beta)}{\gamma b + \beta(b+1)}\,, \quad y_0 = \frac{(b-a)\gamma}{\gamma b + \beta(b+1)}\,, \quad z_0 = -\frac{(b-a)\beta}{\gamma b + \beta(b+1)}\,.$$

As shown in Sect. 4.2.1, the "outer" states P^- and P^+ are locally asymptotically stable, while the origin O is unstable. For some class of excitations on these states, the array can support stable localized solutions (fronts, pulses, pulse trains) traveling in space with finite velocity, as we now show. Note that the terms soliton, solitary wave or pulse would be used according to the context.

Traveling Waves. Properties of the Phase Space. As done in Sect. 4.2.2 here we study the existence of traveling waves using the continuum approximation. Let us look for a solution of (4.4) in the following form:

$$\begin{aligned} x_j(t) &= x(\xi)\,, \\ y_j(t) &= y(\xi)\,, \\ z_j(t) &= z(\xi)\,, \end{aligned}$$

where $\xi = t + jh$ is a coordinate moving along the array with a constant velocity $c = 1/h$. In this case, as γ has an arbitrary though finite value, we have the following fourth-order system of ODE [compare to (4.23)]:

$$\begin{cases} \dot{x} &= u\,, \\ k\dot{u} &= u - \alpha[y - x - f(x)]\,, \\ \dot{y} &= x - y + z\,, \\ \dot{z} &= -\beta y - \gamma z\,, \end{cases} \tag{4.31}$$

where $k = d/c^2$ is a parameter characterizing the dependence of the velocity of the waves on the magnitude of the diffusion d.

Any bounded trajectory of (4.31) determines the possible profile of a traveling wave which uniformly translates along the array with constant velocity. The trivial translating solutions, homogeneous steady states O and $P^{\pm}$ of the array, correspond to the steady points of (4.31), which we denote by the same letters. Nonconstant solutions homoclinic to the points $P^{\pm}$ define solitary pulses propagating relative to the "background" steady states $P^{\pm}$. Orbits that with $\xi \to \pm\infty$ asymptotically tend to different steady points (heteroclinic orbits) correspond to the traveling fronts selecting the terminal homogeneous stable state of the arrays. In this section we show that (4.31) admits various homoclinic and heteroclinic solutions of highly complex profile.

Let us first make some general remarks about the properties of the trajectories of (4.31) which will be helpful in studying the homoclinic and heteroclinic bifurcations.

Note, first, that the symmetry properties of the function $f(x)$ imply that the vector field of (4.31) is invariant under the transformation

$$(x, u, y, z) \to (-x, -u, -y, -z)\,. \tag{4.32}$$

Hence, for any given trajectory of (4.31) another trajectory, defined by (4.32), coexists in the phase space. For instance, if a homoclinic orbit of the steady point P^+ appears, the orbit homoclinic to the point P^- whose profile is defined by (4.32) also appears. Furthermore, due to the piecewise linearity of $f(x)$ (4.3), the 4D phase space of (4.31) can be divided into three regions where motions are governed by *linear* systems. The planes making this division are

$$U^+ : \{x = 1\} \quad \text{and} \quad U^- : \{x = -1\}\,.$$

In each linear region the dynamics of (4.31) is defined *only* by the four eigenvalues of the linear matrix whose corresponding eigenvectors define the manifolds of the steady points O and $P^{\pm}$.

Let us fix the parameters of the unit in the array with the values $\{a = -1.5, b = 2, \beta = 0.5, \gamma = 0.01\}$ and take $\{\alpha, k\}$ as control parameters.

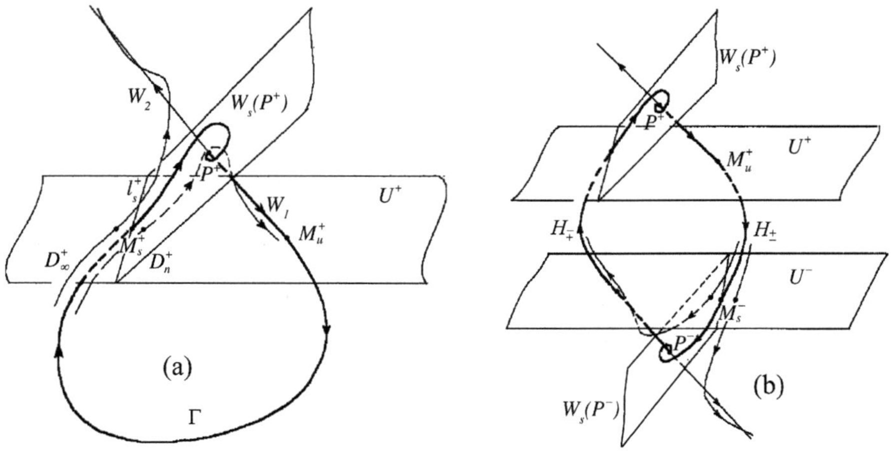

Fig. 4.10. Manifolds of the steady point $P^\pm$ in the phase space of (4.31). (**a**) Qualitative behavior of the trajectory forming a homoclinic loop and neighbouring trajectories. If the point M_u^+ is mapped to a point in region D_∞^+, in region D_n^+ or on line l_s^+, then the trajectory runs to infinity, the trajectory is going to make the next loop or homoclinic bifurcation occurs, respectively. (**b**) Qualitative behavior of the trajectory forming a heteroclinic linkage and its neighboring trajectories

Then for each $\{\alpha > 0, k > 0\}$ the eigenvalues corresponding to the "outer" points $P^\pm$ are

$$\{\lambda_1 > 0, \lambda_2 < 0, \lambda_{3,4} = -h + i\omega \ (h > 0)\} .$$

Accordingly, the points $P\pm$ have a 1D unstable manifold $W_u(P^\pm)$ and a 3D stable manifold $W_s(P^\pm)$. Note that the type of the steady point O depends on the control parameters. It can be either asymptotically stable (all eigenvalues have negative real part) or a saddle (saddle-focus) with two eigenvalues with positive real parts.

Within the corresponding linear regions, $W_s(P^\pm)$ represents the separatrix *plane* in the 4D phase space, and $W_u(P^\pm)$ the *line* containing point $P^\pm$ and two half-lines W_1 and W_2 extending to the different sides of the separatrix plane. Figure 4.10 a qualitatively illustrates the arrangement of the manifolds $W_s(P^+)$ and $W_u(P^+)$ relative to the division plane U^+ in the phase space of (4.31). Let us denote by M_u^+ the intersection point of the half-line W_1 with the plane U^+ and by l_s^+ the line of intersection of the separatrix plane $W_s(P^+)$ with the plane U^+. By the symmetry of (4.31) the corresponding point M_u^- and line l_s^- with analogous properties are well defined in the plane U^-:

$$M_u^\pm : \{W_u(P^\pm) \cap U^\pm\}, \quad l_s^\pm : \{W_s(P^\pm) \cap U^\pm\} .$$

Its neighborhood near the plane $U^\pm$ is divided by the line $l_s^\pm$ in two regions denoted by $D_\infty^\pm$ and $D_n^\pm$ (Fig. 4.10a). When the trajectories intersect ($U^\pm$ within $D_\infty^\pm$), they travel "above" the plane $W_s(P^\pm)$ and run to infinity, while those going through $D_n^{\pm}$ are located "below" the plane $W_s(P^{\pm})$ and after some time should return back to the plane $U^{\pm}$.

Homoclinic and Heteroclinic Orbits.

Homoclinic Orbits. Consider, first, how the orbit homoclinic to the steady point P^+ is formed in (4.31). It corresponds to the nonconstant solution of (4.31) that with $\xi \to \pm\infty$ asymptotically tends to the fixed point. Such a solution is the trajectory simultaneously belonging to the unstable manifold $W_u(P^+)$ and to the stable manifold $W_s(P^+)$ of the fixed point P^+. This trajectory contains the point M_u^+ and intersects, at the same time, the line l_s^+. Therefore, the condition for the existence of homoclinic orbits of P^+ is that if the initial conditions of (4.31) are taken at M_u^+, then, if the flow corresponding to (4.31) maps M_u^+ to some point $M_s^+ \in U^+$ and this point belongs to l_s^+, then a homoclinic loop, Γ, appears in the phase space of (4.31).

Let us change the control parameters, keeping them very near to the point of homoclinicity. Then the mapped point M_s^+ can shift from l_s^+ either to the region D_∞^+ or to the region D_n^+. In the first case a trajectory starting at M_u^+ will tend to infinity as shown in Fig. 4.10, while in the second case this trajectory will go to the nearby neighborhood of P^+, then intersect the plane U^+ near M_u^+, hence giving the possibility for the formation of multiloop homoclinic orbits near a given loop. According to the Shilnikov theorem this possibility occurs if the saddle quantity σ of the saddle-focus P^+,

$$\sigma = \lambda_1 + \max\{\lambda_2, h\},$$

is positive for the parameters of homoclinicity.

By the symmetry earlier mentioned (4.32), with the appearance of a loop homoclinic to the steady point P^+, there also appears a loop homoclinic to the steady point P^-, and the profile of the loop is defined by (4.32).

Heteroclinic Orbits. Let us search for heteroclinic orbits formed by the fixed points P^+ and P^-. Such orbits correspond to a solution of (4.31) which simultaneously belongs to the unstable manifold of P^+ $[W_u(P^+)]$ and to the stable manifold of P^- $[W_s(P^-)]$. Thus the solution which contains M_u^+ intersects l_s^- at some point M_s^-. Accordingly, the parameters for which the flow (4.31) maps M_u^+ to $M_s^- \in U^-$ on l_s^- also correspond to the appearance of a heteroclinic orbit $H_\pm$ "linking" the fixed points P^+ and P^-.

Applying the inverse transformation (4.32) to $H_\pm$ we obtain the heteroclinic orbit $H_\mp$ that "starts" at P^- ends at P^+. Thus, there exists a *contour* $P^+ \to H_\pm \to P^- \to H_\mp \to P^+$ in the phase space of (4.31) (Fig. 4.10b). We show below that this contour is associated with the existence in the medium of solitary pulses formed of two fronts, each of which is described by $H_\pm$ and $H_\mp$ taken separately.

Bifurcation Set of Homoclinic and Heteroclinic Orbits. To determine the parameter values corresponding to the homoclinic and heteroclinic bifurcations, we calculate, for fixed $k = k^*$, the split function

$$S_\Gamma(\alpha) = \mathrm{dev}(M_s^+, l_s^+)$$

for a homoclinic loop Γ, and the corresponding

$$S_{H_\pm}(\alpha) = \mathrm{dev}(M_s^-, l_s^-)$$

for a heteroclinic orbit $H_\pm$; dev denotes the deviation of a point from a line. The parameter α^*, corresponding to a zero of the split function, determines the bifurcation point (α^*, k^*) in the state space (α, k). The location of $l_s^\pm$ and $M_u^\pm$ is calculated in the same manner as in Sect. 4.2.2. The map $M_u^+ \to M_s^\pm$ is determined by numerical integration of (4.31).

Let us now turn to Fig. 4.11, in which the bifurcation diagram is displayed. We denote the bifurcation curves by the letters Γ and H, with

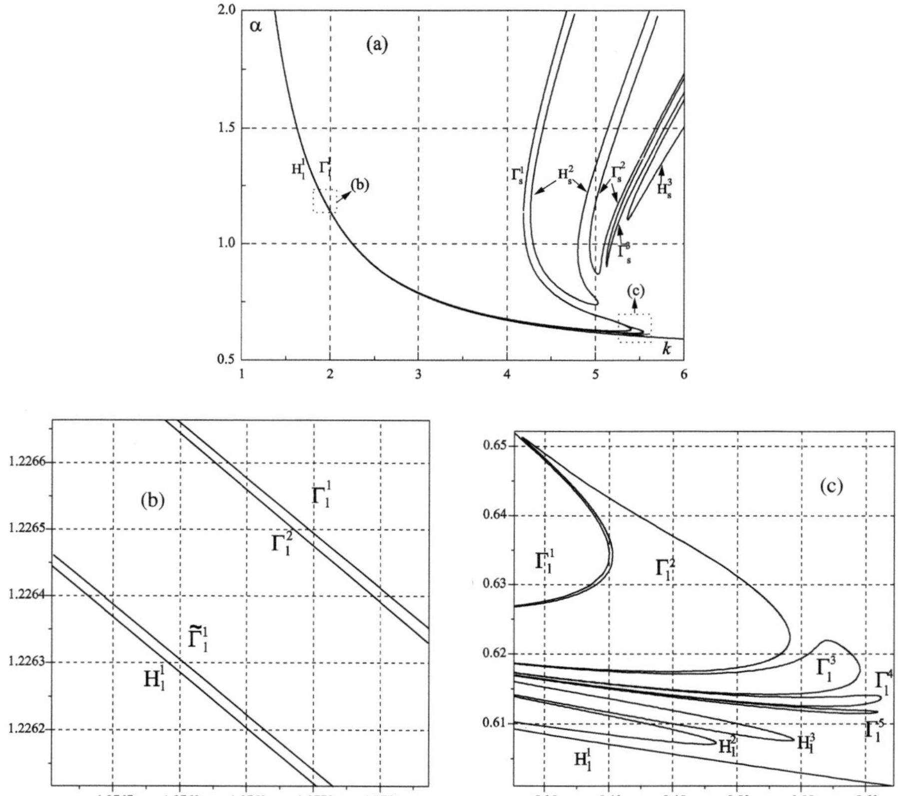

Fig. 4.11. Bifurcation diagram for homoclinic $\Gamma^n_{l,s}$ and heteroclinic $H^n_{s,l}$ orbits in the (k, α) parameter plane. The superscript n indicates the number of rotations or loops made by an orbit in the phase space (or the number of humps along the profile of the orbit). The subscripts s and l distinguish the orbits containing at least one loop of small "magnitude" and those with all loops of large "magnitude", respectively. (**a**) Full diagram; (**b,c**) two enlarged regions

the superscript n indicating the number of loops made by the orbit and a subscript, s or l, characterizing the "magnitude" of the pulse (s – small, l – large). Figure 4.11a shows the diagram in enlarged scale in the (α, k) plane. The two closed curves Γ_l^1 and H_l^1 correspond to the simplest types of homoclinic and heteroclinic orbits, respectively. The shapes of the orbits in the phase space and their profiles $[x(\xi)]$ are displayed in Fig. 4.12 and are marked by the same letters as the corresponding bifurcation curves (see Fig. 4.12a for profiles of the simplest orbits Γ_l^1, H_l^1 and Γ_s^1). For large enough values of k there exists a number of homoclinic and heteroclinic bifurcations corresponding to pulses of small "magnitude" (the curves marked by letters H_s^n and Γ_s^n in Fig. 4.11). Note that we distinguish large and small orbits in the following way: each loop of a large orbit of P^+ extends to the "opposite" outer linear region intersecting both planes U^- and U^+, while the small loops have intersections with the plane U^+ only. The index n indicates the number of rotations made by the orbits around the fixed point O. The dashed curve in Fig. 4.11, restricting the region of existence of these orbits, corresponds to the bifurcation of the steady state O. This point becomes a stable focus (to the right of the dashed curve) with two pairs of complex eigenvalues. Some profiles of multirotated orbits are shown in Fig. 4.12b.

The saddle quantity σ of the saddle-focus P^+ has a positive value for each point of all bifurcation curves associated with single-loop homoclinic bifurcation $(\Gamma_l^1, \Gamma_s^1, \Gamma_s^2, \Gamma_s^3, \ldots)$. Hence, according to the Shilnikov theorem, for the parameters changing very near to each curve, there occurs a number of homoclinic bifurcations, resulting in the appearance of multiloop homoclinic orbits. In addition, the neighborhood of the homoclinicity in the phase space contains a countable set of hyperbolic periodic orbits. Figure 4.11b,c present two enlarged regions taken near the bifurcation curve Γ_l^1. Here different multiloop orbits can be distinguished. The multiloop orbits and their profiles are shown in Fig. 4.12c. Note that due to the stable and unstable manifolds of the hyperbolic fixed point P^+ they have been analytically calculated and extended rather far from P^+, allowing us to clearly identify the homoclinic orbits with a rather high number of loops ($n = 3, 4, 5$) (Fig. 4.11c).

As earlier mentioned, the heteroclinic bifurcation H_l^1 yields the appearance of the heteroclinic contour in the phase space of (4.31). We find that the neighborhood of the curve H_l^1 contains a number of bifurcation points which correspond to the appearance of single- and multiloop homoclinic orbits $\tilde{\Gamma}^n$. The profile of each loop of these orbits is "constructed" from two symmetric heteroclinic orbits taken from the contour (see the simplest one, $\tilde{\Gamma}^1$, in Fig. 4.12a). Moreover, there exist solutions having the form of multiloop heteroclinic orbits. Figure 4.12c illustrates such orbits for $n = 2$ and 3 loops.

In summary, the fourth-order system (4.31) describing the profiles of possible traveling waves in the array (4.4) allows a number of very diverse bounded solutions. These are

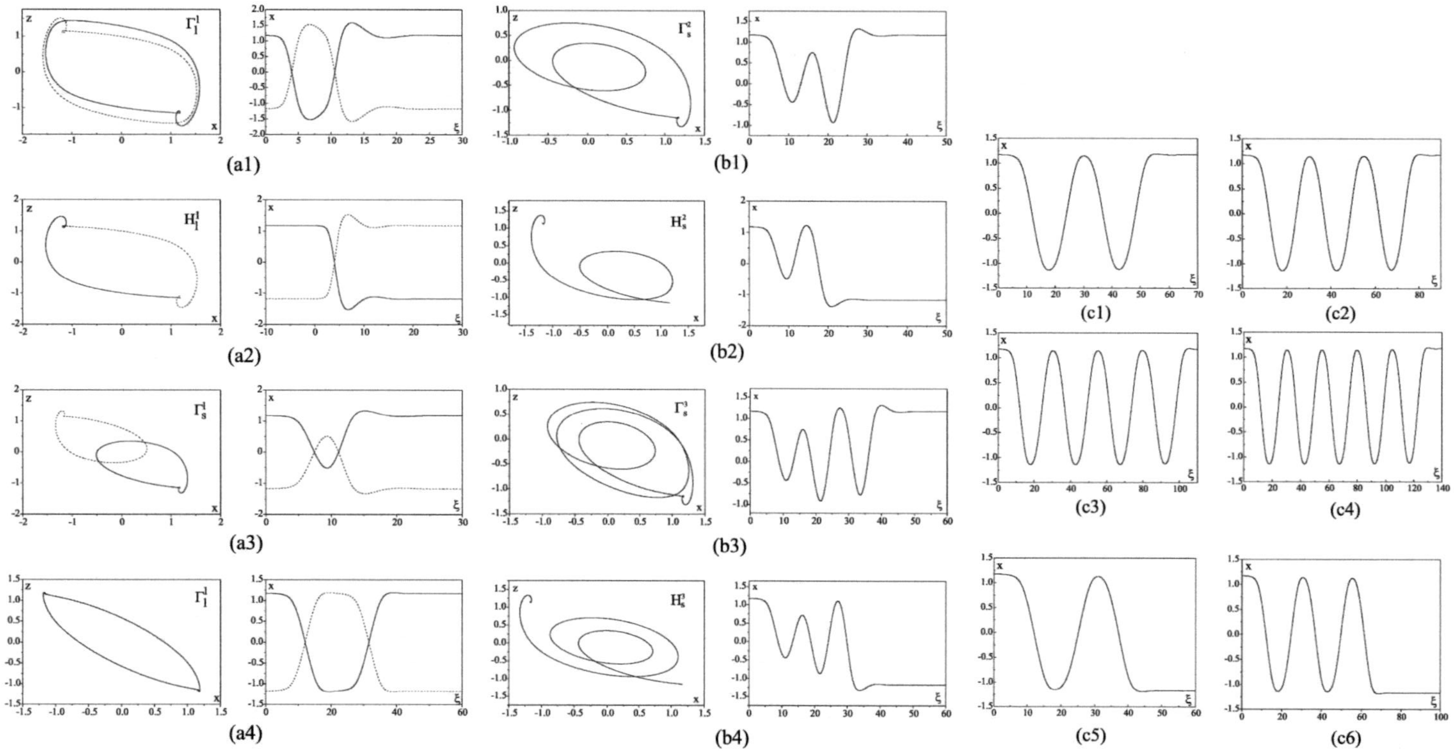

Fig. 4.12. Orbits in phase space and their corresponding profiles for different points of the bifurcation diagram. (**a**) The simplest, single-loop homoclinic and heteroclinic orbits: (**a1**) the orbit Γ_l^1 for $\alpha = 1.500, k = 1.624$; (**a2**) H_l^1, $\alpha = 1.502, k = 1.622$; (**a3**) Γ_s^1; $\alpha = 1.501, k = 4.324$; (**a4**) $\tilde{\Gamma}_l^1$, $\alpha = 0.610, k = 5.300$. *Dashed curves* show the symmmetric orbits that also exist in (4.31). (**b**) Multirotated orbits of small "magnitude" and corresponding wave profiles: (**b1**) Γ_s^2, $\alpha = 1.503, k = 5.576$; (**b2**) H_s^2, $\alpha = 1.502, k = 4.405$; (**b3**) Γ_s^3, $\alpha = 1.502, k = 5.706$; (**b4**) H_s^3, $\alpha = 1.500, k = 5.824$. (**c**) The profiles of multi-looped orbits of large "magnitude": (**c1**) Γ_l^2, $\alpha = 0.620, k = 5.304$; (**c2**) Γ_l^3, $\alpha = 0.618, k = 5.301$; (**c3**) Γ_l^4, $\alpha = 0.618, k = 5.302$; (**c4**) Γ_l^5, $\alpha = 5.301, k = 0.618$; (**c5**) H_l^2, $\alpha = 0.615, k = 5.300$; (**c6**) H_l^3, $\alpha = 0.617, k = 5.301$

- Single- and multiloop homoclinic orbits with different "magnitudes";
- Multirotated homoclinic orbits;
- Heteroclinic orbits of various forms;
- Homoclinic solutions constructed from two symmetric heteroclinic orbits;
- Periodic orbits of hyperbolic type with highly complex profiles (occurring near each point of homoclinicity).

Each of these solutions implies the *existence* of a wave with definite profile propagating along the array with constant velocity, $c = 1/h$. However, the study done of (4.31) does not answer the question of whether or not either type of wave can be excited in the array. Are these possible waves attractors of (4.4)?

Excitation and Evolution of Pulses and Fronts. Let us assume that the discrete medium modeled by (4.4) has initially been in a stable steady state associated with the homogeneous state P^+. How should we perturb this state to observe, for example, traveling pulses associated with an orbit of (4.31) homoclinic to P^+? As earlier shown, near a given loop there are multiloop orbits whose profiles determine the multihump pulses. Thus, for a given parameter set a number of different traveling waves can propagate. To distinguish each type of wave, we need to construct the initial excitation of P^+ as close as possible to the profile of a given homoclinic orbit. If the pulse is stable, we will observe in the numerical study the uniform translation of the initial profile along the array. If, however, this pulse is unstable it will go to another attractor of (4.4).

Construction of Wave Profiles. Let us take a point (α^*, k^*) of the bifurcation curve corresponding, for instance, to a homoclinic orbit Γ_l^1. The profile of the solitary pulse associated with this orbit can be constructed from the profile of the homoclinic orbit $\big(x(\xi), y(\xi), z(\xi)\big)$. To satisfy the quasicontinuum approximation, used to obtain (4.31), the parameter h should be chosen smaller than a value h_0, hence providing the necessary smoothness of the wave profile. The value of the diffusion coefficient d for a given h is $d = k^*/h^2 > d_0 = k^*/h_0^2$. Let us take the core of the profile of the homoclinic orbit of "length" T ($\xi \in [\xi_0, \xi_0 + T]$) satisfying the condition

$$|x(\xi) - x_0| \ll 1\,, \quad |y(\xi) - y_0| \ll 1\,, \quad |z(\xi) - z_0| \ll 1\,,$$
$$\forall \xi < \xi_0\,, \quad \xi > T + \xi_0\,;$$

hence the "tails" of the pulse are very close to the steady state P^+. Let us perturb the homogeneous steady state P^+ of the array by "forcing" N_1 elements ($N_1 = T/h$) starting from the element k_0 mimicking the profile of a homoclinic orbit:

$$x_{k+k_0}(t = 0) = x(\xi_0 + kh)\,,$$
$$y_{k+k_0}(t = 0) = y(\xi_0 + kh)\,,$$
$$z_{k+k_0}(t = 0) = z(\xi_0 + kh)\,,$$

where $k = 1, \ldots, N_1$. We use such a space distribution as the initial condition for the numerical integration of (4.4) to study the evolution of the traveling waves.

Stable and Unstable Pulses. Figure 4.13 illustrate the evolution of the solitary pulses associated with the homoclinic orbits of large "magnitude" (obtained from the curves Γ_l^n of the bifurcation diagram). Figure 4.13a shows a single-hump pulse (orbit Γ_l^1), Fig. 4.13b a 2-hump pulse (orbit Γ_l^2) and Fig. 4.13c a 5-hump pulse (orbit Γ_l^5). Choosing the initial distributions close to the homoclinic orbit profile, as earlier described, we observe the solutions traveling along the array with constant velocity and with no apparent change of form. For zero-flux boundary conditions the pulses are absorbed by the boundary and the terminal state of the array becomes the homogeneous state P^+. The array exhibits stable pulses propagating around the circle. These pulses represent waves of cnoidal type whose images in the phase space of (4.31) are hyperbolic periodic orbits which exist near each homoclinic loop. Note that by symmetry (4.32) solitary pulses of inverse profiles can also propagate relative to the corresponding "background" homogeneous state P^-.

Consider, now, the behavior of pulses associated with homoclinic orbits of small "magnitude" (curves Γ_s^n in Fig. 4.11). All such pulses appear unstable in the numerical study of (4.4). Figure 4.14 illustrates the diverse evolution of a single pulse of small amplitude Γ_s^1 for two parameter sets. The first case provides the decay of a small pulse to the "background" homogeneous

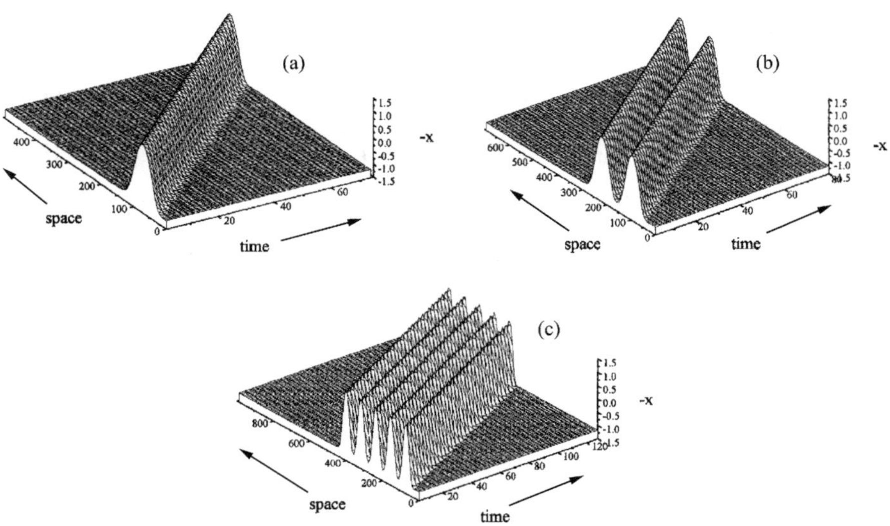

Fig. 4.13. Space–time evolution of stable, large "magnitude" pulses. For the numerical simulation of the discrete medium, the array has been initially forced by a disturbance close to the profile of the homoclinic orbits of the system (4.31). (**a**) Single-hump pulse, $\alpha = 0.627, d = 133$; (**b**) 2-hump pulse, $\alpha = 0.620, d = 133$; (**c**) 5-hump pulse, $\alpha = 5.301, d = 59$

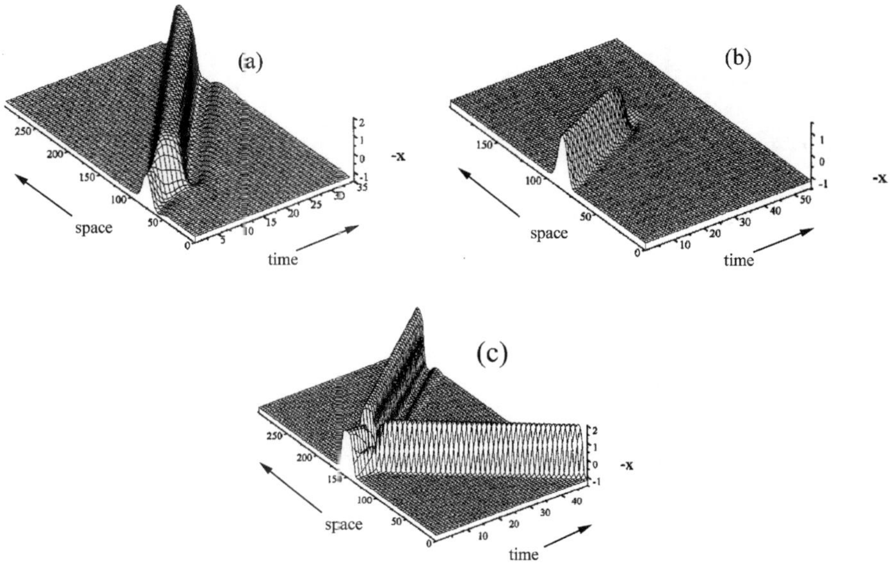

Fig. 4.14. Evolution of the initial condition derived from the homoclinic orbits of small "magnitude". (**a**) Decay of the pulse to the "background" steady state. Parameter values: $\alpha = 0.36$, $d = 6.5$. (**b**) Transformation of a small "magnitude" pulse to a stable pulse large "magnitude". Parameter values: $\alpha = 2.00, d = 52$. (**c**) Onset of two stable pulses travelling in opposite directions from initial conditions taken in the form of a pedestal. Parameter values: $\alpha = 3.74, d = 10$

state P^+ (Fig. 4.14a), while in the second one the pulse evolves to the stable wave of large "magnitude" (Fig. 4.14b) associated with the homoclinic orbit Γ_l^1. Thus, depending on the parameter values, the initial excitation can fall into the basin of attraction of either P^+ or the stable pulse. Physically, we may consider P^+ as the "excitable" one. Thus there exists a finite threshold for perturbations capable of exciting the stable pulse. Finally, pulses of large "magnitude" can be excited using other initial perturbations. Figure 4.14c illustrates the appearance of two pulses from initial conditions taken in the form of a "pedestal" on P^+.

Wave Fronts and Pulses Constructed from Two Fronts. Together with the single- and multihump pulses, (4.4) can exhibit different types of traveling wave fronts associated with the heteroclinic orbits of (4.31). These solutions can be interpreted as a transient process of "switching" between the two stable homogeneous states of the discrete medium, P^+ and P^-. The evolution of single and 2-hump fronts is illustrated in Fig. 4.15a,b. The profiles of these fronts are derived from the heteroclinic orbits H_l^1 and H_l^2. The fronts whose profiles contain humps of small "magnitude" (curves H_s^n of Fig. 4.11) become unstable and evolve to the simplest front (H_l^1), and, as occurs to the pulses of small "magnitude", they are unstable (Fig. 4.15c).

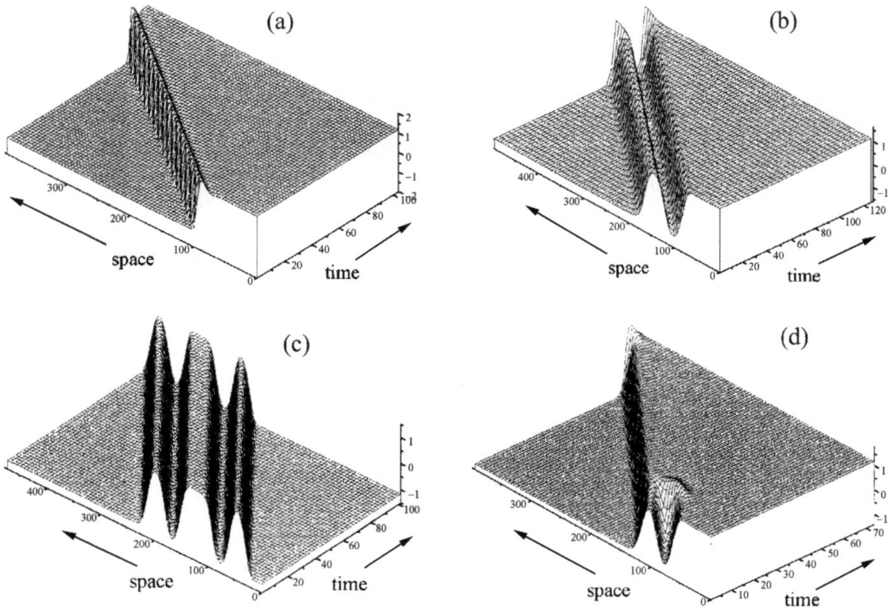

Fig. 4.15. Space–time evolution of wave fronts whose profiles have been derived from the heteroclinic orbits of (4.31). The fronts provide the "selection" of the terminal ($t \to \infty$) homogeneous steady state of the array P^+ or P^-. (**a**) Simple wave front associated with the orbit H_l^1. Parameter values: $\alpha = 2.00, d = 15.2$. (**b**) Stable multihump wave front (orbit H_l^2). Parameter values: $\alpha = 0.615, d = 59$. (**c**) Transformation of the unstable wave front associated with the orbit H_s^2 to the simplest front (orbit H_l^1). Parameter values: $\alpha = 0.755, d = 56$. (**d**) Stable localized structure consisting of two stable multihump fronts. It can be interpreted as the multihump pulse associated with a homoclinic orbit from the family $\tilde{\Gamma}_l^n$. Parameter values: $\alpha = 0.615, d = 33$

As was predicted in the previous section, pulses whose profiles are composed of two symmetric fronts can also propagate. The initial conditions for such waves can be derived from a homoclinic orbit $\tilde{\Gamma}_l^1$ or, in a simpler way, by taking the two symmetric kinks with a finite delay. Such fronts are stable if the constituting fronts are stable. Figure 4.15d illustrates the evolution of a pulse constructed with two stable 2-hump fronts.

Chaotic Wave Trains. Let us first note that the "tails" of all stable pulses and fronts decay rather quickly along the spatial coordinate. As earlier mentioned the velocities of all large "magnitude" waves are close to each other. Then, if we put into the array two different pulses (for example, a single-hump pulse and a 2-hump pulse) with a finite but rather long delay, they can form a *bound state* due to the weak interaction between the initial pulses. This state represents a 3-hump pulse traveling with constant velocity (different, but close to the velocities of the initial pulses). The image of this state in the phase space of (4.31) is a 3-loop homoclinic orbit (or "3-loop" periodic orbit for the circular array). Its existence near the loops, with a positive saddle quantity, is ensured by the Shilnikov theorem. The construction and evolution of complex wave trains composed of single-, 2- and 3-hump pulses is illustrated in Fig. 4.16a. The numerical integration of the array (4.4) shows the stability of this train, as expected, because the wave consists of stable pulses.

In a similar way, more complex wave trains can be constructed. For instance, consider a rather long array with a large number of elements. Building the train from a large number of stable pulses and fronts, which can propagate separately, we obtain a wave of rather complex profile. Note that the delay or the distance in space between neighboring components could vary in very wide limits. Extending the boundaries of the array to infinity brings the possibility of exciting *chaotic wave trains*, as the profile consists of a disordered sequence of pulses and fronts. For (4.31) such a sequence can

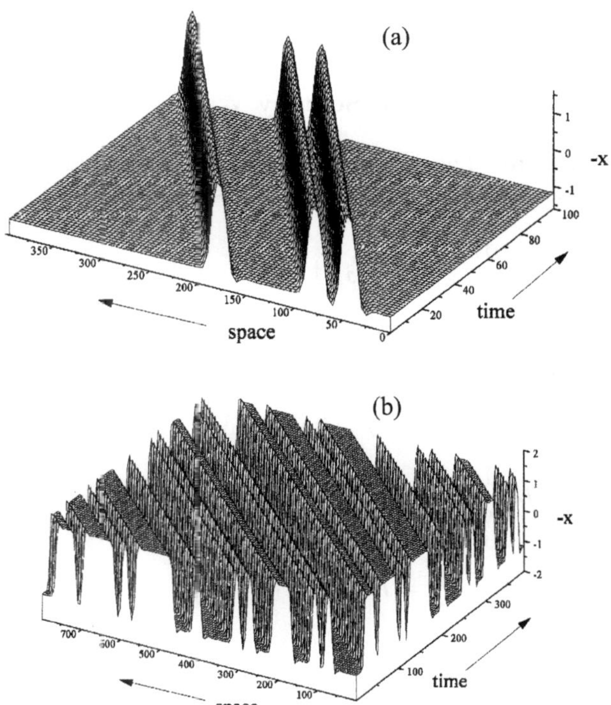

Fig. 4.16. Wave trains composed of a sequence of stable pulses and fronts for the parameters taken near $\alpha = 0.8, d = 11.75$. (**a**) The interaction of single- and 2-hump pulses ends up in a stable bound state. (**b**) Complex wave train composed of a *disordered* sequence of stable pulses and fronts. Its evolution corresponds to the numerical simulation in the circular array

be associated with a homoclinic (heteroclinic) orbit of an infinite number of loops.

To investigate the evolution of chaotic trains, we must consider a large enough number of elements in the array. In this case the image of the disordered train in the phase space of (4.31) is a complex periodic orbit from the countable set (see Sect. 4.2). Taking as the initial condition a structure with an arbitrary sequence of pulses and fronts we find that it tends to the stable solution uniformly translating along the array (Fig. 4.16b). It represents a stationary wave of a given complex profile. Accordingly, the discrete medium modeled by (4.4) is capable of sustaining chaotic wave trains.

4.3 Spatio-Temporal Dynamics of Arrays with Inductively Coupled Units

In this section we consider arrays with inductively coupled identical elements [4.5].

4.3.1 Homoclinic Orbits and Solitary Waves

Dynamical System. The 1D array of inductively coupled Chua's circuits (Fig. 4.17) can be described by the following set of equations:

$$\begin{cases} C_1 \dfrac{\mathrm{d}V_j}{\mathrm{d}t} = -g(V_j) + G(U_j - V_j)\,, \\ C_2 \dfrac{\mathrm{d}U_j}{\mathrm{d}t} = G(V_j - U_j) + I_{L_j} + I_{j-1} - I_j\,, \\ L \dfrac{\mathrm{d}I_{L_j}}{\mathrm{d}t} = -U_j - R_0 I_{L_j}\,, \\ L_0 \dfrac{\mathrm{d}I_j}{\mathrm{d}t} = U_j - U_{j+1}\,, \end{cases} \tag{4.33}$$

$$I_0 = I_1\,,\ \ U_{N+1} = U_N\,,$$

$$j = 1, 2, \ldots, N\,,$$

where V_j, U_j and I_{L_j} denote, respectively, voltage across the capacitor C_1, voltage across the capacitor C_2 and current through the inductor L. I_j is the current through the inductor L_0. As in Sect. 4.1 $g(V)$ is the current–voltage $(I - V)$ characteristics of the nonlinear resistor shown in Fig. 4.1b. Index j stands for the variables of the j-th element of the array. N is the number of elements in the array. In dimensionless form (4.33) become

$$\begin{cases} \dfrac{dx_j}{d\tau} = \alpha[y_j - x_j - f(x_j)] , \\ \dfrac{dy_j}{d\tau} = x_j - y_j + z_j + w_{j-1} - w_j , \\ \dfrac{dz_j}{d\tau} = -\beta y_j - \gamma z_j , \\ \dfrac{dw_j}{d\tau} = d(y_j - y_{j+1}) , \end{cases} \tag{4.34}$$

$$j = 1, 2, \ldots, N ,$$

with

$$\tau = \frac{G}{C_2} t, \quad x_j = \frac{V_j}{B_p}, \quad y_j = \frac{U_j}{B_p}, \quad z_j = \frac{I_{L_j}}{B_p G}, \quad w_j = \frac{I_j}{B_p G} .$$

The parameters α, β, γ and d have positive values and characterize the dynamics of the unit [see Sect. 4.1 and compare with (4.4), $d = C_2/L_0 G^2$]. The nonlinear function $f(x)$ describes the three-segment, piecewise-linear resistor characteristics $g(V)$ and satisfies (4.3).

As we search for *"traveling" wave solutions* of (4.34) and (4.3) we pose

$$\begin{aligned} x_j(t) &= x(\xi) , \\ y_j(t) &= y(\xi) , \\ z_j(t) &= z(\xi) , \\ w_j(t) &= w(\xi) , \end{aligned} \tag{4.35}$$

where $\xi = t - jh$ is the moving frame coordinate; h is a parameter, $h > 0$, the inverse of wave velocity. Substituting (4.35) into (4.34) we obtain

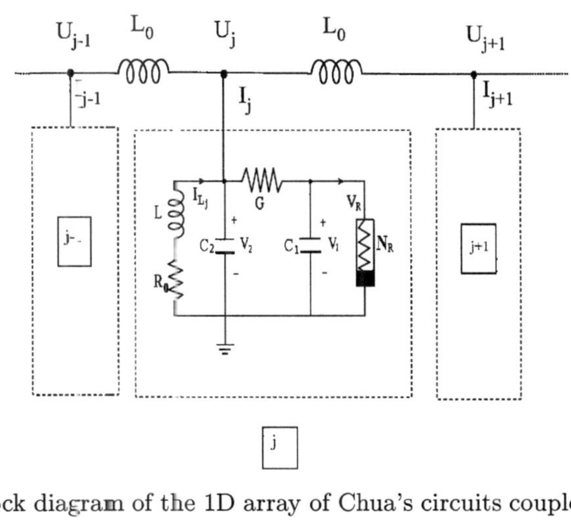

Fig. 4.17. Block diagram of the 1D array of Chua's circuits coupled by the inductors L_0

$$\begin{cases} \dot{x} = \alpha[y - x - f(x)]\,, \\ \dot{y} = x - y + z + w(\xi - h) - w(\xi)\,, \\ \dot{z} = -\beta y - \gamma z\,, \\ \dot{w} = d[y(\xi) - y(\xi + h)]\,, \end{cases} \tag{4.36}$$

where the dot denotes differentiation with respect to ξ. For small enough h (i.e. high enough c) the two difference terms in (4.36) can be replaced by the first derivatives $\dot{w}$ and $-\dot{y}$ (with respect to ξ), respectively. Then we obtain the following system of coupled first-order ODE:

$$\begin{cases} \mu\dot{x} = y - x - f(x)\,, \\ \dot{y} = 1/\delta(x - y + z)\,, \\ \dot{z} = -\beta y - \gamma z \end{cases} \tag{4.37}$$

where $\mu = 1/\alpha$ and $\delta = 1 - d/c^2$. Equations (4.37) describe traveling waves propagating with velocity c along the array.

Solitary waves of (4.34) correspond to nonconstant solutions of (4.37) which satisfy the condition

$$\lim_{|\xi|\to\infty} \big(x(\xi), y(\xi), z(\xi)\big) = 0\,, \tag{4.38}$$

hence corresponding to the homoclinic orbits of (4.37).

Note that for $\delta = 1$ (4.37) coincides with the equations describing a single element of the array. This highly simplifying feature comes with the inductive coupling.

Homoclinic Orbits. Phase Space Analysis. Let us analyze the behavior of trajectories in the phase space of (4.37). For simplicity, we restrict consideration to the following parameter range:

$$\begin{aligned} &\mu \ll 1\,, \quad d > 0\,, \quad \delta > 0\,, \quad \beta > \beta_{ab}\,, \\ &\gamma \geq 0\,, \quad a < -1\,, \quad -1 < b < 0\,, \end{aligned} \tag{4.39}$$

with

$$\beta_{ab} \equiv \max\{\beta_a, \beta_b\}\,,$$

$$\beta_q \equiv \frac{\delta}{4}\left(\gamma - \frac{q}{\delta(q+1)}\right)^2 + \frac{1}{4}\left(\frac{\gamma^2}{(q+1)^2} - \frac{q^2}{\delta^2(q+1)^4}\right)\mu + O(\mu^2)\,,$$

where the index q is either a or b. Then (4.37) has three fixed points:

$$O(0,0,0)\,, \quad P^+(x_0, y_0, z_0)\,, \quad P^-(-x_0, -y_0, -z_0)\,,$$

with

$$\begin{aligned} &x_0 \equiv \frac{(\gamma+\beta)D}{\beta - \gamma B}\,, \quad y_0 \equiv \frac{\gamma + D}{\beta - \gamma B}\,, \quad z_0 \equiv -\frac{\beta D}{\beta - \gamma B}\,, \\ &D \equiv \frac{b-a}{b+1}\,, \quad B \equiv -\frac{b}{(b+1)}\,. \end{aligned} \tag{4.40}$$

Each of these steady states has a pair of complex-conjugate eigenvalues. The point O is a saddle-focus (Appendix A) with a 1D unstable manifold $W^u_\mu(O)$ and a 2D stable manifold $W^s_\mu(O)$. The manifold $W^u_\mu(O)$ consists of O and two outgoing trajectories W^u_1 and W^u_2. In the (x, y, z) phase space the trajectory W^u_1 goes into the region $x > 0$, while W^u_2 goes into the region $x < 0$. The points $P^\pm$ may be either stable foci (in the parameter region G_s, Fig. 4.18) or saddle-foci (in the parameter region G_u, Fig. 4.18) with the corresponding 1D stable manifolds $W^s_\mu(P^\pm)$ and 2D unstable manifolds $W^u_\mu(P^\pm)$. The stability of $P^\pm$ changes at the bifurcation value $\beta = \beta_s$ with

$$\beta_s \equiv - \frac{\left(1 + \dfrac{\delta(b+1)}{\mu}\right)\left[\gamma^2 + \gamma\left(\dfrac{1}{\delta} + \dfrac{b+1}{\mu}\right) + \dfrac{b}{\mu\delta}\right]}{\gamma + \dfrac{1}{\delta}} . \tag{4.41}$$

At $\beta = \beta_s$ the trajectories located on $W^u_\mu(P^\pm)$ are equivalent to trajectories of elliptic points in a 2D manifold.

Let us consider the homoclinic orbits in the phase space of (4.37) with the parameter values taken in the region G_u (Fig. 4.18). Let us examine the homoclinic orbits formed by W^u_1. Since the vector field of (4.37) is invariant under the transformation

$$(x, y, z) \to (-x, -y, -z) , \tag{4.42}$$

the homoclinic orbit formed by W^u_1 coexists with the homoclinic orbit formed by W^u_2.

Consider now the behavior of W^u_1 when $\mu \ll 1$ (4.37). As μ is the coefficient of the highest derivative in the system, we have a singular perturbation

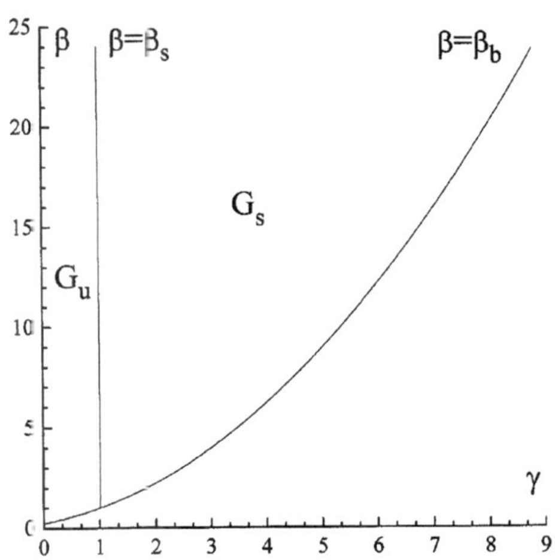

Fig. 4.18. 2D parametric planar representation (γ, β) with $a = -2, b = -1/2$, $\mu = 0.01$, and $\delta = 0.95$. For the parameters values in the region G_s the stationary point P^+ is a stable focus. In the region G_u the point P^+ is a saddle-focus

problem. Accordingly, the motion in the 3D phase space has both fast and slow features.

Let us start our analysis with $\mu = 0$. In this case a 2D manifold W_0 of slow motions exists in the (x, y, z) phase space of (4.37). The shape of the manifold is given by

$$W_0 : \{(x, y, z) \mid y = x + f(x)\} . \tag{4.43}$$

Comparing (4.43), (A.12) and (A.13) we see that W_0 coincides with the 2D manifolds of O and P^+. In the regions $x \geq 1$ and $x \leq -1$, W_0 is given by the planes $W_0^u(P^\pm)$. In the region $| x | \leq 1$, W_0 is the plane $W_0^s(O)$ (Fig. 4.19). Therefore, the character of the slow motions on W_0 is conditioned by the complex-conjugate eigenvalues of $P^\pm$ and O. As shown in Appendix C, the trajectories located on $W_0^u(P^\pm)$ form unstable foci, and the trajectories on $W_0^s(O)$ form a stable focus. Outside W_0 the dynamics of the system generates fast motions with

$$z = \text{const.} , \quad y = \text{const.} \tag{4.44}$$

Then, from (4.37) it follows that $W_0^u(P^\pm)$ are attracting, while $W_0^s(O)$ sends the trajectories to the regions of fast motions. The qualitative behavior of the trajectories corresponding to fast and slow motions is shown in Fig. 4.19.

Now let us consider $0 < \mu \ll 1$. In this case the manifold of slow motions W_μ of (4.37) comprises the 2D manifolds of O and $P^\pm$. Within the region

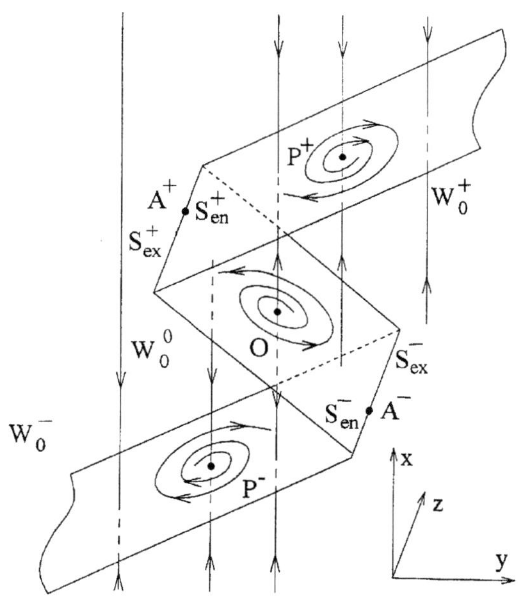

Fig. 4.19. Qualitative picture of the phase portrait of (4.36) in the relaxation regime with $\mu = 0$

$x \geq 1$, W_μ is formed by the plane $W_\mu^u(P^+)$, and in the region $|x| \leq 1$ it is formed by the plane $W_\mu^s(O)$ [(A.6) and (A.12)]. The shape of W_μ is shown in Fig. 4.20a. Then, as shown in Appendix B, $W_\mu^s(O)$ and $W_\mu^u(P^+)$ intersect the plane

$$U_{+1} : \{(x, y, z) \mid x = 1\} \tag{4.45}$$

at l_μ^s and l_μ^u, respectively. The lines l_μ^s and l_μ^u intersect each other at $L(y_l, z_l)$ [Fig. 4.20b and (B.4)].

Since the point L lies in the intersection between $W_\mu^u(P^+)$ and $W_\mu^s(O)$, there is a trajectory which originates from P^+, passes through the point L, and then tends to O. In Fig. 4.20a this trajectory is the solid line marked Γ_l which defines a heteroclinic orbit.

Besides L, the lines l_μ^s and l_μ^u contain the points K and M, which are essential for understanding the dynamics of the system. The locations of these points is shown in Fig. 4.20b, and their coordinates are determined in Appendix C (C.1–2). These points divide the trajectories located in $W_\mu^u(P^+)$ and $W_\mu^s(O)$ into trajectories which leave the planes and trajectories which come to the planes at l_μ^s and l_μ^u. The trajectories passing above M come to $W_\mu^s(O)$. The trajectories passing below K come to $W_\mu^u(P^+)$.

Since $0 < \mu \ll 1$, slow motions occur not only on $W_\mu^s(O)$ and $W_\mu^u(P^+)$, but also in some *thin layers* (thicknesses of order μ) containing these planes. Let us consider W_1^u, which originates from O and evolves in the (x, y, z) phase space. This trajectory has portions of fast and slow motions. Fast motion along the trajectory occurs in the region of the phase space which is located outside the thin layers associated with $W_\mu^s(O)$ and $W_\mu^u(P^+)$ (Fig. 4.20a). In this region of the phase space the shape of W_1^u is close to a straight line $\{z = 0, y = 0\}$. After passing this region W_1^u comes into the thin layer of slow motions associated with $W_\mu^s(P^\pm)$. In this layer the behavior of the trajectory is qualitatively similar to the behavior of the nearest trajectory located on

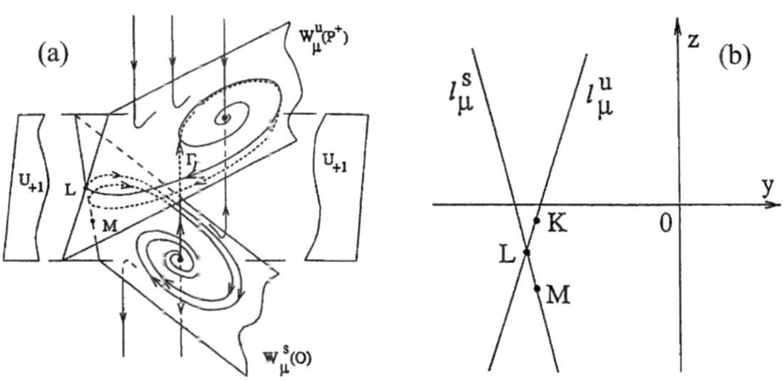

Fig. 4.20. (**a**) A portion of the qualitative picture of the phase portrait of (4.36) in the relaxation regime with $0 < \mu \ll 1$. (**b**) Relative location of cross-sections of the manifolds $W_\mu^u(P^+)$ and $W_\mu^s(O)$ at the plane $x = 1$

$W^s_\mu(P^\pm)$. As (4.39) impose that the trajectories in $W^u_\mu(P^\pm)$ have a spiral shape, W^u_1 will also have an interval of the form of an spiral (Fig. 4.20a). Therefore, being in the thin layer of slow motions W^u_1 makes one or several turns under $W^u_\mu(P^+)$ and then intersects the plane U_{+1}. Let M^u_1 be the point of the first intersection of W^u_1 with U_{+1} when it leaves the domain $x > 1$. Then as W^u_1 passes the region of slow motion under $W^u_\mu(P^+)$, M^u_1 is located to the right of l^u_μ, and M^u_1 is located in a μ-neighborhood of l^u_μ.

Now, let us consider l^s_μ, which is the intersection of $W^s_\mu(O)$ and U_{+1} (Fig. 4.20b and Appendix B). In U_{+1} this line divides the right μ-neighborhood of l^u_μ into two domains, U^s_μ and U^u_μ, located, respectively, to the right and to the left of l^u_μ. M^u_1 can belong to either one of these domains or to l^s_μ. If M^u_1 is in U^s_μ then after intersecting U_{+1} the trajectory W^u_1 goes into the region of fast motions *under the plane* $W^s_\mu(O)$. If M^u_1 is in U^u_μ then after intersecting U_{+1} the trajectory W^u_1 stays in the layer of slow motions in a μ-neighborhood of $W^u_\mu(P^+)$, and therefore, *above the plane* $W^s_\mu(O)$. Finally, if $M^u_1 \in l^s_\mu$, then W^u_1 belongs to $W^s_\mu(O)$, and as $\xi \to +\infty$ it tends to O. Since W^u_1 originates at O, the latter case corresponds, indeed, to the existence of a homoclinic orbit in the phase space of (4.37).

Similar arguments can be used for the analysis of homoclinic bifurcations in the parameter region G_s. The main difference from the case just considered is that in G_s only bifurcations of the homoclinic orbit with a single rotation in the slow motion layer are found. A detailed description of bifurcations of this type is presented in Sect. 4.3.1.

To obtain the parameter values corresponding to homoclinicity, we use the piecewise linearity of $f(x)$ (4.3). The results are correct for any value of μ. First, we search for the point M^u_0 where the trajectory W^u_1 first intersects the plane U_{+1}. It follows from (A.11) that the coordinates of this point are

$$x = 1\,, \quad y = y^u_0\,, \quad z = z^u_0\,, \qquad (4.46)$$
$$y^u_0 \equiv -\frac{1}{k^a_4}\,, \quad z^u_0 \equiv -\frac{1 + k^a_4 - k^a_3}{k^a_4}\,,$$

with k^a_3 and k^a_4 given by (A.9). In the regions $x > 1$, the solution of (4.37), corresponding to the trajectory passing through M^u_0, can be written in the form

$$x = \varphi_1(\xi, C_u)\,, \quad y = \varphi_2(\xi, C_u)\,, \quad z = \varphi_3(\xi, C_u)\,. \qquad (4.47)$$

Since, for $x > 1$, (4.37) is linear, the functions φ may be easily obtained. The equation

$$\varphi_1(\xi, C_u) = 1 \qquad (4.48)$$

yields the interval of "time", $\xi = \tau$, required for the trajectory starting at M^u_0 to reach M^u_1. The existence of M^u_1 is guaranteed if the parameter values belong to the region G_u and, consequently, the solution of (4.48) exists. The coordinates of M^u_1 are

$$x = 1\,, \quad y = y_1^u\,, \quad z = z_1^u\,, \tag{4.49}$$

with

$$y_1^u \equiv \varphi_2(\tau, C_u)\,, \quad z_1^u \equiv \varphi_3(\tau, C_u)\,.$$

It follows from (4.49) and (B.1) that $M_1^u \in l_\mu^s$ if the parameters of (4.37) satisfy

$$k_1^a \varphi_2(\tau, C_u) + k_2^a \varphi_3(\tau, C_u) + 1 = 0\,. \tag{4.50}$$

Equation (4.49) defines the bifurcation set, Π, corresponding to the appearance of a single-loop homoclinic orbit of (4.37) associated with the stationary state O. Multiloop homoclinic orbits will be discussed later. Note that the single-loop orbits can have a rather complicated shape because such orbits may rotate many times in a thin layer near the plane $W_\mu^u(P^+)$.

As, unfortunately, (4.48) cannot be solved exactly by analytical methods, we have solved it numerically. The results of the numerical analysis of Π, using (4.48), will be discussed in Sect. 4.3.1. Π can also be characterized using an approximate description of the behavior of the trajectory W_1^u when it can be split into portions of fast and slow motions. The fast part of the motion of W_1^u is close to the line $y = z = 0$.

Let us consider in detail the slow motion part of W_1^u. It has been earlier shown that if the homoclinic orbit exists, then $M_1^u \in l_\mu^s$ is located between L and M. First, let us see how W_1^u approaches M_1^u. As it follows from (B.1), when $0 < \mu \ll 1$, l_μ is close to the line $y = a+1$. Therefore, when W_1^u comes close to M_1^u, its motion satisfies the condition $\dot{y} \approx 0$. On other hand, in order to remain in the thin layer of slow motions, W_1^u must satisfy the condition $\dot{x} \approx 0$ in the vicinity of M_1^u. Taking into account both these facts we find that M_1^u, with such properties in the plane U_{+1}, is located in the vicinity of the point

$$y = y_a\,, \quad z = z_a\,,$$

with

$$y_a \equiv a + 1\,, \quad z_a \equiv a\,.$$

Besides, in the layer of slow motions W_1^u moves very close to the plane $W_\mu^u(P^+)$. Therefore, during the slow motion, W_1^u may be approximated by a trajectory located in $W_\mu^u(P^+)$. Taking into account the properties of W_1^u near M_1^u we choose the trajectory from $W_\mu^u(P^+)$ passing through the point

$$y = y_a\,, \quad z = z_a\,.$$

Let Γ_a denote the chosen trajectory from $W_\mu^u(P^+)$. From (B.1) it follows that in the (x, y, z) phase space the trajectory Γ_a passes through the point $A(x_a, y_a, z_a)$, with

$$x_a \equiv 1 - \frac{[a\gamma + \beta(a+1)]}{\delta(b+1)^3}\mu^2 + O(\mu^3)\,.$$

To show that A is close to l^s_μ we estimate their relative distance. From (B.1) it follows that the distance, R, between A and l^s_μ is

$$R = \frac{|\gamma a + \beta(a+1)|}{\delta}\sqrt{\frac{1}{(a+1)^4} + \frac{1}{(b+1)^6}}\mu^2 + O(\mu^3)\,,$$

which is of order μ^2, and hence a small quantity. Thus, the set of the parameter values, Π_{approx}, approximating Π may be obtained by analyzing the conditions in which Γ_a is the best approximation of W^u_1. These conditions may be considered as the boundary problem for the 2D system (C.3). The solution of the boundary problem gives the set of the parameter values Π_{approx}. This solution takes the form

$$\begin{aligned} \frac{D_\mu\sqrt{\beta I_\mu}}{\sqrt{\delta}(\beta I_\mu - \gamma B_\mu)} &= -\left(a + 1 - \frac{\gamma D_\mu}{\beta I_\mu - \gamma B_\mu}\right) \\ &\quad \times \exp\left[\frac{h_b}{\omega_b}\left(2\pi n - \arctan\frac{2\omega_b}{\gamma + \frac{B_\mu}{\delta}}\right)\right], \\ n &= 1, 2, \ldots\,, \end{aligned} \tag{4.51}$$

with $B_\mu, I_\mu, D_\mu, \omega_b$ and h_b given by (C.3) and (C.4). The index n characterizes the number of turns made by W^u_1 when moving in the layer of slow motions it unwinds around the stationary point P^+. The solutions of (4.51) obtained for $n = 1, 2, 3$ and 4 with the fixed parameter values $a = -2,\ b = -1/2,\ \mu = 0.01$ are shown in Fig. 4.21 by dashed curves in the parameter plane (β, γ).

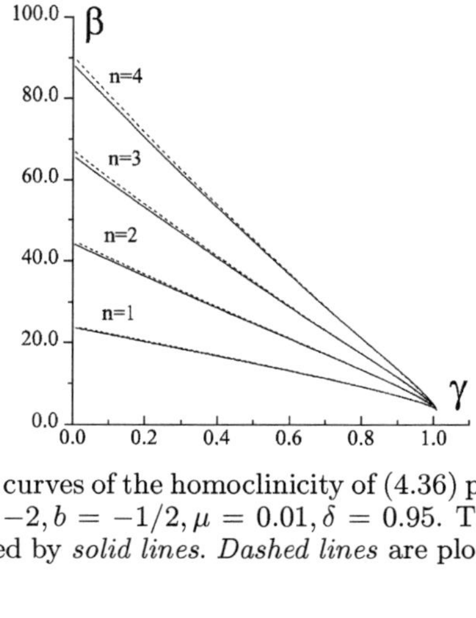

Fig. 4.21. Bifurcation curves of the homoclinicity of (4.36) plotted in the parameter plane (γ, β) with $a = -2, b = -1/2, \mu = 0.01, \delta = 0.95$. The results of numerical simulation are presented by *solid lines*. *Dashed lines* are plotted using approximate formula (4.20)

Note that all the curves defined by (4.51) start from the same point Π_0 of coordinates

$$\gamma = \frac{B_\mu}{\delta}, \quad \beta = \frac{[D_\mu + (a+1)B_\mu]^2}{\delta(a+1)^2}. \tag{4.52}$$

The behavior of these curves near Π_0 can be explained as follows: At Π_0 (C.3) is conservative and, therefore, the trajectories on $W^u_\mu(P^+)$ given by this system are equivalent to the trajectories near an elliptic point, and hence Γ_a is closed and has the form of an ellipse. A is the extreme left point of the ellipse. The ellipse is "nearly" tangential to l^s_μ. Tangency occurs at the "zeroth"-order in μ. Therefore, we use the contour Γ_0 as an interval of the trajectories which approximate a homoclinic orbit. Γ_0 goes along the ellipse which is tangent to l^s_μ and intersects the line $\{y = 0,\ z = 0\}$ originating from O, simultaneously. Then Γ_0 can approximate the homoclinic orbits making any number of rotations n in a neighborhood of P^+.

Homoclinic Orbits. Numerical Results. Π can be analyzed in two ways. One way is to use the results of Sect. 4.3.1 It suffices to numerically integrate (4.48), together with (4.37), to obtain the parameter values of the bifurcation set. The second approach consists in numerically integrating (4.37) in the region $x \geq 1$ with initial conditions at M^u_0. The integration is stopped when M^u_1 is reached. This is the first intersection of the trajectory with the plane U_{+1}. Locating M^u_1 for different parameter values, we find the set of parameter values which satisfy the condition $M^u_1 \in l^s_\mu$ (Sect. 4.3.1). Both methods practically give the same values.

In this subsection we only provide the results obtained from the numerical analysis of the homoclinic orbits with the direct integration (i.e. using the second method). To integrate (4.37) a fourth-order Runge–Kutta routine was used. The absolute and relative errors of numerical integration did not exceed 10^{-6} and 10^{-8}, respectively. The system (4.37) was integrated in the region $x > 1$ from M^u_0 to the first intersection of the trajectory W^u_1 with U_{+1}. This intersection gives M^u_1. Then we obtain the deviation, d_M, of M^u_1 from l^s_μ located at the intersection between the planes $W^a_\mu(O)$ and U_{+1}. Depending upon the location of M^u_1 in U_{+1}, d_M can be either positive or negative. If $M^u_1 \in U^u_\mu$, then d_M is negative, while if $M^u_1 \in U^s_\mu$, d_M is positive. Then we vary one of the parameters of the system, for example β, and obtain the *splitting function* $d_M(\beta)$ (dev, Fig. 4.22). The discontinuity of this splitting function (dashed line in Fig. 4.22) corresponds to the approach of the separatrix W^u_1 to U_{+1} at a point (from very near to K), where the vector field of (4.37) is tangent to U_{+1}. The parameter value, where the splitting function crosses zero, corresponds to the case $M^u_1 \in l^u_\mu$ and, therefore, to the existence of a homoclinic orbit in the phase space of (4.37).

The solid lines in Fig. 4.21 depict the Π obtained in the numerical simulations of the bifurcation values in the parameter plane (γ, β) for fixed $\mu = 0.01$, $\delta = 0.95$, $a = -2$ and $b = -1/2$. The index n characterizes the

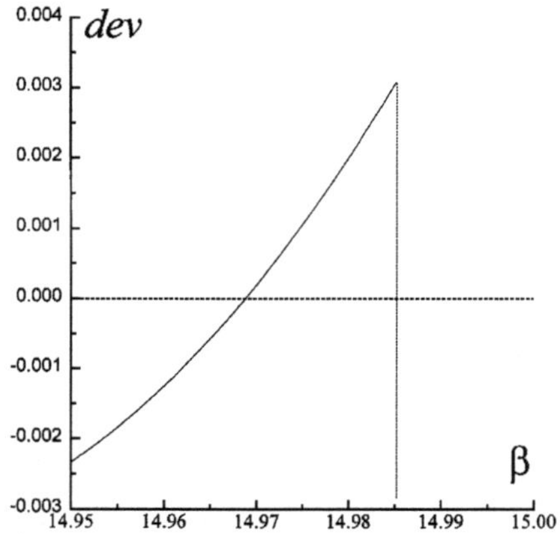

Fig. 4.22. The splitting function $d_M = d_M(\beta) \equiv$ dev calculated with $a = -2$, $b = -1/2, \mu = 0.01, \delta = 0.95, \gamma = 0.5$. The value of β where $d_M = 0$ corresponds to the appearance of homoclinic orbit with $n = 1$. For the homoclinic bifurcations with $n = 2, 3, \ldots$ the splitting functions exhibit a similar behavior

number of turns made by the homoclinic orbit around P^+. We denote by h_n the parameter values from Π corresponding to the homoclinic orbit marked by n. The bifurcation values obtained in the numerical simulations coincide with the approximate values of the corresponding bifurcation parameters given by (4.51) (dashed lines in the Fig. 4.21). Equation (4.51) gives the best approximation for the homoclinic orbits with small n. The shapes of the homoclinic orbits with different n, for the parameter values from Π, are shown in Fig. 4.23. The orbits correspond to $\gamma = 0.5$ and β values taken within the bifurcation set Π (see Fig. 4.21).

As mentioned in Sect. 4.2.1, Π does not exhaust the whole bifurcation set of homoclinic orbits of (4.37). This fact can be confirmed with the analysis of the saddle-focus value, σ_{sf}, of the stationary point O. This saddle-focus value is

$$\sigma_{sf} = \lambda_a + h_a \,,$$

where λ_a is the positive eigenvalue of the fixed point and h_a is the real part of the complex-conjugate eigenvalues. Here λ_a is given by (B.3) and h_a by (C.3). When $0 < \mu \ll 1$, the eigenvalue $\lambda_a \gg 1$ and, therefore, $\sigma_{sf} > 0$. According to the Shilnikov theorem, if the saddle-focus value is positive, then other bifurcation curves corresponding to multiloop homoclinic orbits will exist in the neighborhood of the curves of Π. However, they are hardly observable in numerical simulations. Indeed, when $\lambda_a \gg 1$ the plane $W^s_\mu(O)$ is strongly unstable, and any incoming trajectory to the slow motion layer near $W^s_\mu(O)$ rapidly leaves this layer. This instability in the transitions from slow to fast motions causes the stiffness of (4.37). When $\mu = 0.1$ the situation is easier and multiloop homoclinic orbits may be observed in the numerical simulations. Such homoclinic orbits are generated by the trajectory W^u_1. At

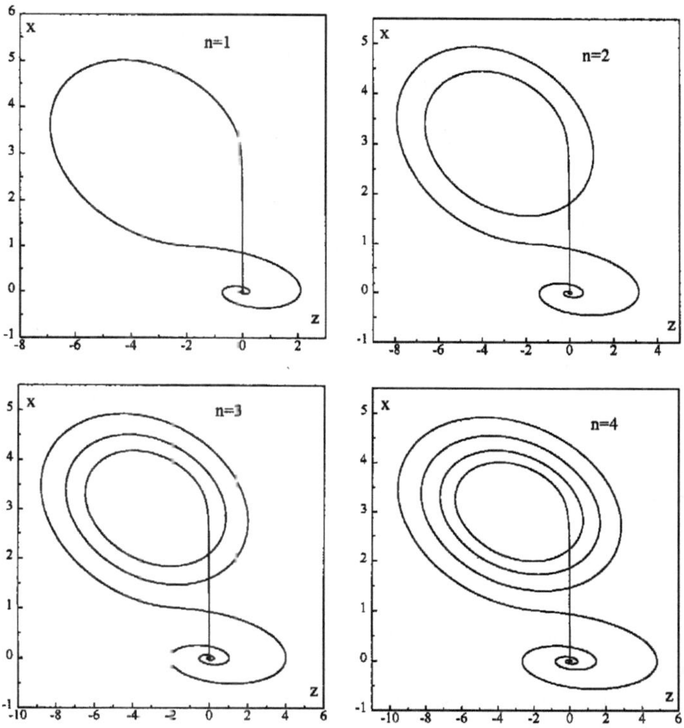

Fig. 4.23. Various types of the homoclinic orbits of (4.36) associated with the stationary point $O(0,0,0)$. (**a–d**) provide the homoclinic orbits which, respectively, make $n = 1, 2, 3$, and 4 rotations in the slow motions layer near the manifold $W^u_\mu(P+)$, respectively. The parameter values are the following: $a = -2, b = -1/2$, $\mu = 0.01, \delta = 0.95, \gamma = 0.5$ and β corresponding to the solid curves of the bifurcation diagram (Fig. 4.21) labelled $n = 1, 2, 3$, and 4, respectively

the first intersection of the plane U_{+1} (from the region $x > 1$), the trajectory gets into the region U^ε_μ (Fig. 4.20), then it moves in the slow motion layer above $W^s_\mu(O)$, and at the second intersection with U_{+1} (from the region $x > 1$) the trajectory gets into l^s_μ and forms the homoclinic orbit.

Homoclinic Orbits. Results from an Experiment. Homoclinic orbits of (4.33) have been observed in experiments with electronic circuits (Fig. 4.24) [4.11]. For illustration we concentrate on the (v_1, v_2) projections of the trajectories of the circuit, hence voltages v_1 and v_2 are applied to the "X" and "Y" terminals of the oscilloscope. A periodic pulse generated by a function generator is used to periodically set the initial state of the circuit near the steady point O by short-circuiting with a relay the nonlinear active element N_R. This short-circuiting makes the origin, in the resulting system, asymptotically stable. The pulses from the function generator are also used for intensity modulation (via the "Z" terminal of the oscilloscope) to display the intervals of the trajectories starting from the vicinity of O

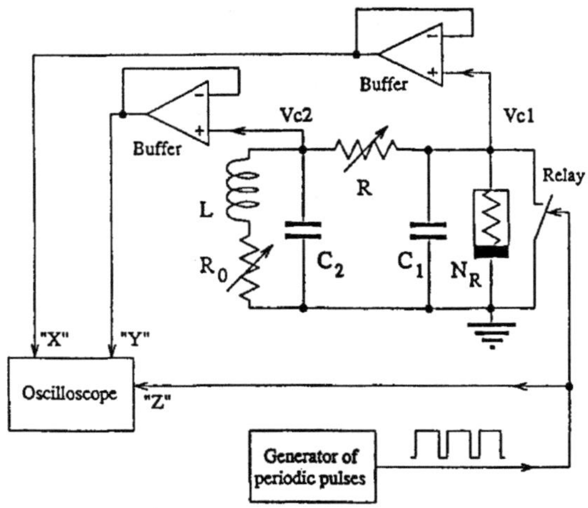

Fig. 4.24. Schematic of the experimental setup for generating pictures of homoclinic orbits

and ending when the relay is switched on. As it follows from Sect. 4.3.1 the behavior of the trajectories originating from O changes qualitatively when the parameter value of the circuit crosses the bifurcation values where the system has homoclinic orbits. This qualitative change of the trajectory is used to detect the transition of the circuit through homoclinicity.

In the experiments the evolution equations of the circuit are

$$\begin{cases} C_1 \dfrac{dv_1}{dt} = G(v_2 - v_1) - g(v_1)\,, \\[2ex] C_2 \dfrac{dv_2}{dt} = G(v_1 - v_2) + i_L\,, \\[2ex] L \dfrac{di_L}{dt} = -v_2 - R_0 i_L\,, \end{cases} \tag{4.53}$$

where $G = 1/R$ and the nonlinear function $g(v_1)$, which defines the $I - V$ characteristics of the nonlinear active element, N_R, is described by the piecewise-linear function

$$g(v_1) = m_0 v_1 + \frac{1}{2}(m_1 - m_0)\big(|v_1 + B_p| - |v_1 - B_p|\big)\,. \tag{4.54}$$

In the experimental setup, the values $m_1 = -0{,}5\,\mathrm{mS}$, $m_0 = -0{,}11\,\mathrm{mS}$ and $B_p = 0{,}5\,\mathrm{V}$ were held fixed. Also fixed were the linear elements of the circuit at $L = 129\,\mathrm{mH}$, $C_1 = 1\,\mathrm{nF}$ and $C_2 = 47\,\mathrm{nF}$. The values of the resistors R and R_0 are the control parameters of the dynamics of the circuit. The bifurcations found associated with homoclinic orbits are given in Fig. 4.25. These bifurcation curves are coded by indices h_n. When the parameters of

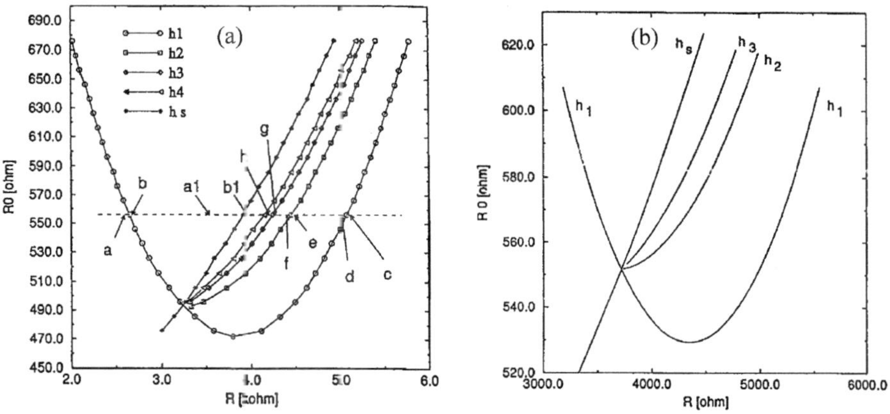

Fig. 4.25. The bifurcation diagram in the plane of physical parameters (R, R_0). (**a**) Bifurcation curves measured in an experiment with the circuit. (**b**) Bifurcation curves obtained in numerical simulation of (4.36) when the parameters of the system are the same as in the experiment (**a**)

the circuit are chosen on h_n, the homoclinic trajectory starts at the unstable stationary point O, goes to the plane of slow motions, makes n turns around P^+ and then returns back to O.

As earlier explained, the behavior of a trajectory originating at the stationary hyperbolic point changes qualitatively when the parameter value of the circuit crosses the bifurcation values associated with the homoclinicity. Consider the trajectory studied in the experiment with $R_0 = 556\,\Omega$. Two different types of bifurcation associated with the homoclinic orbit h_1 can be observed. The first type appears in the parameter range where the stationary point P^+ is stable. Figure 4.26b shows the trajectory obtained when the parameter values of the circuit are chosen in the region above the left branch of h_1, where P^+ is stable. The trajectory starts at O, goes to the manifold of slow motions and then is attracted by P^+. After bifurcation, when the parameter values of the circuit are below h_1 the trajectory starting at O goes to the manifold of slow motions, makes one turn around P^+, then falls off the plane of slow motions and travels fast to the other branch of the manifold of slow motions associated with P^-. This trajectory is shown in Fig. 4.26a.

The second type of bifurcation is observed in the parameter range where P^+ and P^- are unstable. After the homoclinic bifurcation the trajectory behaves like the trajectory observed after the bifurcation of the first type, as a comparison of Fig. 4.26a and c shows. However, prior to the bifurcation the behavior of the trajectory is different from the trajectory observed before the bifurcation of the first type. Now it is not attracted by P^+ because this fixed point is no longer stable. Then the trajectory makes a second turn in the slow motion layer near the manifold $W^u_\mu(P+)$, moves away from P^+ and falls off the manifold, as illustrated in Fig. 4.26d.

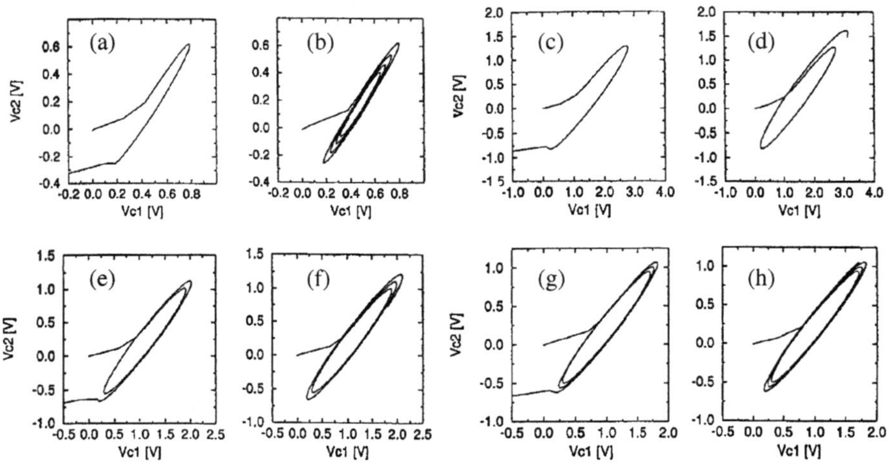

Fig. 4.26. The behavior of the trajectory originating from the stationary point O measured in the experiment with the circuit. The parameters of the circuit were taken close to the homoclinic bifurcation. The trajectories shown in (**a**–**h**) were measured with the parameter values used for Fig. 4.25 a identified by arrows with corresponding labels

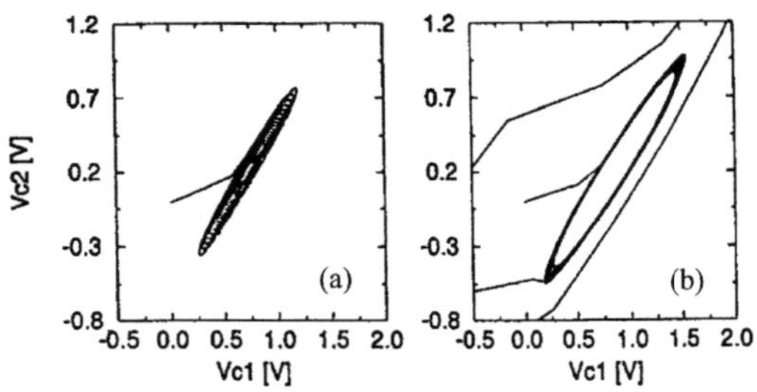

Fig. 4.27. The behavior of the trajectory originating from the stationary point O. The parameters of the measurements (**a**) and (**b**) are those used for Fig. 4.25a identified by arrows with the labels $a1$ and $b1$, respectively

In the experiment it was observed that the homoclinic orbits, h_n, with $n > 1$ appear only with the bifurcation of the second type. Trajectories observed in the vicinity of higher-order homoclinicity are shown in Fig. 4.26e–h.

As earlier mentioned, the existence of two different types of homoclinic bifurcation, h_1, originates from the different stability properties of P^+. The parameter range corresponding to different types of bifurcation is divided by the bifurcation curve h_s where P^+ loses stability. Figure 4.27 shows the behavior of the trajectories originating from O, observed before and after P^+ loses stability.

Solitary Waves. As shown earlier, the appearance of homoclinic orbits in the phase space of (4.37) indicates the existence of *solitary waves* of (4.34) with the boundary conditions

$$\begin{cases} y_{N+1} = y_N \,, \\ w_0 = w_1 \,. \end{cases} \tag{4.55}$$

The characteristics (velocity, etc.) of the solitary waves depend on the parameter values on the bifurcation set Π. The wave profiles correspond to the forms of the homoclinic orbits as discussed in Sects. 4.3.1–3. The number of humps in the solitary wave profile is determined by the number of rotations of the homoclinic orbit both around P^+ and around O (Fig. 4.23). Since O is of a saddle-focus type, the profiles of the solitary waves contain oscillating, wavy tails.

Let us see the characteristics of the possible solitary wave solutions in (4.34) with (4.9), as a function of the coupling parameter. The dependence of their velocity upon the coupling parameter is shown in Fig. 4.28a,b. Values were obtained for two different parameter sets (β, γ, μ). To find the values of δ corresponding to the existence of solitary waves we examined the splitting function, described when earlier in this Section discussing numerical results for homoclinic orbits, with argument δ. We have

$$c^2 = \frac{d}{1-\delta_n} \,,$$

where δ_n are the values of δ corresponding to the appearance of homoclinic orbits of (4.37) with index n. The propagation of such solitary waves in the array depends on their stability. Two different approaches have been taken in studying stability. In the numerical simulations we investigated the initial value problem with data close to the solitary wave solution (Fig. 4.29). The other approach follows the criteria used in the theory of continuously distributed systems, where the stability of the spatially homogeneous state associated with the solitary wave may be used as one of the conditions for stability of the solitary wave. For an array (i.e. discrete medium) this condition should be satisfied too. It is shown in Appendix D that all homogeneous

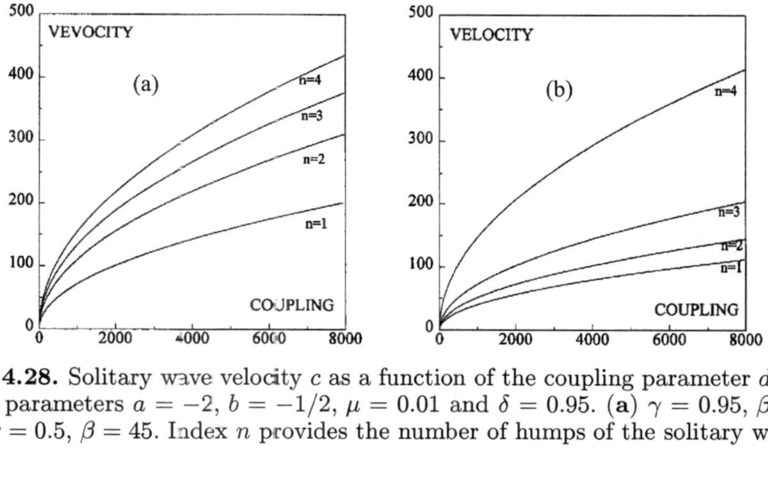

Fig. 4.28. Solitary wave velocity c as a function of the coupling parameter d with fixed parameters $a = -2$, $b = -1/2$, $\mu = 0.01$ and $\delta = 0.95$. (**a**) $\gamma = 0.95$, $\beta = 9$; (**b**) $\gamma = 0.5$, $\beta = 45$. Index n provides the number of humps of the solitary wave

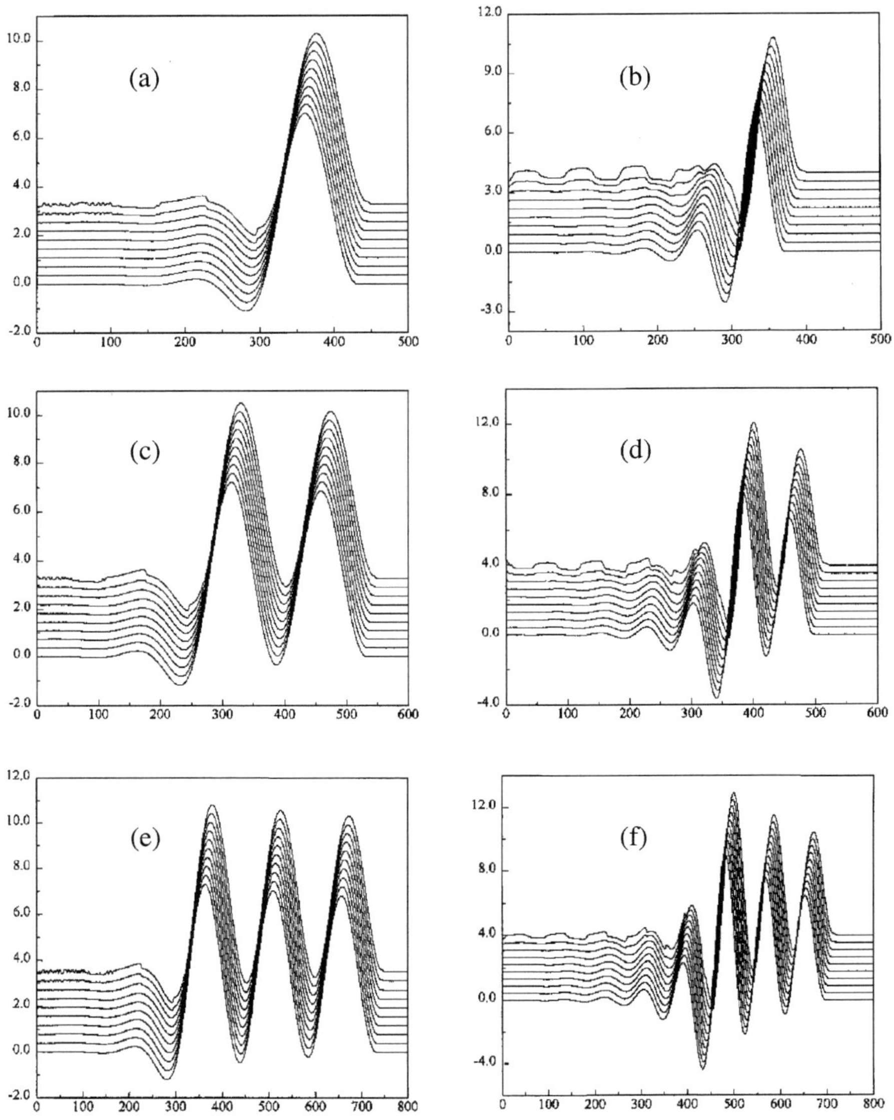

Fig. 4.29. Evolution of solitary waves. The x-coordinate corresponds to the "spatial" coordinate j, the y-coordinate to the value $-z + s\tau$, with z being the dimensionless coordinate proportional to the current through the inductor L, τ the dimensionless time and s a scaling coefficient. The parameter values in the array are $a = -2, b = -1/2, \mu = 0.01$. (**a**) $\gamma = 0.95, \beta = 9, d = 662.067, c = 66.667, s = 15$. The final time of integration $t_{end} = 0.234$. (**b**) $\gamma = 0.95, \beta = 9, d = 208.937$, $c = 50, s = 15, t_{end} = 0.234$. (**c**) $\gamma = 0.95, \beta = 9, d = 252.996, c = 66.667$, $s = 15, t_{end} = 0.234$. (**d**) $\gamma = 0.5, \beta = 45, d = 9995, c = 125, s = 20, t_{end} = 0.18$. (**e**) $\gamma = 0.5, \beta = 45, d = 3876.1, c = 100, s = 20, t_{end} = 0.18$. (**f**) $\gamma = 0.5, \beta = 45$, $d = 1971.5, c = 100, s = 20, t_{end} = 0.18$

stationary states of (4.34) and (4.55) are unstable. Therefore, strictly speaking, all solitary waves in (4.34) and (4.55) are unstable. However, if instability does not set in too fast, or too strongly, the solitary wave may actually occur and propagate for quite a long time before any appreciable change is seen. Then, as for the dissipation-modified Boussinesq–Korteweg–de Vries equation treated in Chap. 2, we may very well speak about the long-term "practical" stability of the waves which show slow enough "aging" effects. In numerical simulations of (4.34) we have found that for some parameter values solitary waves can propagate for quite some time with only slight changes in their profile.

The evolution of initial wave profiles chosen close to the solitary-wave solutions is illustrated in Fig. 4.29. Figure 4.29a–c show the evolutions of the solitary waves propagating with the velocities given in Fig. 4.28a. Figure 4.29d–f correspond to the velocities given in Fig. 4.28b. The x-axis in Fig. 4.29 corresponds to the "spatial" coordinate j, while the y-axis corresponds to $-z + s\tau$, where τ is a dimensionless time and s is a scaling coefficient. Figure 4.29 shows to what extent solitary waves of different shape can propagate with no appreciable changes. However, their oscillating tails affect the unstable stationary state ($x_j = y_j = z_j = w_j = 0$), and, indeed, after some time small perturbations grow in the background of the solitary wave (Fig. 4.29). These perturbations grow high enough to finally destroy the solitary waves. Note that the scale of the instability is different for different parameter sets. In particular, if the value of the coupling parameter is large enough (Fig. 4.29d–f), then the scale of the instability also widens.

4.3.2 Periodic Waves in a Circular Array

A Circular Array. Let us now focus attention on a circular, periodic ring of inductively coupled Chua's circuits [4.10]. Thus, we consider (4.34) and (4.3) with the boundary condition

$$\begin{cases} y_{N+1} = y_1\,, \\ w_0 \quad = w_{N+1}\,. \end{cases} \tag{4.56}$$

We again search for "traveling" wave solutions of (4.34) and (4.3) in the form of (4.35). Due to (4.56) we should look for only periodic solutions of (4.34). Note, that for a solution of period T, the quantity

$$h = \frac{Tm}{N}\,,$$

$$m = 1, 2, 3, \ldots N\,,$$

may be interpretated as a "phase" difference between two nearest neighbors in the array. If $h \ll 1$ the characteristic space scale of the traveling wave solutions (4.35) is significantly greater than the corresponding "discrete" spacing of (4.34). Thus, we can approximately define the variables $(x(\xi), y(\xi), z(\xi), w(\xi))$ not only in the array junctions j but between them, and hence use the continuum approximation to (4.34). It allows us to define a derivative with respect to the moving frame coordinate, ξ, leading to the simpler system (4.37). Hence periodic wave trains of the original model problem (4.34) correspond to periodic orbits of (4.37). The following three subsections are devoted to the study of these periodic orbits.

Phase-Space Analysis and a Suitable Poincaré Return Map. Again, to simplify the analytical description we start with the case in which $\mu = 0$. Needless to say, the singular perturbation analysis in this case yields 3D phase-space motions with fast and slow features. We denote by W_0 the 2D manifold of slow motions in the (x, y, z) phase space of (4.37). The shape of this manifold is

$$W_0 : \{(x, y, z) \mid y = x + f(x)\} .$$

Outside the surface W_0, the dynamics of (4.37) generates fast motions which may be described by

$$y = \text{const.}, \quad z = \text{const.} \tag{4.57}$$

It has been shown in Sect. 4.3.1 that (4.37) has three steady states:

$$O(0,0,0), \quad P^+(x_0, y_0, z_0), \quad P^-(-x_0, -y_0, -z_0) .$$

$P^\pm$ are stable foci corresponding to the pair of complex-conjugate eigenvalues

$$\lambda_{1,2} = -h \pm \mathrm{i}\omega , \tag{4.58}$$

with

$$h \equiv \frac{\gamma - \dfrac{B}{\delta}}{2} ,$$

$$\omega \equiv \sqrt{\frac{\beta}{\delta} - \frac{1}{4}\left(\gamma + \frac{B}{\delta}\right)^2} .$$

We consider (4.37) for the parameter range defined by the inequalities

$$\begin{gathered} a < -1 , \quad -1 < b < 0 , \quad \mu = 0 , \\ \delta > 0 , \quad \beta > \beta_b , \quad \gamma > B/\delta , \end{gathered} \tag{4.59}$$

with

$$\beta_b \equiv \frac{1}{4}\left(\gamma + \frac{B}{\delta}\right)^2$$

and B defined as in (4.40). In the (β, γ) plane, this region (denoted by G_s) is shown in Fig. 4.30. The 2D stable manifolds $W_0^s(P^{\pm})$ associated with the eigenvalues (4.58) coincide with the parts of the surface W_0 for $x > 1$ and $x < -1$, respectively. The part of W_0 restricted to $|x| < 1$ coincides with the stable manifold $W_0^s(O)$ of the saddle fixed point O. $W_0^s(P^+)$ attract trajectories while $W_0^s(O)$ ejects them to the region of fast motions. The qualitative behavior of the trajectory intervals corresponding to the fast and slow motions is illustrated in Fig. 4.31.

The evolution of (4.37) in $W_0^s(P^+)$, $W_0^s(P^-)$ and $W_0^s(O)$ is, respectively, defined by the following linear subsystems:

$$\begin{cases} \dot{y} = \frac{1}{\delta}(By + z + D)\,, \\ \dot{z} = -\beta y - \gamma z\,, \end{cases} \tag{4.60}$$

$$\begin{cases} \dot{y} = \frac{1}{\delta}(By + z - D)\,, \\ \dot{z} = -\beta y - \gamma z\,, \end{cases} \tag{4.61}$$

$$\begin{cases} \dot{y} = \frac{1}{\delta}(Ay + z)\,, \\ \dot{z} = -\beta y - \gamma z\,, \end{cases} \tag{4.62}$$

with $A \equiv -a/1 + a$. We denote by S^+ and S^- the lines of intersection of the manifold of slow motions W_0 with the planes $x = 1$ and $x = -1$, i.e.

$$S^+ : W_0 \cap x = 1 : \{(x, y, z) : x = 1,\ y = (a + 1)\}\,,$$
$$S^- : W_0 \cap x = -1 : \{(x, y, z) : x = -1,\ y = -(a + 1)\}\,.$$

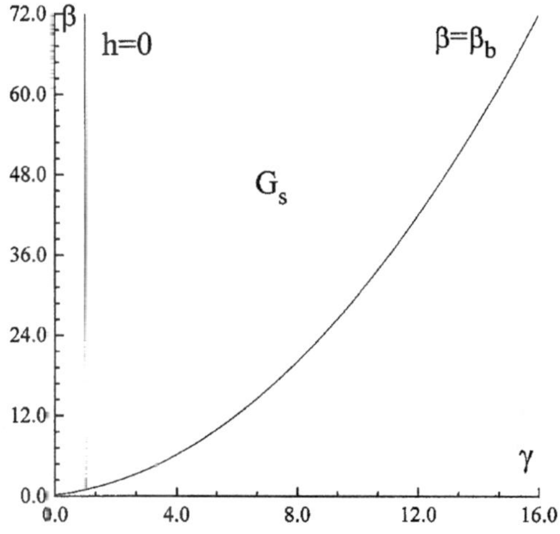

Fig. 4.30. The G_s region on the (β, γ) plane with $a = -2, b = -1/2, \delta = 1$, and $\mu = 0$. Here the stationary points P^+ and P^- are stable foci

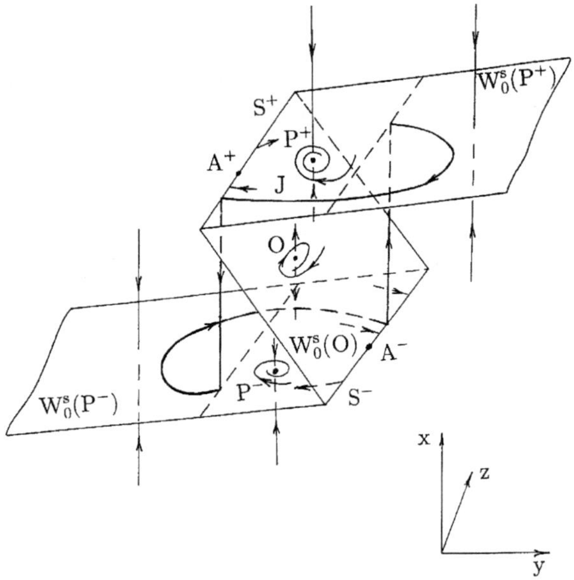

Fig. 4.31. The phase portrait (4.36) for $\mu = 0$ and the qualitative form of a periodic orbit

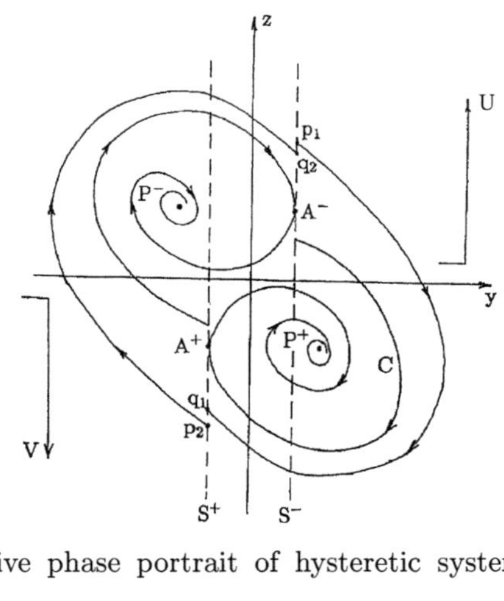

Fig. 4.32. Qualitative phase portrait of hysteretic system (4.32) in the state space (y, z)

The periodic orbits may be described in the following way: Let a trajectory start at some point of the "switch" line S^+ (Fig. 4.32) such that under the influence of the fast motions it falls down onto the stable slow-motions plane $W_0^s(P^-)$. After some time the trajectory intersects S^- and leaves $W_0^s(P^+)$. It goes through the region of fast motions and comes back to $W_0^s(P^+)$. Moving on this plane the trajectory finally returns to the initial point, hence delineating the periodic orbit (Fig. 4.31).

Point $A^+(y_a, z_a)$ of S^+ divides the flows of trajectories into incoming ($z > z_a$) and outgoing ($z < z_a$) from $W_0^s(P^+)$ along S^+. Point $A^-(-y_a, -z_a)$ of S^- is symmetric to A^+ and also divides the trajectory flows in incoming and outgoing from $W_0^s(P^-)$. The coordinates of these points are

$$y_a = (a+1)\,, \quad z_a = a\,.$$

Accordingly, the periodic orbits should intersect S^+ with $z < z_a$, and the line S^- with $z > -z_a$, respectively. These orbits are described, in fact, by the following *hysteretic* system

$$\begin{cases} \dot{y} = \frac{1}{\delta}[By + z + g(y)]\,, \\ \dot{z} = -\beta y - \gamma z\,, \end{cases} \tag{4.63}$$

with an hysteresis function

$$g(y) = \begin{cases} +D & \text{if} \quad y > -(a+1)\,, \\ \pm D & \text{if} \quad (a+1) < y < -(a+1) \\ & \qquad \text{and the last switch is } \mp D \text{ to} \pm D\,, \\ -D & \text{if} \quad y < (a+1)\,. \end{cases}$$

This system is constituted by two linear systems (4.60–61) which are symmetric about the origin. Figure 4.32 illustrates the phase portrait of (4.63) in the state space (y, z).

The periodic orbits of (4.63) can be described by a 1D half-return map. Let p_1 be a point of S^- with $z > -z_a$. The trajectory of (4.60) maps p_1 onto a point $q_1 \in S^+$ with $(z < z_c)$, where q_1 is the first intersection point of the trajectory with S^+. Define the half-return map, r_1, as

$$r_1 : p_1 \to q_1\,,$$

i.e. $r_1(p_1) = q_1$ for all points $p_1(y, z) : \{p_1 \in S^-,\ z > -z_a\}$. By the symmetry already described, for all points $p_2(y, z) : \{p_2 \in S^+,\ z < z_a\}$, the half-return map

$$r_2 : p_2 \to q_2 \ (q_2 \in S^-)$$

is also well defined. It is

$$r_2 = I \circ r_1 \circ I\,,$$

where $\circ$ denotes the sequential action of the operators (from right to left). I is the inverse map $I : R^2 \to R^2$ with $I \circ (y, z) = (-y, -z)$. The full-return map H mapping S^- onto itself is

$$H = r_2 \circ r_1 = I \circ r_1 \circ I \circ r_1\,.$$

Then, studying all fixed points of the 1D map, H, and their stability is equivalent to studying the fixed points of $r : S^- \to S^-$.

$$r = I \circ r_1\,.$$

Thus in what follows we consider only the r map generated by the trajectories of (4.60).

A solution of the linear system (4.60) is

$$\begin{aligned} y(t) &= \mathrm{e}^{-ht}(c_1 \cos \omega t + c_2 \sin \omega t) + y_0 \,, \\ z(t) &= \mathrm{e}^{-ht}\{[c_2 \omega \delta - c_1(\delta h + B)] \cos \omega t \\ &\quad -[c_1 \omega \delta + c_2(\delta h + B)] \sin \omega t\} + z_0 \,. \end{aligned} \tag{4.64}$$

To determine r, (4.60) is integrated with the conditions

$$\begin{aligned} t = 0 \to y &= -(a+1)\,, \\ z &= U\,, \\ t = \tau \to y &= (a+1)\,, \\ z &= -V\,, \end{aligned} \tag{4.65}$$

where $U > 0$ and $-V < 0$ are the z-coordinates of p_1 and q_1, respectively. After some simple transformations we obtain

$$\begin{aligned} U(\tau) =& -\frac{\delta\omega}{\sin \omega\tau}\left[a^+ \mathrm{e}^{h\tau} + a^- \left(\cos \omega\tau + \frac{h}{\omega} \sin \omega\tau\right)\right] \\ & + B(a+1) - D\,, \\ V(\tau) =& \frac{\delta\omega}{\sin \omega\tau}\left[a^- \mathrm{e}^{-h\tau} + a^+ \left(\cos \omega\tau - \frac{h}{\omega} \sin \omega\tau\right)\right] \\ & + B(a+1) + D\,, \end{aligned} \tag{4.66}$$

with

$$\begin{aligned} a^+ &= -(a+1) + y_0\,, \\ a^- &= -(a+1) - y_0\,. \end{aligned}$$

The parameter τ is varied from τ_∞ to τ_0, with

$$\tau_\infty = \frac{\pi}{\omega} : \{U(\tau_\infty) = V(\tau_\infty) = +\infty\}\,.$$

and τ_0 is defined by the relation

$$U(\tau_0) = -a, V(\tau_0) > -a \quad \text{or} \quad V(\tau_0) = -a, U(\tau_0) > -a\,,$$

with $\pi/\omega < \tau_0 < 2\pi/\omega$.

Let us denote by S^n and S^{n+1} the points in S^- and let $S^{n+1} = r \circ S^n$. The "time" label $n = 0, 1, 2, \ldots$ indicates the number of iterations of some initial point S^0 by r [$m = 2n$ is the number of crossings of S^- with a trajectory of (4.37)]. r is well defined by the function $V = r(U)$; hence in what follows we concentrate on the properties of this function in the parameter region G_s.

Periodic Orbits. Analytical Results. The first and the second derivatives of the function $V = r(U)$ are, respectively,

$$\frac{\mathrm{d}V}{\mathrm{d}U} = \mathrm{e}^{-2h\tau} \frac{U - B(a+1) + D}{V - B(a+1) - D}, \tag{4.67}$$

$$\frac{\mathrm{d}^2V}{\mathrm{d}U^2} = -\frac{\delta\,\mathrm{e}^{-3h\tau}(\omega^2 + h^2)\sin\omega\tau}{\omega(V - B(a+1) + D)^3} \times \left\{a^-[V - B(a+1) - D] - a^+[U - B(a+1) + D]\right\}. \tag{4.68}$$

When $U \to \infty$ the function $V = r(U)$ asymptotically tends to the line a_1

$$V = \exp(-h\pi/\omega)U - 2h\delta\left[a^+ - a^- \exp(-h\pi/\omega)\right] + \exp(h\pi/\omega)\left[D - B(a+1)\right] + \left[B(a+1) + D\right]. \tag{4.69}$$

First, we show that crossing the $h = 0$ boundary of G_s a region of stable periodic orbits appears (near this boundary outside the G_s region such orbit does not appear).

When $h = 0$ the asymptotic line a_1 is

$$V = U + 2D. \tag{4.70}$$

It is located above the line $V = U$, denoted by b_1. How the curve $r(U)$ approaches a_1 is determined by the sign of the second derivative (4.68). Since, when $U \to \infty$, we have

$$a^-[V - B(a+1) - D] - a^+[U - B(a+1) + D] = -2y_0(U + a + 2D) < 0, \tag{4.71}$$

the sign is negative and $r(U)$ approaches a_1 from below. As a_1 and b_1 are parallel there are no fixed points of r with $U \to \infty$. Figure 4.33a illustrates $r(U)$ obtained using (4.66). All trajectories of (4.37) starting at an initial point S^0 go transitively to infinity.

Inside the region G_s, where h has small positive values, a_1 must intersect b_1 $[\exp(-h\pi/\omega) < 1]$. Therefore, $r(U)$ also intersects b_1 at some point, J_s, which corresponds to the appearance at J_s $(r'(U) < 1)$ of a stable periodic orbit of an infinitely large size (Fig. 4.33b). Note that when h is small and negative such an orbit does not appear [because $\exp(-h\pi/\omega) > 1$], and there is no intersection point J_s with $U \to \infty$.

Increasing $h > 0$, i.e. proceeding further inside G_s, we have the bifurcation of an unstable periodic orbit. It occurs when the critical trajectory C (Fig. 4.32) passes through both A^+ and A^-. The parameter set C_u associated with this bifurcation is determined by

$$2D = -\frac{\delta\omega}{\sin\omega\tau^*}\left[a^+\mathrm{e}^{h\tau^*} + a^-\left(\cos\omega\tau^* + \frac{h}{\omega}\sin\omega\tau^*\right)\right], \tag{4.72}$$

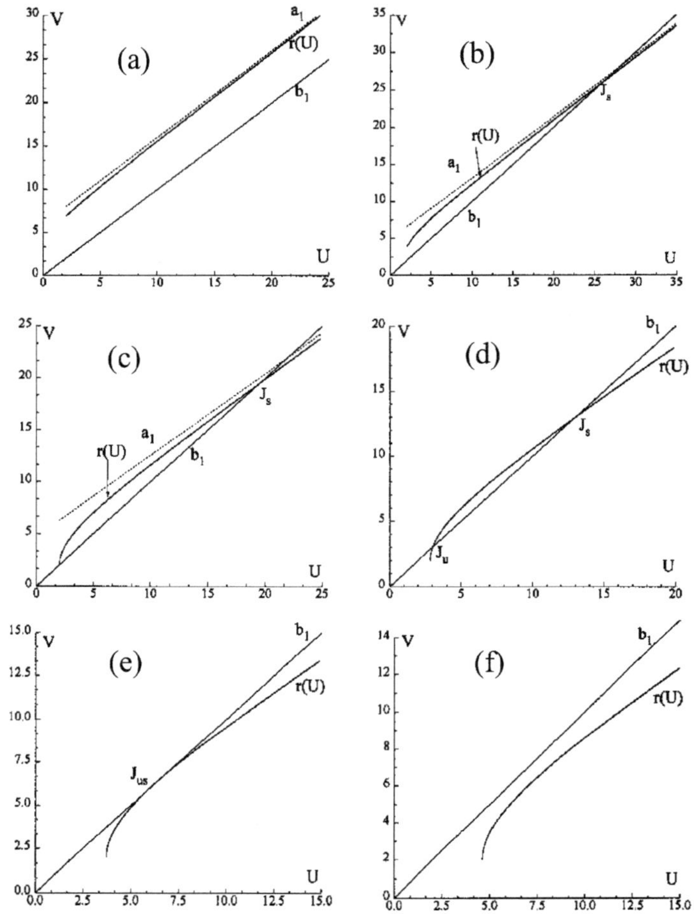

Fig. 4.33. 1D half-return map r for $\mu = 0$ with $a = -2, b = -0.5, \delta = 1$, and (β, γ) in the G_s region. (**a**) $\beta = 12, \gamma = 1$; (**b**) $\beta = 12, \gamma = 1.4$; (**d**) $\beta = 12, \gamma = 1.492$; (**c**) $\beta = 12, \gamma = 1.62$; (**e**) $\beta = 12, \gamma = 1.74$; (**f**) $\beta = 12, \gamma = 1.85$

where τ^* is defined using the equation

$$a^- \mathrm{e}^{-h\tau^*} + a^+ \left(\cos \omega\tau^* - \frac{h}{\omega} \sin \omega\tau^* \right) = 0\,, \tag{4.73}$$

and $\pi/\omega < \tau^* < 2\pi/\omega$. r in this case is shown in Fig. 4.33c.

With a further increase in h inside G_s the size of the stable orbit J_s decreases and the unstable orbit J_u grows (Fig. 4.33d). Finally, for some parameter values (we denote this set by C_c), J_s and J_u merge in a saddle-node bifurcation, J_{us}. C_c is determined by

$$2D = -\frac{\delta\omega}{\sin \omega\tau_2} \left(a^+ \mathrm{e}^{h\tau_2} + a^- \mathrm{e}^{-h\tau_2} - (a+1) \cos \omega\tau_2 - \frac{2y_0 h}{\omega} \sin \omega\tau_2 \right), \tag{4.74}$$

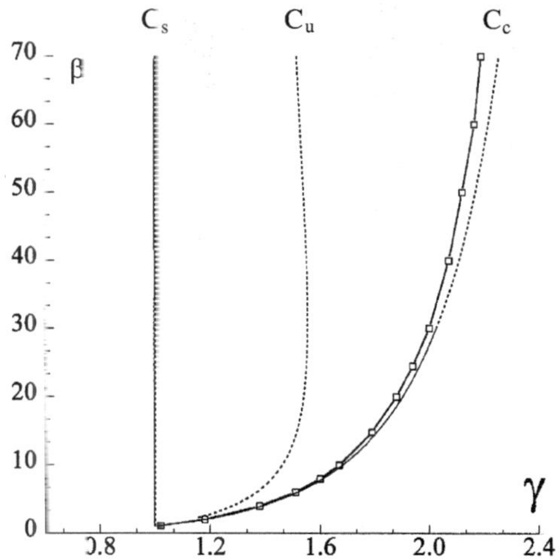

Fig. 4.34. Bifurcation boundaries in the (β, γ) plane for $a = -2, b = -1/2, \delta = 1$. *Dashed lines* correspond to $\mu = 0$ and *solid lines* to $\mu = 0.01$

where τ_2 is defined by

$$e^{-2h\tau_2} \frac{U(\tau_2) - B(a+1) + D}{V(\tau_2) - B(a+1) - D} = 1 , \tag{4.75}$$

with $\pi/\omega < \tau_2 < 2\pi/\omega$. r in this case is shown in Fig. 4.33e.

Outside the parameter region restricted by the curves C_s and C_c, r has no fixed point (Fig. 4.33f) and all the trajectories of (4.37) are attracted by one of the stable steady states, either P^+ or P^-. The relative location of C_s, C_u, C_c in the (β, γ) parameter plane is illustrated in Fig. 4.34 with the dashed lines corresponding to the following parameter values: $a = -2, b = -0.5$ and $\delta = 1$.

Periodic Orbits. Numerical Study. To make this study complete, a numerical exploration of the results earlier obtained has been carried out with $\mu \neq 0$ but $0 < \mu \ll 1$. As already discussed in Sect. 4.3.1, in this case the manifolds of steady states are close to the corresponding part of the slow-motions surface W_0. The fast motions are close to the lines $y =$ const., $z =$ const. and are inside the manifolds in a (order μ) neighborhood of the thin slow-motions layers. The periodic orbits generated by (4.37) may be described in a similar way as done in the subsection on phase-space analysis and suitable Poincaré return maps. However, the trajectory formed by this orbit does not come to the manifolds $W_s(P^+)$ and $W_s(P^-)$ of the steady states P^+ and P^-. It goes to the slow-motions layers near these manifolds. Then for small μ, the periodic orbits can also be described by a 1D Poincaré return map. Due to the symmetry properties of (4.37), the half-return map describes these orbits well.

To integrate (4.37), a fourth-order Runge–Kutta routine was used. An initial point was taken in the switch line $S^- : \{x = -1 \cap W_s(P^-)\}$ with

$y > -(a+1)$. Integration was stopped when the second intersection of the trajectory with the plane $x = 1$ was attained (the first intersection takes place in the fast-motions region). The point given by this intersection is very close to $S^+ : \{x = 1 \cap W_s(P^+)\}$. Note that the distance between this point and S^+ does not exceed μ. By moving the initial point along S^-, a 1D half-return map is obtained.

The bifurcation diagram in the (β, γ) plane for $\mu = 0.01$ is given in Fig. 4.34 by the solid lines. The curve C_s corresponding to the appearance of the stable periodic orbits is given by $h_b = 0$. The real part, h_b, of the complex-conjugate eigenvalues when $\mu \neq 0$, $0 < \mu \ll 1$, may be numerically calculated or obtained with good accuracy in an asymptotic approach (C.3–4).

Stability of the Periodic Waves. The traveling waves in the array described in the previous subsections may be interpreted as closed trajectories in the $4N$ phase space of the system of ODE (4.34). Therefore, to investigate their local stability we must analyze the eigenvalue Lyapunov spectrum associated with (4.34). The stable solutions all have eigenvalues with negative real parts (here we do not count the zero eigenvalue which always exists for a closed trajectory and corresponding time invariance). A single eigenvalue with a positive real part makes the solutions unstable. To overcome the difficulty of having a nonautonomous system when linearizing (4.34) around the periodic wave solution, we use an approximate scheme to locate the eigenvalue Lyapunov spectrum.

First, we consider the stability properties of the periodic orbits of the 3D system (4.37) describing traveling waves. It has earlier been done by means of the half-return map. Here we use an alternative approach. The periodic orbits consist of two parts of slow motions ($|x| > 1$) and two parts of fast motions ($|x| < 1$) (Fig. 4.31) and are described by the hysteretic system (4.63). Let us consider the transformation of the small volume V of the phase space (y, z) along a periodic orbit. As an initial value problem, here the decrease in this volume along eigenvector directions with $t \to \infty$ ensures the stability of the orbit (the flow has negative divergence). Since the orbit is periodic and symmetric it is enough to consider only a half-period containing one part of both slow and fast motions. Let $V_0 = dy_0 dz_0$ be a small phase volume near the point

$$
\begin{aligned}
y &= -(a+1)\,,\\
z &= R_p\,,
\end{aligned}
$$

where R_p is the fixed point coordinate of the half-return map r. Since in the region $y > -(a+1)$ (4.63) is linear, the transformation of V_0, up to the intersection with S^+, is defined only by the eigenvalues (4.58) and the time of the transformation. We consider the infinitesimal volume V_0 (dy_0, dz_0) and hence time is equivalent for all parts of the volume and V_0 can be considered as a "point" of the phase space. We have

$$V_2 = V_0 \exp\left(-2h\frac{T}{2}\right) \tag{4.76}$$

where T is the period of the orbit and h is the real part of the eigenvalue (4.55). The transformation of V_1, on S^+, into $V(T/2)$ is defined by

$$\begin{aligned} y &= (a+1)\,, \\ z &= -R_p\,. \end{aligned} \tag{4.77}$$

Let $d\tau$ be the time of the transformation. The phase volumes V_1 and $V(T/2)$ are approximately

$$V_1 = \mathrm{d}y_1\,\mathrm{d}z_1 \approx \dot{y}(1)\dot{z}(1)\,\mathrm{d}\tau^2\,,$$

$$V(T/2) = \mathrm{d}y_2\,\mathrm{d}z_2 \approx \dot{y}(2)\dot{z}(2)\,\mathrm{d}\tau^2\,,$$

where $\big(\dot{y}(1), \dot{z}(1)\big)$ and $\big(\dot{y}(2), \dot{z}(2)\big)$ are the vector-field directions of (4.60–1), respectively, at the point (4.77). The relation between V_1 and $V(T/2)$ is

$$V(T/2) = \frac{B(a+1) - D - R_p}{B(a+1) + D - R_p} V_1\,. \tag{4.78}$$

Substituting (4.76) into (4.78) we obtain the transformation of V during a half period:

$$V(T/2) = K(R_p)V_0\,, \tag{4.79}$$

with

$$K(R_p) = \frac{B(a+1) - D - R_p}{B(a+1) + D - R_p} \exp\left(-2h\frac{T}{2}\right)\,.$$

Denote by R_p^* the "size" of the orbit for which $K(R_p^*) = 1$. It follows from (4.79) that if $R_p > R_p^*$ then $K(R_p) < 1$ and the orbit is stable, while if $-a < R_p < R_p^*$ then $K(R_p) > 1$ and the orbit is unstable. Accordingly, if the size of the periodic orbit is larger than some critical size R_p^*, the "compression" of the phase volume during the slow-motion period is greater than the "expansion" given by the intersection with the switch lines. Thus the criterion for linear stability of the orbits are as follows:

- Within the region $x > 1\,(x < -1)$, all Lyapunov eigenvalues of the linear system (4.37) have negative real parts.
- The geometric condition $R_p > R_p^*$ is satisfied.

This approach to stability can be applied to the traveling waves representing the periodic trajectories in the $4N$ phase space of (4.34). Let μ be infinitesimal. The dynamics of (4.34) may be described in terms of slow and fast motions (Sect. 3.1). The slow-motions surface W_N in the $4N$ phase space is determined by

$$y_j - x_j - f(x_j) = 0\,,$$

$$j = 1, 2, 3, \ldots, N\,.$$

The "middle" part ($|x_j| \leq 1$) of the surface expels all trajectories, while the "outer" parts ($x_j > 1$ and $x_j < 1$) attract them. In the fast region (out of W_N) motions are close to

$$\begin{aligned} y_j &= \text{const.}\,, \\ z_j &= \text{const.}\,, \\ w_j &= \text{const.}\,, \\ j &= 1, 2, 3, \ldots, N\,. \end{aligned} \tag{4.80}$$

The trajectories which are close to a periodic wave are in the stable parts of the $4N$ surface W_N. For fixed "j", these parts are

$$x_j = By_j + D\,, \quad x_j > 1 \quad \text{and} \quad x_j = By_j - D\,, \quad x_j < -1\,.$$

During evolution the corresponding trajectory in the j-th direction of the $4N$ phase space falls down from one stable part of W_N to another passing through the fast region $|x_j| < 1$ in negligible time. Hence what matters is the transformation of the $4N$ phase volume V^{4N} of the small disturbances in the slow-motions part governed by (4.34) in the regions $x_j > 1$ and $x_j < -1$. Equations (4.34) in these regions represent linear systems with equivalent Lyapunov eigenvalue spectra, L. The compression of V^{4N} along $4N$ eigenvector directions demands that all eigenvalues of L have negative real parts. The second condition of the stability is geometric. The expansion of the phase volume in the j-th direction on S_j^+ must be smaller than the compression in this direction. It follows from (4.35) that for fixed j, the equations describing the orbit j-th element of the array coincide with (4.37). The geometric condition for stability of the periodic orbits of this system is $R_p > R_p^*$. In particular, traveling waves corresponding to the unstable periodic orbits of (4.37) are always unstable as $R_p < R_p^*$. The condition of negativeness of the L-eigenvalue real parts suffices to ensure local stability. Note that this stability analysis carries over to $\mu \neq 0$ with $0 < \mu \ll 1$. In this case the slow motions are restricted to thin layers (of order μ) near the surface W_N. The fast motions are close to the lines (4.80). Accordingly, the $4N$ orbit corresponding to the traveling wave may be described as done for $\mu = 0$.

The spectrum L is defined by the following linear system:

$$\begin{cases} \dot{x}_j &= \alpha(y_j - x_j - bx_j)\,, \\ \dot{y}_j &= x_j - y_j + z_j + w_{j-1} - w_j\,, \\ \dot{z}_j &= -\beta y_j - \gamma z_j\,, \\ \dot{w}_j &= d(y_j - y_{j+1})\,, \end{cases} \tag{4.81}$$

$$j = 1, 2, \ldots, N\,,$$

where (x_j, y_j, z_j, w_j) may be interpreted as small disturbances of the periodic wave. With periodic boundary conditions, the solution of (4.81) can be written as follows:

$$\begin{aligned} x_j &= \sum_{s=0}^{N-1} \tilde{x}_s \exp\left(\mathrm{i}\frac{2\pi s}{N} j\right), \\ y_j &= \sum_{s=0}^{N-1} \tilde{y}_s \exp\left(\mathrm{i}\frac{2\pi s}{N} j\right), \\ z_j &= \sum_{s=0}^{N-1} \tilde{z}_s \exp\left(\mathrm{i}\frac{2\pi s}{N} j\right), \\ w_j &= \sum_{s=0}^{N-1} \tilde{w}_s \exp\left(\mathrm{i}\frac{2\pi s}{N} j\right). \end{aligned} \tag{4.82}$$

Substituting (4.82) into (4.81) we obtain

$$\begin{cases} \dot{\tilde{x}}_s &= \alpha[\tilde{y}_s - \tilde{x}_s - f(\tilde{x}_s)], \\ \dot{\tilde{y}}_s &= \tilde{x}_s - \tilde{y}_s + \tilde{z}_s + [\exp(-\mathrm{i}2\pi s/N) - 1]\,\tilde{w}_s, \\ \dot{\tilde{z}}_s &= -\beta\tilde{y}_s - \gamma\tilde{z}_s, \\ \dot{\tilde{w}}_s &= d\,[1 - \exp(\mathrm{i}2\pi s/N)]\,\tilde{w}_s. \end{cases} \tag{4.83}$$

The characteristic equation for this system is

$$\begin{aligned} \lambda^4 + [1 + \gamma + \alpha(b+1)]\lambda^3 + [\gamma + \beta + \alpha b + \alpha(b+1)\gamma + 2d(1 - \cos 2\pi s/N)]\lambda^2 \\ + \{\alpha b\gamma + \alpha(1+b)\beta + 2d[\gamma + \alpha(b+1)](1 - \cos 2\pi s/N)\}\,\lambda \\ + 2d\alpha(1+b)\gamma(1 - \cos 2\pi s/N) = 0\,, \end{aligned} \tag{4.84}$$

$$s = 0, 1, 2, \ldots, N-1\,,$$

which defines the set of $4N$ eigenvalues of (4.81). The parameter values such that all $\{\lambda_i\}_{i=1}^{N}$ have negative real parts correspond to the stable traveling wave if, in turn, this wave corresponds to the stable periodic orbit ($R_p > R_P^*$)

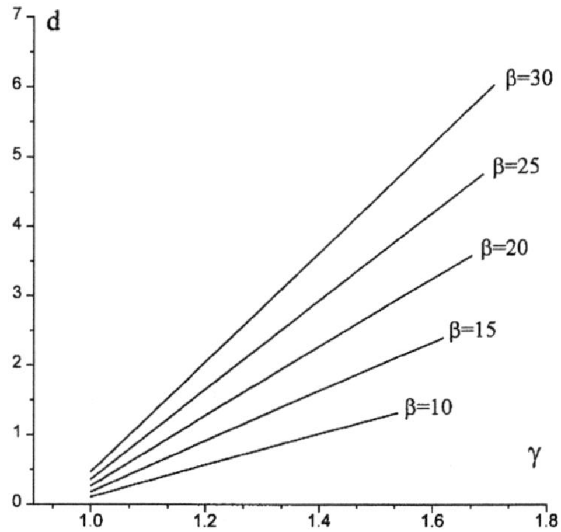

Fig. 4.35. Stability boundaries for the travelling waves in the (γ, d) plane for different values of β. For $d < d^*(\gamma, \beta)$ the waves are stable

of (4.37). For fixed values of the parameters α, β, γ, a and b of the unit in the lattice, (4.84) provides the critical value of the coupling, $d = d^*$. Then from (4.84) when

- $d < d^*$, all eigenvalues $\{\lambda_i\}_{i=1}^N$ have negative real parts and the wave is stable,
- $d > d^*$, a pair of complex-conjugate eigenvalues has a positive real part and the wave becomes unstable.

The dependence $d^* = d^*(\gamma)$, for different but fixed values of β, is shown in Fig. 4.35. This curve is restricted by the boundaries of the region of existence of the waves,

$$\gamma_{min} < \gamma < \gamma_{max} ,$$

where γ_{min} and γ_{max} are associated with the curves C_s and C_c, respectively (Fig. 4.34).

Periodic Waves. The periodic orbits of (4.37) determine the shape of the waves which can propagate in the ring (4.34). The dependence of the velocity of these waves on the coupling parameter, d, is

$$c^2 = \frac{d}{1 - \delta} . \tag{4.85}$$

For illustration we give an initial distribution which is close to a periodic wave of (4.34) with (4.56). The values of variables (x_j, y_j, z_j, w_j) in a junction (site) of the array are defined by the trajectory (x, y, z) of (4.37), which is close to the stable periodic orbit

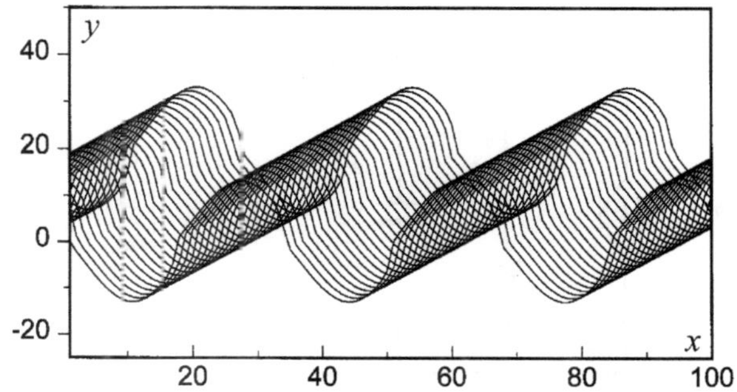

Fig. 4.36. Evolution of a periodic wave for $\beta = 30, \gamma = 1.5, \alpha = 100$. The abscissa is a spatial coordinate while the ordinate is defined by $y = x + s\tau$, where x is the variable defined by (4.33) and (4.9), τ is the time in (4.33) and (4.9), and $s = 20$ is a scaling coefficient

$$\begin{aligned} x_j(t=0) &= x(-jh)\,, \\ y_j(t=0) &= y(-jh)\,, \\ z_j(t=0) &= z(-jh)\,, \\ w_j(t=0) &= -dhy(-jh)\,. \end{aligned}$$

The integration of (4.34) and (4.56) describing the ring chain was carried out using a fourth-order Runge–Kutta integration routine (with absolute and relative numerical error 10^{-7}). The values of the parameters β, γ, α, a and b of the unit were held fixed. The evolution of the periodic waves, and hence their stability, was examined as a function of the coupling parameter d. Well-defined waveforms were observed still after a long propagation time (about 30 turns the ring). Traveling waves which did not change their shapes with respect to the initial distributions were considered as practically stable. If disturbances grew the wave was considered unstable. For low values of the coupling, $d < d^*$, waves were stable. Increasing the coupling, $d > d^*$, makes the waves unstable. The evolution of a periodic wave is illustrated in Fig. 4.36.

4.4 Chaotic Attractors and Waves in a One-Dimensional Array of Modified Chua's Circuits

Let us now investigate the propagation of traveling waves of aperiodic or with a seemingly chaotic profile in an unbounded 1D array of inductively coupled *modified* Chua's circuits [4.9].

4.4.1 Modified Chua's Circuit

Let us consider the circuit shown in Fig. 4.37, where a nonlinear voltage-controlled current source $i_b = g_b(V_1)$ is added across the capacitor C_2 of the earlier studied Chua's circuit. The two nonlinear functions $g_a(V_1)$ and

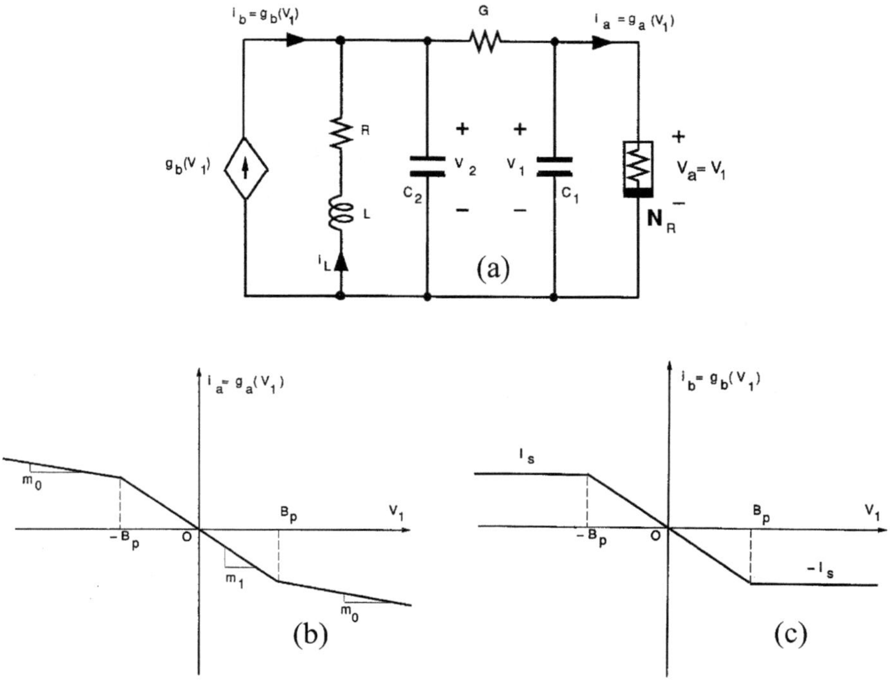

Fig. 4.37. Modified Chua's circuit. (**a**) Circuitry. (**b**) Characteristics of Chua's diode $i_a = i_a(V_1)$. (**c**) Characteristics of the voltage-controlled current source $i_b = g_b(V_1)$

$g_b(V_1)$ are shown in Fig. 4.37b,c, respectively. The evolution equations of this *modified* Chua unit are

$$\begin{cases} C_1 \dfrac{\mathrm{d}V_1}{\mathrm{d}t} = G(V_2 - V_1) - g_a(V_1)\,, \\ C_2 \dfrac{\mathrm{d}V_2}{\mathrm{d}t} = G(V_1 - V_2) + I_L + g_b(V_1)\,, \\ L \dfrac{\mathrm{d}I_L}{\mathrm{d}t} = -V_2 - RI_L\,, \end{cases} \tag{4.86}$$

where V_1, V_2 and I_L are the voltage across the capacitor C_1, the voltage across the capacitor C_2, and the current through the inductor L, respectively. By choosing suitable scales, (4.86) becomes in dimensionless form

$$\begin{cases} \dfrac{\mathrm{d}x}{\mathrm{d}\tau} = \alpha[y - x - g(x)]\,, \\ \dfrac{\mathrm{d}y}{\mathrm{d}\tau} = x - y + z + k(x)\,, \\ \dfrac{\mathrm{d}z}{\mathrm{d}\tau} = -\beta y - \gamma z\,, \end{cases} \tag{4.87}$$

with

$$x = \frac{V_1}{B_p}, \quad y = \frac{V_2}{B_p} \quad z = \frac{I_L}{B_p G}, \quad \tau = \frac{tG}{C_2}.$$

$g(x)$ and $k(x)$ are the dimensionless forms of $g_a(V_1)$ and $g_b(V_1)$:

$$\begin{cases} g(x) = bx + \frac{1}{2}(a-b)\left(|x+1| - |x-1|\right), \\ k(x) = \frac{1}{2}k\left(|x+1| - |x-1|\right). \end{cases} \tag{4.88}$$

The following dimensionless parameters have been introduced:

$$\alpha = \frac{C_2}{C_1}, \quad \beta = \frac{C_2}{LG^2}, \quad \gamma = \frac{RC_2}{GL}, \quad a = \frac{m_1}{B_p}, \quad b = \frac{m_0}{B_p}, \quad k = \frac{I_s}{GB_p}.$$

4.4.2 One-Dimensional Array

Now, as in earlier sections of this chapter, we consider a 1D array of inductively coupled (modified) Chua's circuits, where the inductance L_c connects the upper node of the capacitor C_2 between adjacent units. The state equations of the resulting 1D array are

$$\begin{cases} \dot{x}_j = \alpha[y_j - x_j - g(x_j)], \\ \dot{y}_j = x_j - y_j + z_j + k(x_j) + w_{j-1} - w_j, \\ \dot{z}_j = -\beta y_j - \gamma z_j, \\ \dot{w}_j = d(y_j - y_{j+1}), \end{cases} \tag{4.89}$$

$$j = 1, 2, \ldots, N,$$

where d denotes the inductive coupling between the units. Let us first explore some features of this modified Chua's circuit.

4.4.3 Chaotic Attractors

Phase-Space Analysis. Consider (4.87) when $\mu = 1/\alpha \ll 1$ is infinitesimal small. Again, in this case the motions of the system have both fast and slow features over thin layers, whose thicknesses are of the order of μ. Recall that systems exhibiting fast and slow motions are systems with relaxation oscillations.

Once more, for simplicity let us start with $\mu = 0$. Then the slow motions can be approximately described by the system

$$y - x - g(x) = 0, \tag{4.90}$$

$$\begin{cases} \dot{y} = x - y + z + k(x), \\ \dot{z} = -\beta y - \gamma z. \end{cases} \tag{4.91}$$

Equation (4.90) defines the surface of slow motions W_0, and (4.91) are the equations of the slow motions on W_0. Then the approximate equations of the fast motions of (4.87) are

$$\begin{cases} \mu \dot{x} = y - x - g(x)\,, \\ y \quad = \text{const.}\,, \\ z \quad = \text{const.} \end{cases} \tag{4.92}$$

Thus (4.87) has one fast variable x and two slow variables (y, z). We impose on the parameters a and b of $g(x)$ suitable conditions defining the Z shape of the surface W_0 as shown in Fig. 4.38. These conditions are

$$a < -1\,, \quad -1 < b < 0\,. \tag{4.93}$$

We assume that the other parameters of (4.87) satisfy the conditions

$$\gamma < 0\,, \quad \beta > 0\,, \quad k > D\,, \quad D \equiv \frac{b-a}{1+b}\,. \tag{4.94}$$

Since the functions $g(x)$ and $k(x)$ are piecewise linear, the phase space of (4.87) consists of three regions. Within each of the regions, motions are described by a corresponding linear system. These regions are divided by the planes U_{+1} and U_{-1} with

$$U_{\pm 1} = \{(x, y, z) : x = \pm 1\}\,.$$

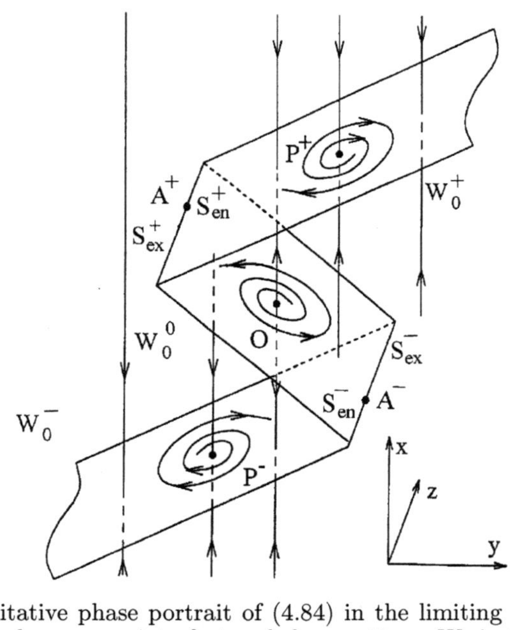

Fig. 4.38. Qualitative phase portrait of (4.84) in the limiting case $\mu = 0$ shows "fibering" of the phase space into fast and slow motions. W_0 is the surface of slow motions. The fast motions are vertical

The system (4.87) has three steady states: $O(0,0,0)$, $P^+(x_0, y_0, z_0)$ and $P^-(-x_0, -y_0, -z_0)$, with

$$x_0 \equiv \frac{\gamma(-k+D) - B\gamma k + \beta D}{\beta - \gamma B},$$
$$y_0 \equiv \frac{\gamma(-k+D)}{\beta - \gamma B}, \quad z_0 \equiv -\frac{\beta(-k+D)}{\beta - \gamma B},$$

where

$$B \equiv \frac{b}{1+b}.$$

All steady states lie on W_0. The point O lies in the middle part, W_0^0, of W_0 bounded by U_{+1} and U_{-1}; P^+ lies in the part W_0^+ of W_0 above U_{+1} (i.e. in the region $x > 1$); and P^- lies in the part W_0^- of W_0 under U_{-1} (i.e. in the region $x < -1$). Accordingly, on W_0 we find the Lyapunov characteristic exponents defining the character of the motion at W_0. They are

$$\lambda_{1,2}(b) = -h_b \pm \mathrm{i}\omega_b \quad \text{at} \quad W_0^+, W_0^-,$$
$$\lambda_{1,2}(a) = -h_a \pm \mathrm{i}\omega_a \quad \text{at} \quad W_0^0,$$

with

$$h_b \equiv \frac{\gamma - B}{2}, \qquad h_a \equiv \frac{\gamma + \frac{k+a}{1+a}}{2},$$

$$\omega_b \equiv \sqrt{\beta - \frac{(\gamma+B)^2}{4}}, \quad \omega_a \equiv \sqrt{\beta - \frac{\left(\gamma - \frac{k+a}{1+a}\right)^2}{4}}.$$

Let us add to (4.93–94) the following inequalities

$$\gamma < B, \quad \beta > (\gamma+B)^2/4, \qquad \gamma + k + a/1 + a < 0. \tag{4.95}$$

When the conditions (4.95) are satisfied, all three steady states are unstable in W_0: P^+, P^- are foci and O is either a focus or a node.

Let us now consider the stability of the slow motions relative to the fast ones. It follows from (4.92) that W_0^+, W_0^- are stable and W_0^0 is unstable with regard to fast motions. Since all steady states are unstable in W_0, after a finite time the trajectories of the slow motions will reach the boundary of the region of slow motions. These trajectories are assumed to "jump" from W_0. Then, the loci of points where the jump occurs are

$$S^{\pm} = W_0^{\pm} \cap U_{\pm 1}.$$

The fibered structure of the (x, y, z) phase space of the systems (4.91–92) on the fast and slow motions is shown in Fig. 4.38. Since O and W_0^0 are unstable, all trajectories of (4.91–92) must leave the region $|x| < 1$ of the phase space. Therefore, nontrivial stationary motions of (4.91–92) can be formed only when the fast–slow motions "connect" the stable planes W_0^+ and W_0^-.

When μ, $\mu > 0$, is small enough, the structure of the partitioning of the phase space into trajectories does not significantly change with respect to the case of (4.91–92). In the phase space, layers of slow motions appear, containing the slow-motions planes W_μ^+, W_μ^- and W_μ^0, which are close to the corresponding planes of (4.91–92) ($W_\mu^\pm \to W_0^\pm, W_\mu^0 \to W_0^0$ with $\mu \to 0$). The trajectories of (4.87) are close to the trajectories of (4.91) within the layers of the slow motions and to the trajectories of (4.92) outside these layers. Therefore, the motions of (4.87) are also formed by the fast–slow motions "connecting" the stable slow layers of the neighborhoods of W_μ^+ and W_μ^-.

For completeness, let us study the behavior of the system when $\mu \to 0$. In this limit the fast motions occur almost instantaneously and the motions of (4.87) may be described as piecewise discontinuous (x is the discontinuous variable). Therefore, the trajectories of (4.87) reside most of the time on the stable planes W_0^+ and W_0^-. Note that the projection of W_0^+, W_0^- onto the plane (y, z) is not one to one as shown in Fig. 4.38. Thus, when $\mu \to 0$ the fast motions can be *assumed* to make an instantaneous jump from one of these planes to another without changing the coordinates (y, z). These jumps proceed when the trajectories on W_0^+ and W_0^- arrive at the impasse lines S^+ and S^-. Thus, in the limiting case in which $\mu \to 0$, (4.87) can be modeled by a 2D system with hysteresis.

Two-Dimensional System with Hysteresis. Consider the following hysteretic system

$$\begin{cases} \dot{y} = \frac{1}{\delta}[By + z + \Gamma(y)]\,, \\ \dot{z} = -\beta y - \gamma z\,, \end{cases} \tag{4.96}$$

where $\Gamma(y)$ is a hysteresis delay function defined by

$$\Gamma(y) = \begin{cases} +(D-k) & \text{if} \quad y > -(a+1)\,, \\ \pm(D-k) & \text{if} \quad (a+1) < y < -(a+1) \\ & \qquad \text{and the last switch is } \mp(D-k) \text{ to } \pm(D-k)\,, \\ -(D-k) & \text{if} \quad y < (a+1)\,. \end{cases}$$

The motions of (4.95) are located on the half-planes W_0^+ or W_0^- and are governed by two linear systems with unstable foci. Then all the (nontrivial) trajectories of (4.96), after a finite time, must reach S^+ and S^- and leave the half-planes W_0^+ and W_0^-, respectively. Analyzing the behavior of (4.96) on W_0^- we find that S^- is divided by the point $A^-(-y_a, -z_a)$ in two parts, S_{ex}^- and S_{en}^- (Fig. 4.38), with

$$y_a \equiv a + 1 < 0\,, \quad z_a \equiv k + a > 0\,.$$

Along the trajectories of (4.96) the coordinate y decreases down to the half-line

$$S_{en}^{-} = \{(y,z) : y = -y_a, z > -z_a\}$$

and increases over the half-line

$$S_{ex}^{-} = \{(y,z) : y = -y_a, z < -z_a\}\,.$$

Hence, the trajectories enter W_0^- through S_{en}^- and leave this half-plane through S_{ex}^-. The trajectories of (4.96) have similar properties in W_0^+. S^+ is divided by $A^+(y_a, z_a)$ into the following two parts:

$$S_{en}^{+} = \{(y,z) : y = y_a\,,\ z > z_a\}\,,$$

$$S_{ex}^{+} = \{(y,z) : y = y_a\,,\ z < z_a\}\,.$$

The vector field of (4.96) on W_0^+ and W_0^- is tangent to S^+ and S^- at A^+ and A^-, respectively. Let us denote by C^+ the part of the trajectory lying on W_0^+ and containing A^+, and by C^- the part of the trajectory lying on W_0^- and containing A^-. The qualitative picture of C^+ is shown in Fig. 4.38. We shall see that C^+ and C^- play an important role in the bifurcation analysis.

Poincaré Return Map. It follows from the properties of the trajectories of (4.96) that

- any point on the half-line S_{ex}^- is, in finite time, mapped by a trajectory of (4.96) at W_0^+ to a point on the half-line S_{ex}^+;
- any point on the half-line S_{ex}^+ is, in finite time, mapped by a trajectory of (4.96) at W_0^- to a point on the half-line S_{ex}^-.

Hence, putting together the two maps

$$r^- : S_{ex}^- \to S_{ex}^+ \quad \text{and} \quad r^+ : S_{ex}^+ \to S_{ex}^-\,,$$

we find a map $f : S_{ex}^- \to S_{ex}^-$, with $f = r^+ \circ r^-$. Since (4.96) is invariant under the transformation $J : (y,z) \to (-y,-z)$ the trajectories of the maps r^- and r^+ are characterized by the property of mutual symmetry; hence r^+ and r^- become

$$r^- = J \circ r \quad \text{and} \quad r^+ = r \circ J\,,$$

where r is a *half-return* map. The "full"-return map f describing the dynamics of (4.96) may be obtained as $f = r^+ \circ r^- \equiv r \circ r$. $r : S_{ex}^- \to S_{ex}^-$ is defined in the following way (Sect. 4.3.2): Let p_1 be a point on S_{ex}^- (Fig. 4.39). A trajectory of (4.96) on W_0^+ maps p_1 onto a point q_1 in S_{ex}^+, where q_1 is the first intersection point of the trajectory with the half-line S_{ex}^-. Inverting p_1, with the map J, to q_1, we define the half-return map as

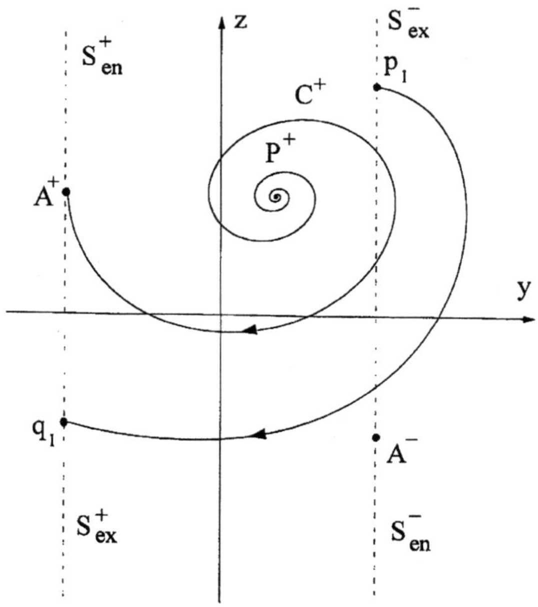

Fig. 4.39. Qualitative phase portrait of the hysteretic system (4.93). S^+ and S^- are the impasse lines where the motions jump from the half-plane W_0^+ to the half-plane W_0^-, and vice versa

$$r : p_1 \to q_1 ,$$

for all $p_1 \in S_{ex}^-$.

r defines a function $V = r(U)$, where U is the "z"-coordinate of p_1 and V is the absolute value of the "z"-coordinate of q_1. Then the "full" return map, defined by a function $f(U) = r\big(r(U)\big) \equiv r^2(U)$, is

$$\begin{aligned} &U(n+1) = f\big(U(n)\big) = r^2\big(U(n)\big) , \\ &n = 0, 1, 2, 3, \ \ldots , \end{aligned} \tag{4.97}$$

where n indicates the number of intersections of the trajectory starting from p_1 with S_{ex}^- (with switching, i.e. the trajectory "jumps" from W_0^- to W_0^+). To derive the function $r(U)$, we integrate the linear system on W_0^- with the initial conditions

$$\begin{aligned} &y(t=0) = -(a+1) , \\ &y(t=0) = U , \end{aligned} \tag{4.98}$$

until the first intersection with S_{ex}^+:

$$\begin{aligned} &y(t=\tau) = (a+1) , \\ &y(t=\tau) = -V . \end{aligned} \tag{4.99}$$

The general solution of (4.96) on W_0^- is

$$\begin{aligned} y(t) &= e^{-h_b t}(c_1 \cos\omega_b t + c_2 \sin\omega_b t) + y_0 , \\ z(t) &= e^{-h_b t}\{[c_2\omega_b - c_1(h_b + B)] \cos\omega_b t \\ &\quad -[-c_1\omega_b - c_2(h_b + B)] \sin\omega_b t\} + z_0 . \end{aligned}$$

Using (4.98–99) we obtain

$$\begin{aligned} c_1 &= -(a+1) - y_0 \\ c_2 &= 1/\omega_b\{V - [(a+1) + y_0](h_b + B) - z_0\} . \end{aligned}$$

Thus $r(U)$ is implicitly defined by

$$\begin{aligned} V = -e^{-h_b\tau}\Big(U \cos\omega_b\tau - \tfrac{1}{\omega_b}\{(U - z_0)(h_b + B) \\ -\beta[(a+1) + y_0]\} \sin\omega_b\tau\Big) - z_0 , \end{aligned} \qquad (4.100)$$

where τ is the lowest positive root of

$$e^{-h_b t}(c_1 \cos\omega_b\tau + c_2 \sin\omega_b\tau) + y_0 = -(a+1) .$$

Dynamics of the Continuous Map and Attractors of the Hysteretic System. Consider the dynamics of the map f defined by (4.97–100). Let us fix the parameter values

$$a = -2 , \quad b = -0.1 , \quad \beta = 4 .$$

We calculate the topological entropy and the Lyapunov exponent (Appendix E) of the map f by using an iteration scheme. We fix the parameter $\gamma = -0.6$ and vary the parameter k. Then:

(i) When $0 \le k \le k_{sn}$, $k_{sn} = 2.88$, the map has no bounded trajectory. For all initial points U^0 the map trajectories go to infinity as shown in Fig. 4.40a. When $k = k_{sn}$ a *saddle-node* bifurcation appears. The behavior of the map near this bifurcation is shown in Fig. 4.40a–c. As a consequence of this bifurcation both a stable and an unstable fixed point, U^s and U^u, respectively, appear. These points correspond to the stable and unstable periodic orbits of (4.96) in the phase plane (y, z), as shown in Fig. 4.40d.

(ii) The deformation of the map for $k_{sn} < k \le k_{pf}$, $k_{pf} = 3.56$ is illustrated in Fig. 4.41a. The coordinate of the unstable periodic point U^u increases and the map becomes nonmonotonic. For $k = k_{pf}$ the graph of the map and the diagonal are tangent at U^s, hence a *pitchfork* bifurcation. The variation of the map in the neighborhood of this bifurcation is shown in Fig. 4.41b,c. For $k > k_{pf}$ the point U^s loses its stability and two stable periodic points, U^1 and U^2, appear. Figure 4.41d shows the stable periodic orbits of (4.96) corresponding to U^1 and U^2.

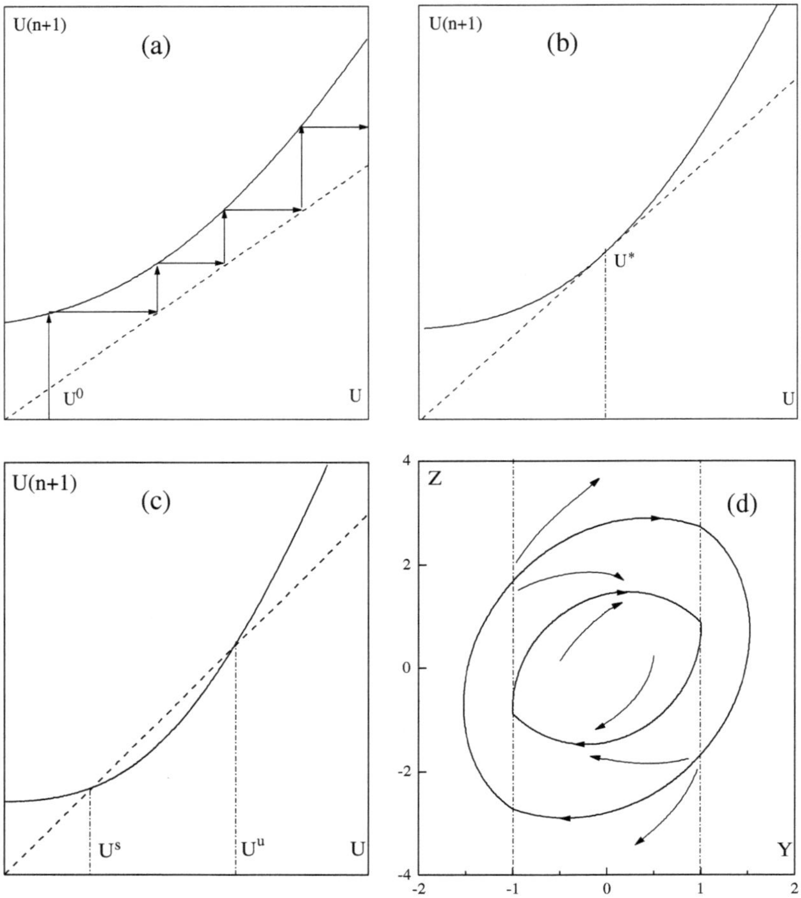

Fig. 4.40. The saddle-node bifurcation. Changing the map in a neighborhood of the bifurcation. (**a**) $k < k_{sn} = 2.88$. (**b**) $k = k_{sn}$. (**c**) $k > k_{sn}$. (**d**) The qualitative phase portrait of the periodic orbits in the (y, z) phase plane of the hysteresis system (4.93). The stable and unstable orbits are associated, respectively, with the stable fixed point U^s and unstable fixed point U^u of the map for $k > k_{sn}$

(iii) For $k_{pf} < k < k_0$, $k_0 = 3.965$, there exist two intervals, I_1 and I_2, invariant under the action of the map, as shown in Fig. 4.42a. These two regions attract all the trajectories of the map, and the dynamics of (4.96) is defined only by the action of the map within I_1 and I_2. Then f is a unimodal or a single-hump map of the intervals. Let us consider the interval I_2. Due to the symmetry of the trajectories of the hysteretic system, the attractor described by the map of I_1 is identical to the attractor of I_2 and is located symmetrically to it, relative to the origin. By increasing k from $k = k_{pf}$, the map of I_2 bifurcates in accordance with the Sharkovskii ordering. Figure 4.43 shows the bifurcation diagram numerically obtained by choosing many values

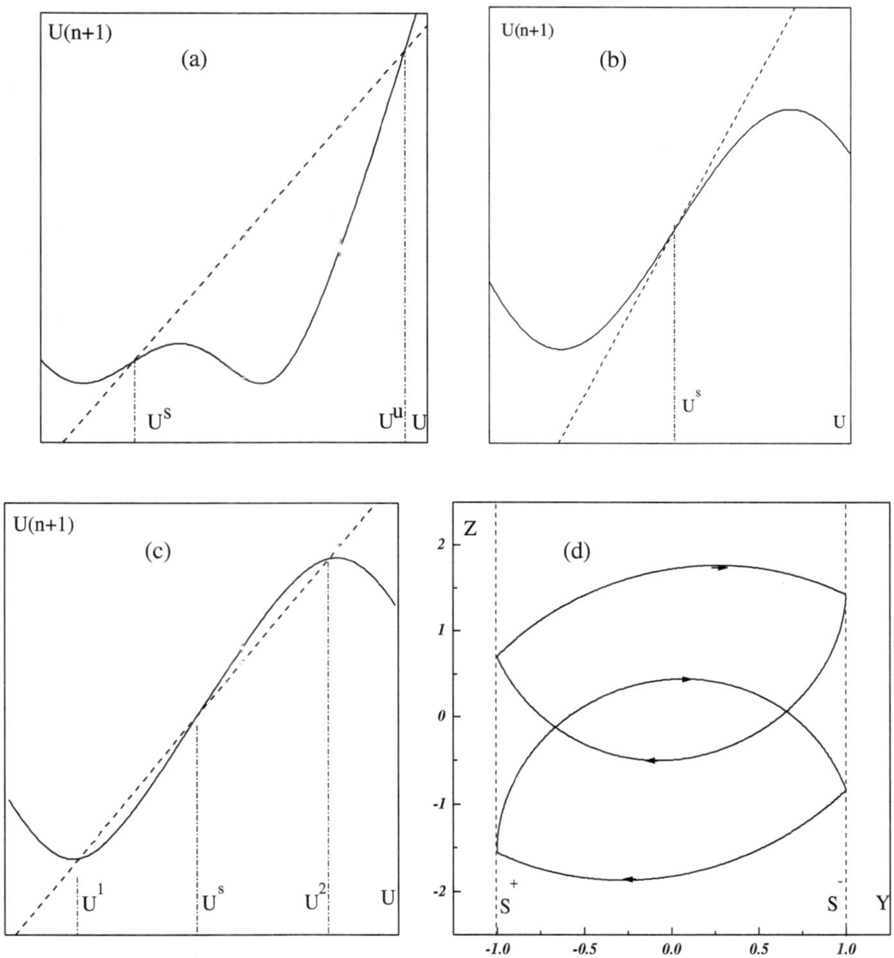

Fig. 4.41. (**a**) The map becomes nonmonotonic for $k = 3.5$. (**b,c**) The map f in a neighborhood of the pitchfork bifurcation $k_{pf} = 3.56$: (**b**) $k < k_{pf}$; (**c**) $k > k_{pf}$. (**d**) The stable periodic orbits in the phase plane of the hysteresis system for $k > k_{pf}$

of the parameter k between k_{pf} and k_0 and iterating an initial point 100 times (not displayed, awaiting transient behavior to die out) and then printing the next 100 values of U. A period-doubling cascade occurs from k_d to k_∞. At each bifurcation a stable fixed point of period 2^{n+1} appears and a fixed point of period 2^n loses its stability. This sequence of bifurcations accumulates at k_∞, which is called the *boundary of chaos* [4.1, 4.3]. In summary,

- for $k < k_\infty$ the map f of the interval I_2 is not chaotic;
- for $k > k_\infty$ the map is chaotic and $h_{top}(f) > 0$;
- for $k = k_\infty$ the map has unstable periodic orbits with periods of all the powers of 2. Other trajectories of the map are quasiperiodic.

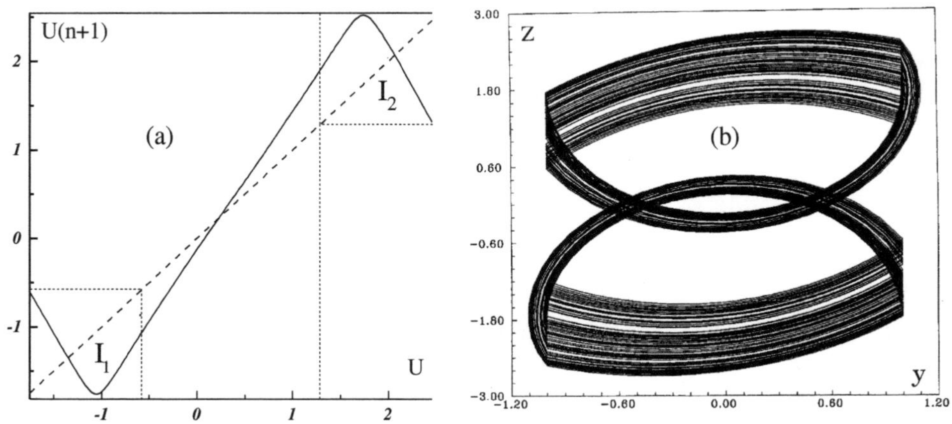

Fig. 4.42. (**a**) The map f for $k = 3.962$. I_1 and I_2 are two intervals which are invariant under the map action. Within each of the intervals the map is unimodal (single-hump) and describes two identical chaotic attractor with the characteristics $h_{top}(f) = 0.481$ and $\lambda \approx 0.18$. (**b**) Phase portrait of the chaotic attractors described by the map shown in (a). The attractors are identical and are located symmetrically with respect to the origin

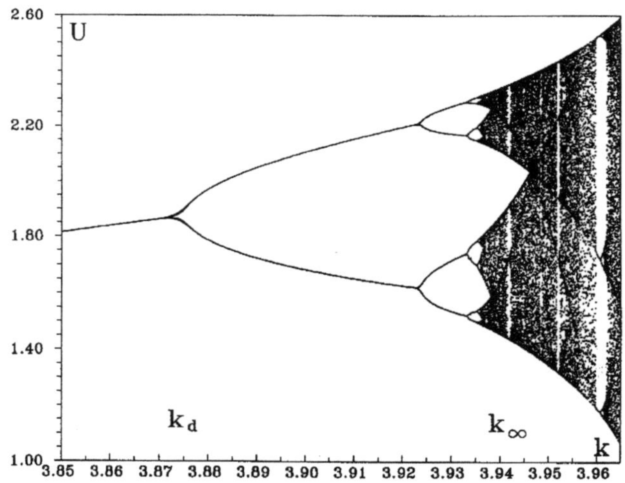

Fig. 4.43. Bifurcation diagram of the continuous map f as a function of the parameter k. $k_{pf} < k < k_0$

(iv) The chaotic map f, for $k_\infty < k < k_0$, can have both attracting periodic points of a period different from a power of 2 and an attracting chaotic interval (or a sequence of intervals) contained in I_2, the chaotic attractor. In the latter case the map of I_2 is characterized by a positive topological entropy and a positive Lyapunov exponent. The chaotic map of I_2 shown in Fig. 4.42a corresponds to $k = 3.962$ with $h_{top}(f) = 0.481$

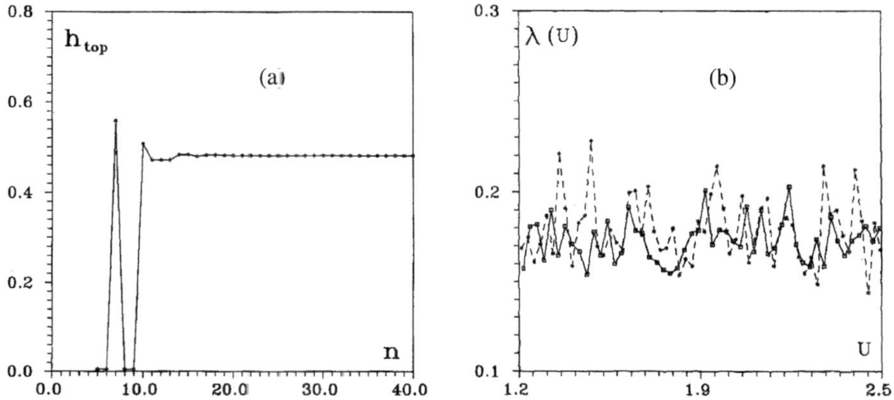

Fig. 4.44. Characteristics of the chaotic map of the interval I_2 for $k = 3.962$ (the map is shown in Fig. 4.43a). (**a**) Dependence of the calculated values of the topological entropy $h_{top}(f)$ on the time length n of the map trajectories. (**b**) The Lyapunov exponent as a function of the initial points U. U varies within the invariant interval I_2. The *dashed curve* shows the calculated value of λ for trajectories with a time length $n = 400$, and the *solid curve* the same but for $n = 800$

and $\lambda \approx 0.18$. The dependence of the entropy on the time length n of the map trajectory is shown in Fig. 4.44a and is typical of all the chaotic maps analyzed later on. The symbol λ denotes an average value of the Lyapunov exponent $\lambda(x)$ for $x \in I_2$ for the time duration of the map trajectories, $n = 800$. The graph of $\lambda(x)$ for the interval I_2 calculated for $n = 400$ (dashed curve) and for $n = 800$ (solid curve) is shown in Fig. 4.44b. The scattering of the calculated values $\lambda(x)$ within the interval decreases as the time duration of the trajectory increases. This situation is also typical of all the chaotic attractors considered below. Figure 4.42b illustrates the phase portrait of the hysteresis system (4.96) in this case. In the phase plane (y, z) there exist two identical chaotic attractors described by the map action in the intervals I_1 and I_2. The parameter value k_0 corresponds to the appearance of the first discontinuity set and will be analyzed later on.

(v) Consider now the transformation of the map as we change γ. For different but fixed values of $\gamma < 0$, and varying k, from $k = 0$ to $k = k_\infty$, there exist bifurcations $k_{sn}, k_{pf}, k_d - k_\infty$ and $k_\infty - k_0$. However, the "location" of k_0 in the Sharkovskii ordering changes with changing γ. Figure 4.45a illustrates the map f of I_1 and I_2 for $\gamma = -0.8$ near the bifurcation boundary $k_0 = 3.89$. In this case the map of I_2 acts like a "tent" map, and the whole interval I_2 is the chaotic attractor The topological entropy and the Lyapunov exponent are, respectively, $h_{top}(f) = 0.684$ and $\lambda \approx 0.65$. The phase portrait of the chaotic attractors in the phase plane (y, z) is shown in Fig. 4.45b. Note that I_1 and I_2 touch each other, with their identical attractors being very close to each other.

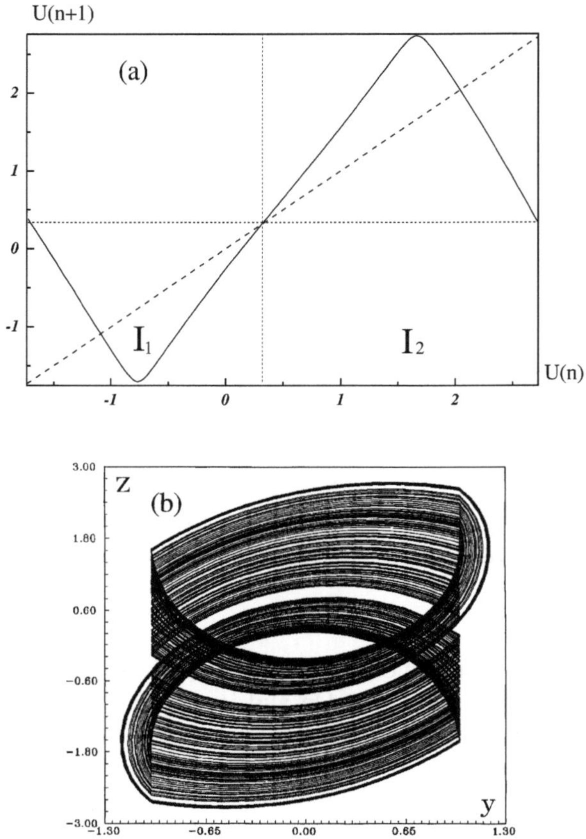

Fig. 4.45. Chaotic attractors for $\gamma = -0.8$, $k = 3.885$. (**a**) The map of the intervals I_1 and I_2 acts like the well-known "tent" map. The map describes two independent identical chaotic attractors with the characteristics $h_{top}(f) = 0.684$, and $\lambda \approx 0.65$. (**b**) The phase portrait of these attractors in the (y, z) phase plane of the hysteretic system. The attractors are very close to each other

(vi) The map which is shown in Fig. 4.46a corresponds to $\gamma = -0.95$ and $k = 3.845$. It describes a symmetric (around the origin) chaotic attractor with $h_{top}(f) = 0.863$ and $\lambda \approx 0.84$. This attractor can be considered as the "interaction" of two identical chaotic attractors described by the map of I_1 and I_2. These intervals are chaotic but are not invariant under the map action. There is a region contained in I_2 which is mapped into I_1, and a region of I_1 which is mapped into I_2. The "strength" of the "interaction" can be characterized by the value of $I_{12} = f(I_1) \cap f(I_2)$. When I_{12} is small the "interaction" is "weak", i.e. the map trajectories are iterated within I_1 (or I_2) for a long time. The phase portrait of the symmetric attractor corresponding to the map of Fig. 4.46a is shown in Fig. 4.46b.

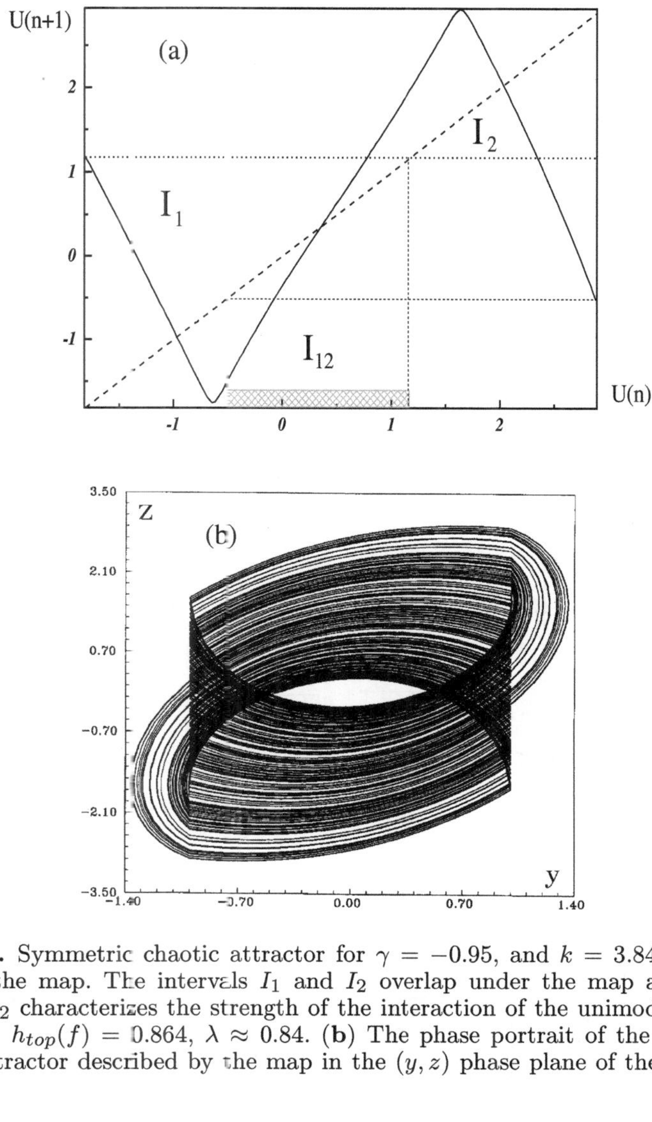

Fig. 4.46. Symmetric chaotic attractor for $\gamma = -0.95$, and $k = 3.845$. (**a**) The graph of the map. The intervals I_1 and I_2 overlap under the map action. The interval I_{12} characterizes the strength of the interaction of the unimodal maps of I_1 and I_2: $h_{top}(f) = 0.864$, $\lambda \approx 0.84$. (**b**) The phase portrait of the symmetric chaotic attractor described by the map in the (y, z) phase plane of the hysteretic system

Dynamics of the Discontinuous Map and Attractors of the Hysteretic System. In this subsection we give examples of the map f in the discontinuous case and consider some typical features of its chaotic attractors.

"River" of Trajectories. The dynamics of (4.96) essentially depends on the arrangement of the *critical* trajectories C^+ and C^-, which have been defined in Sect. 4.3.2 in the phase plane (y, z) (Fig. 4.39). Due to symmetry it suffices to study only how the critical trajectory C^+ appears (the trajectory C^- is located symmetrically with respect the origin).

Consider the situation when C^+ does not intersect the line S^- as shown in Fig. 4.47a. In this case, all trajectories of (4.96) on W_0^+ starting from the points of S^- arrive at S^+ during the first turn around the focus P^+. Then the half-return map $r(U)$ and, hence, the map $f(U) = r^2(U)$ are both continuous for all U. The attractors of f in this case have been considered in the previous subsection and obtained for $0 < k < k_0$.

Let C^+ intersect S^- at two points, a and b, as shown in Fig. 4.47c. Then, a and b are the discontinuity points of $r(U)$. Consider, for example, b. For $U \to b + 0$ (i.e. $U > b$) the trajectory starting from U intersects S^+ for the first time in a neighborhood of the point A^+. For $U \to b - 0$ (i.e. $U < b$) the map trajectory makes one complete turn around P^+ before reaching, for the first time, S^+. Thus the right-hand and left-hand limits of $r(U)$, at b,

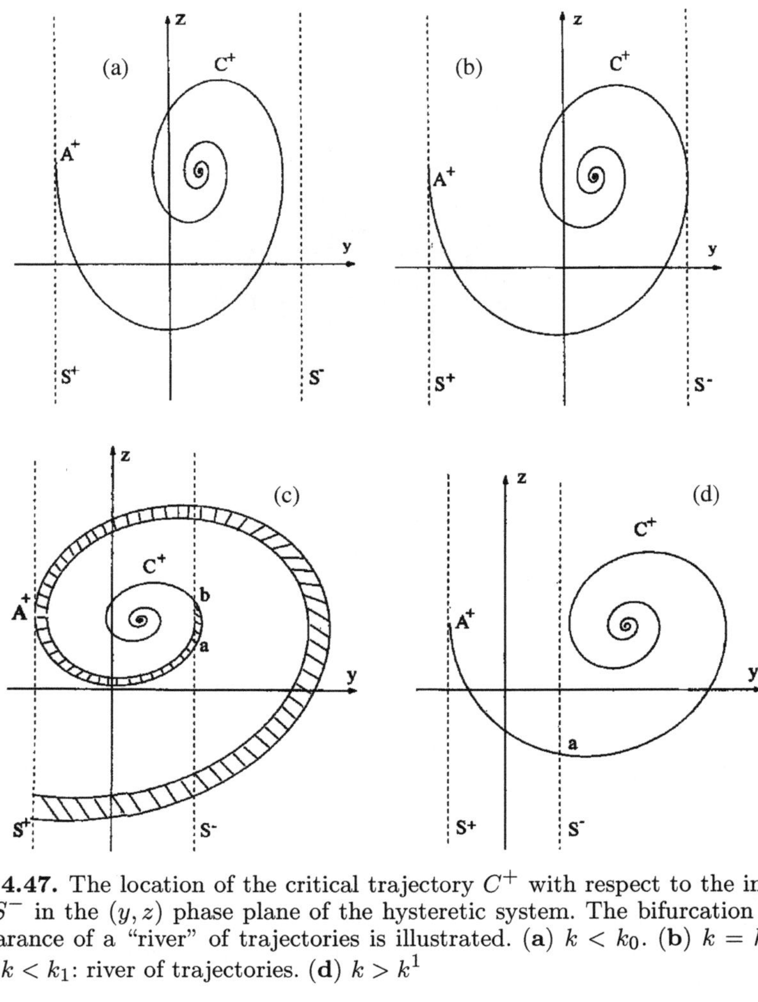

Fig. 4.47. The location of the critical trajectory C^+ with respect to the impasse line S^- in the (y, z) phase plane of the hysteretic system. The bifurcation of the appearance of a "river" of trajectories is illustrated. (**a**) $k < k_0$. (**b**) $k = k_0$, (**c**) $k_0 < k < k_1$: river of trajectories. (**d**) $k > k^1$

are different, i.e. $r(b-0) \neq r(b+0)$, and, hence, r is discontinuous. Similar results apply a. Since f is defined as $r\big(r(U)\big)$ it also has the discontinuities at a and b, given by

$$r(U) = a \quad \text{and} \quad r(U) = b\,.$$

Consider how the trajectories of (4.96) on W_0^+ map the interval (a,b) into the impasse line S^+. The trajectories starting from this interval make one complete turn around the focus until they first intersect with S^+, and then create a "river" of trajectories as shown in Fig. 4.47c. Figure 4.47b illustrates the bifurcation when the critical trajectory C^+ and S^- are tangential. The first discontinuity set of f appears near this bifurcation parameter $k = k_0$.

Bifurcation Order. Note that C^+ shown in Fig. 4.47c can have tangency with S^- during the "previous" turn around the focus, i.e. from the point of tangency to A^+, the trajectory can make $m = 1, 2, 3, \ldots$ complete turns. As a consequence of such bifurcations a new discontinuity appears in f. The mechanism of appearance of discontinuities can be explained in a way similar to what has been earlier done for the first tangency $(m = 0)$ of C^+ and S^-. Let us fix γ and increase k from k_0 up. If we label the bifurcations of the tangency by the subscript m, we obtain the sequence $\{k_m\}_{m=0}^{\infty}$ which has

$$k_f = (a+1)\left(1 - \frac{\beta}{\gamma B}\right) B - D$$

as its accumulation point. When $k = k_f$, P^+ is on S^-. Then, there is a countable set of intersections between C^+ and S^-, and, consequently, there exists a countable set of discontinuities of the map.

Further increasing $k > k_f$, P^+ lies to the right of S^-. The number of intersection points between C^+ and S^- decreases through the bifurcations at the tangency, and the discontinuity sets begin to disappear. We also obtain a sequence of bifurcations

$$k_f \ldots k^m, k^{m-1}, \ldots, k^2, k^1\,,$$

where the index m denotes the number of complete turns of C^+ around the focus from the tangency point to A^+. The location of C^+ in the phase plane of the hysteretic system for $k > k^1$ is shown in Fig. 4.47d. $r(U)$ in this case has a discontinuity at a. The discontinuity set of $f = r\big(r(U)\big)$ consists of a and the points defined by $r(U) = a$.

Attractors with a "River" of Trajectories. As it follows from earlier given arguments, f in the continuous case can have two identical, independent attractors, or a single symmetric chaotic attractor. Let us now consider different possible transformations of these attractors when the map becomes discontinuous for different values of γ and k.

When $\gamma = -0.6$ and $k_0 < k < k_1$, $k = 3.99$, f describes two identical, though independent, attractors with a river of trajectories (Fig. 4.48a). Note that I_2 is invariant with respect to the map and contains the set r_2, which

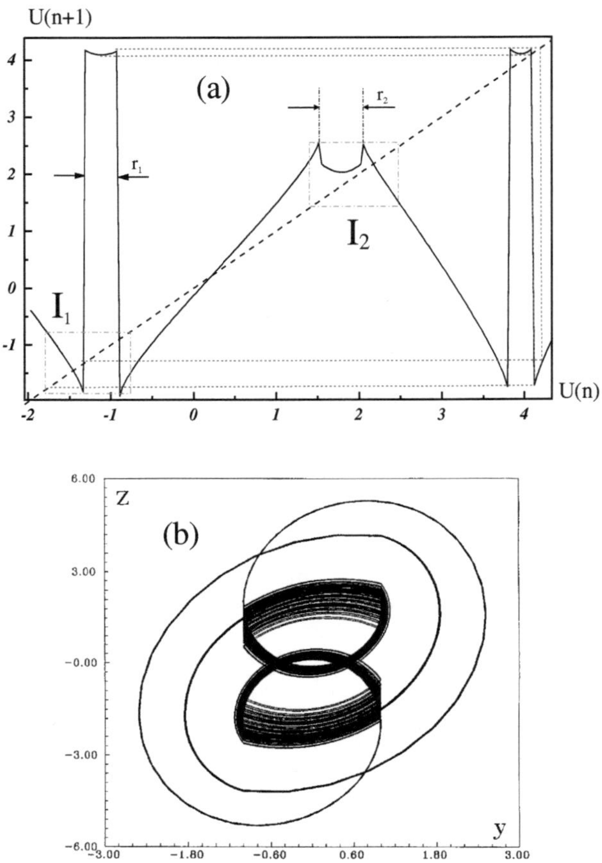

Fig. 4.48. Independent identical chaotic attractors with a river of trajectories: $\gamma = -0.6$, $k = 3.99$. (**a**) Graph of the map f for the invariant interval I_2, $h_{top} = 0.591$, $\lambda \approx 0.35$. (**b**) Phase portrait of the attractors in the (y, z) phase plane of the hysteretic system

corresponds to a river of trajectories [r_2 coincides with the interval (a, b) shown in Fig. 4.47c]. The map of this interval describes a chaotic attractor with $h_{top}(f) = 0.591$ and $\lambda \approx 0.35$. An identical chaotic attractor is described by I_1. The river of trajectories of I_1 is defined by the set r_1. This set is iterated twice and mapped back into I_1 as shown with dashed lines in Fig. 4.48a. Figure 4.48b illustrates the phase portrait of these two chaotic attractors.

Consider next the situation in which two independent chaotic attractors begin to "interact" through the river of trajectories, forming a single symmetric chaotic attractor. This occurs when the river starting from one attractor "flows" into the other attractor. The map obtained for $\gamma = -0.4$ and $k = 4.135$ is shown in Fig. 4.49a. The dashed lines show how the river sets r_1 and r_2 of the intervals I_1 and I_2 map into I_2 and I_1 in one step,

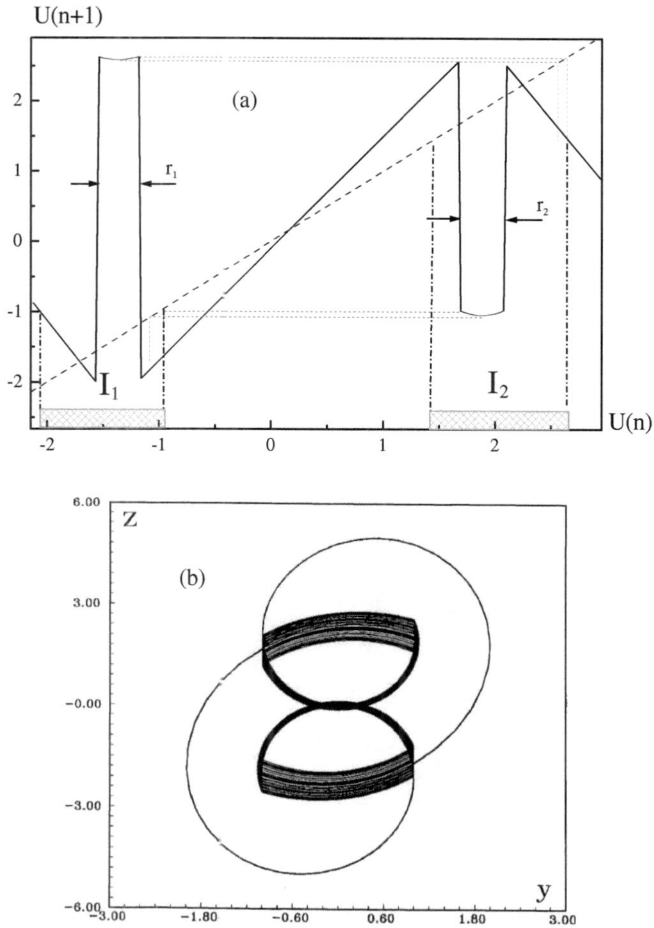

Fig. 4.49. Symmetric chaotic attractor with a simple river form: $\gamma = -0.4$, $k = 4.135$. (**a**) Graph of the map f with $h_{top} = 0.382$ and $\lambda \approx 0.21$. (**b**) Phase portrait of the attractors in the (y, z) phase plane of the hysteretic system

respectively. The topological entropy and the Lyapunov exponent for the map are $h_{top}(f) = 0.382$ and $\lambda \approx 0.21$, respectively. Accordingly, the symmetric attractor described by this map with two rivers is, indeed, chaotic. Its phase portrait is shown in Fig. 4.49b.

The river of trajectories can have either simple or complex forms. Consider the graph of the map for $\gamma = -0.67$ and $k = 4.04$ shown in Fig. 4.50a. The interaction between I_1 and I_2 is produced by the complicated rivers which are described by the map action on r_1 and r_2. This action is shown in Fig. 4.50a (dashed lines). After two iterations, r_1 maps into I_2, and r_2 into I_1. This attractor is characterized by $h_{top} = 0.848$ and $\lambda \approx 0.65$. Figure 4.50b illustrates the phase portrait of the chaotic attractor with two complex rivers.

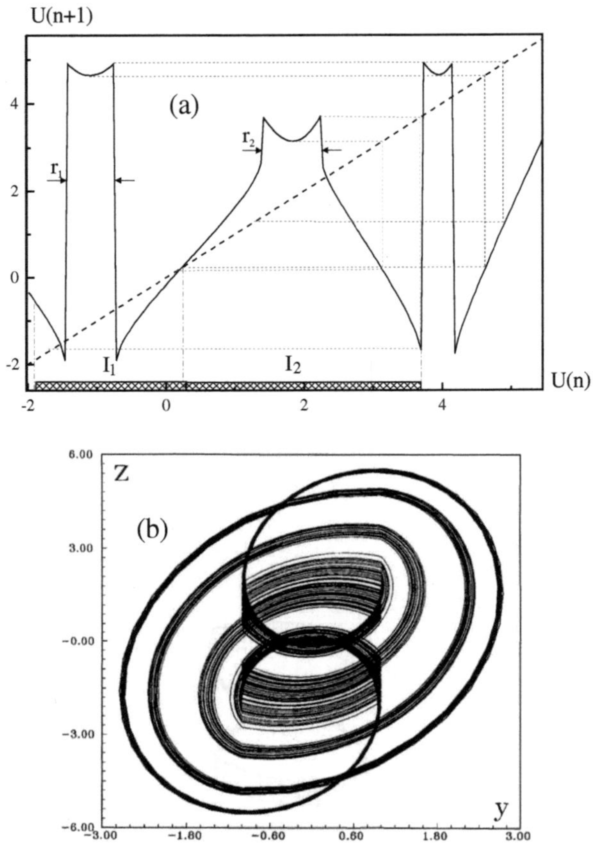

Fig. 4.50. Symmetric chaotic attractor with the complicated river form: $\gamma = -0.67$, $k = 4.04$. (**a**) Graph of the map f. $h_{top} = 0.848$, $\lambda \approx 0.65$. The map action on the river sets is shown by *dashed lines*. (**b**) Phase portrait of the attractor in the (y, z) phase plane of the hysteretic system

Attractors of the Map for Sufficiently Large k. Let us fix $\gamma = -0.6$. Consider the possible attractors of the map for large enough values of k (e.g. $k > k^1$, $k^1 = 15.3$). For $k = 20$, f has two invariant intervals, I_1 and I_2, and describes two independent identical attractors, as shown in Fig. 4.51a. Since the entropy and the Lyapunov exponent are both positive, $h_{top}(f) = 0.479$ and $\lambda = 0.37$, respectively, these attractors are chaotic. The phase portrait of the attractors is shown in Fig. 4.51b.

The map shown in Fig. 4.52a corresponds to $k = 15.7$ and describes the symmetric attractor generated from the interaction between I_1 and I_2. The "strength" of the interaction is characterized by the size of the interval $I_{12} = f(I_1) \cap f(I_2)$. This interval is very small. Hence, for a long time interval the map trajectories are confined within the interval I_1 (or I_2). The characteristics of the map are $h_{top} = 0.518$ and $\lambda \approx 0.43$, and hence the attractor

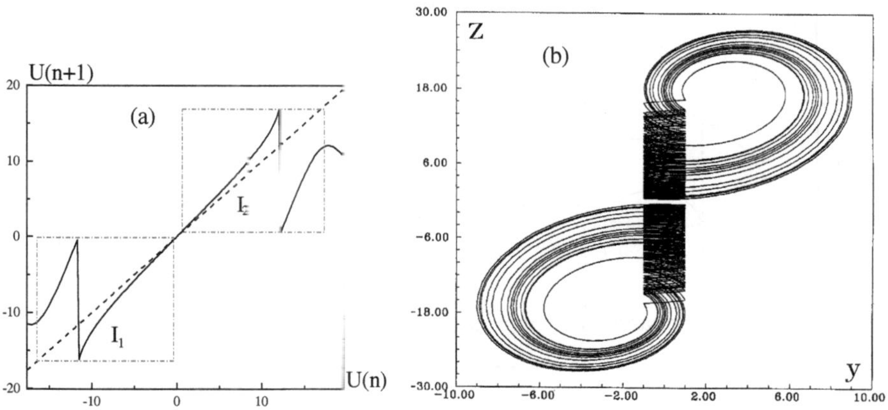

Fig. 4.51. Two independent chaotic attractors for $\gamma = -0.6$ and $k = 20$. (**a**) Graph of the map f. For the map of each interval $h_{top} = 0.479$ and $\lambda \approx 0.37$. (**b**) Phase portrait of the attractors in the (y, z) phase plane of the hysteretic system

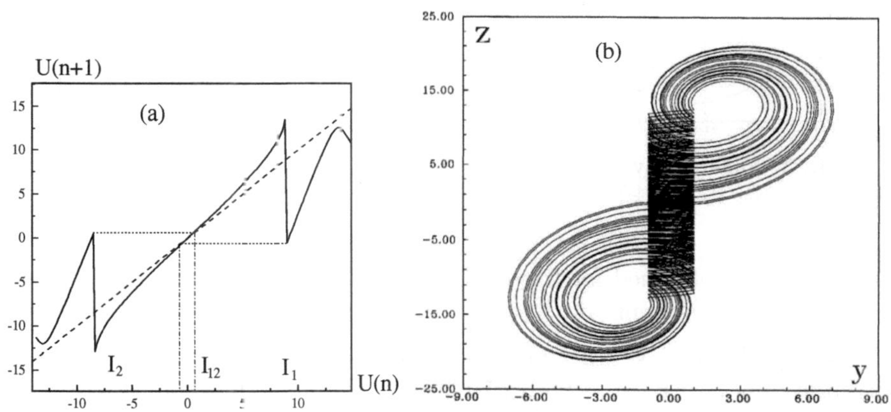

Fig. 4.52. Symmetric chaotic attractor realized for $\gamma = -0.6$ and $k = 15.7$. (**a**) Graph of the map f. The interval of overlapping I_{12} characterizing the strength of the "interaction" between the intervals I_1 and I_2: $h_{top} = 0.518$, $\lambda \approx 0.43$. (**b**) Phase portrait of the attractor in the (y, z) phase plane of the hysteretic system

is chaotic. Figure 4.52b illustrates the phase portrait of the attractor in the phase plane (y, z).

Periodic Orbits. Along with chaotic dynamics, (4.96) also exhibits regular behavior. The map f can have stable periodic points with different periods. These points attract the map trajectories from some interval of U. Let us consider the simplest periodic points and the corresponding orbits of the hysteretic system for a continuous map. When the map has discontinuity sets, then the form of the orbits can be complex. Figure 4.53 shows the

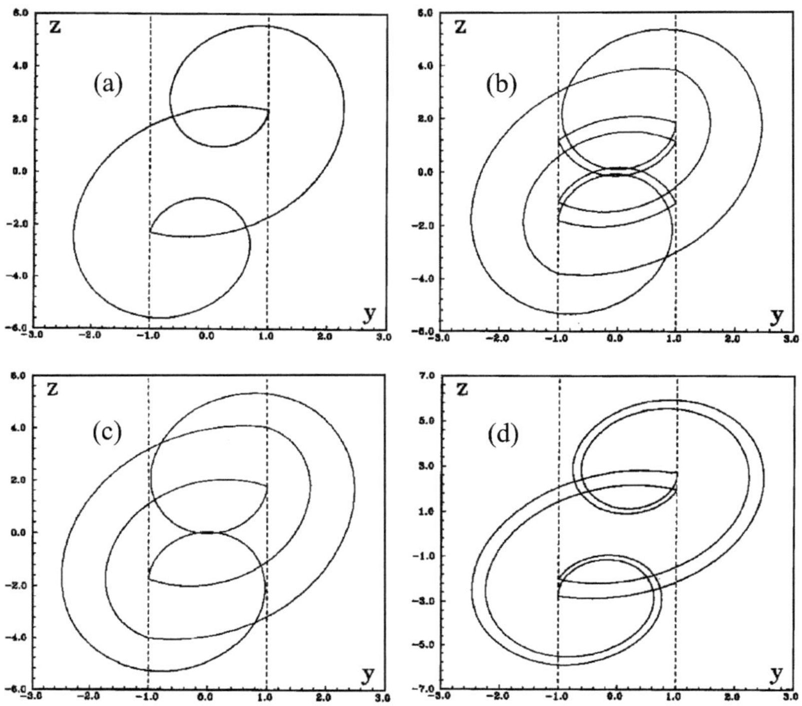

Fig. 4.53. Different types of periodic orbits for fixed $\gamma = -0.6$. (**a**) $k = 4.05$; (**b**) $k = 4.135$; (**c**) $k = 4.7$; (d) $k = 4.9$

periodic orbits obtained for fixed γ but variable k in the (y, z) phase plane of the hysteretic system.

Traveling Waves. A "traveling" wave solution of (4.89) can be described in the form of (4.34). Each of the variables [for example, $z(\xi)$ defined by (4.35) in the space (j, t, z)] has the form of a wave which propagates in the array from left to right, with velocity $c = 1/h$. Let us consider the solution (4.35) using the *continuum approximation* (Sect. 4.2.2). This approach implies some conditions on (4.35). In particular, the parameter h must have values suitably small, i.e. $h \ll 1$. If this is satisfied the spatial grid scale of (4.35) is much finer than the spatial grid of the "discrete" array (4.89). Again, we can, approximately, define the variables $\big(x(\xi), y(\xi), z(\xi), w(\xi)\big)$ not only at the sites j, but also along the bonds, together with the continuous spatial variables and the continuous spatial derivatives of the variables y and w. Therefore, solutions of the form of (4.35) are described by a system of partial differential equations (PDE).

In the continuum approximation the functions $x(\xi)$, $y(\xi)$, $z(\xi)$ satisfy the following ODE:

$$\begin{cases} \dot{x} = \alpha[y - x - f(x)]\,, \\ \dot{y} = \frac{1}{\delta}[x - y + z + g(x)]\,, \\ \dot{z} = -\beta y - \gamma z\,, \end{cases} \tag{4.101}$$

with $\delta = 1 - d/c^2$ and $c = 1/h$. The dot denotes differentiation with respect to ξ, and $w(\xi)$ is defined by

$$w(\xi) = -\frac{d}{c}y(\xi)\,.$$

Since ξ is a moving coordinate, any bounded trajectory of (4.101) determines a traveling wave along the array. Its profile is in one-to-one correspondence with the trajectory. In particular, if (4.101) has a chaotic attractor, then the trajectory of this attractor defines a traveling wave of chaotic profile.

Let us make the following substitution in (4.101):

$$\frac{\xi}{\delta} = \xi_n\,, \quad \beta\delta = \beta_n\,, \quad \gamma\delta = \gamma_n\,.$$

Then, (4.101) transforms into (4.87). Therefore, when $\delta > 0$ all earlier motions described also appear for (4.101). We have shown that when $\mu = 0$ the dynamics of (4.87) can be either regular or chaotic. However, the trajectories of the system have singularities at the points of the intersection with the lines S^+ and S^-, and the spatial scale near these points is infinitesimally small, thus violating the conditions of the continuum approximation.

Let us consider (4.87) when $\mu \neq 0$, albeit suitably small, $\mu \ll 1$. As mentioned in Sect. 4.4, the motions of (4.87) are smooth forms near the points of switching between the fast and slow motions. However, at other points, they are not qualitatively different from the case in which $\mu = 0$. The chaotic attractors and periodic orbits of the system for $\mu \ll 1$ have forms and characteristics similar to the attractors and orbits earlier studied. For sufficiently large number of elements in (4.89), the conditions for the continuum approximation are applicable, and thus we can consider the bounded trajectories of (4.87) as the profiles of "traveling" waves.

Chaotic Waves. The chaotic profile of traveling waves corresponds to a trajectory which is "contained" in the chaotic attractor. Figure 4.54 illustrates different chaotic profiles, $z(\xi)$, of waves for different chaotic attractors. The profile shown in Fig. 4.54a is associated with the symmetric attractor of Fig. 4.46b. That of Fig. 4.54b corresponds to the attractor with the complex river form (Fig. 4.50b). Finally, that of Fig. 4.54c corresponds to the attractor obtained for a high enough value of k (Fig. 4.51b).

The stability analysis of these waves is difficult. Since the chaotic waves have very long spatial "length", the array must have a huge number of elements. Even numerically observing the evolution of such waves and their stability is difficult. However, we can study some stability properties of wave motions in (4.89) by analyzing the stability of periodic waves. Then, chaotic waves of the array may be interpreted as periodic waves with complex profiles and practically aperiodic. Therefore, the study of evolutionary stability (or metastability) of different periodic waves for sufficiently small couplings is expected to provide information about the behavior of chaotic waves in an unbounded array, i.e. we can assume that the waves also are stable for small couplings.

Periodic Waves. The periodic profile of a "traveling" wave is defined by a periodic trajectory of (4.87). The stability of these waves was investigated by numerical integration using a circular array (4.89), i.e. (4.87) with periodic boundary conditions (4.55). Let us consider the array's reaction to an initial distribution which is close to a periodic wave. As in earlier occasions, if the waves propagate without any appreciable change in their forms for a "long" time, these waves are considered to be, practically, stable. For unstable waves the numerical errors and the errors from the iteration routine grow and after some relatively "short" time destroy the wave profile. To observe the propagation of periodic waves, (4.89) with (4.56), the integration was done using a fourth-order Runge–Kutta routine with relative and absolute errors equal to 10^{-7}. For low enough values of the coupling coefficient, d, the waves propagate for a "long" time without an appreciable change in their forms (we

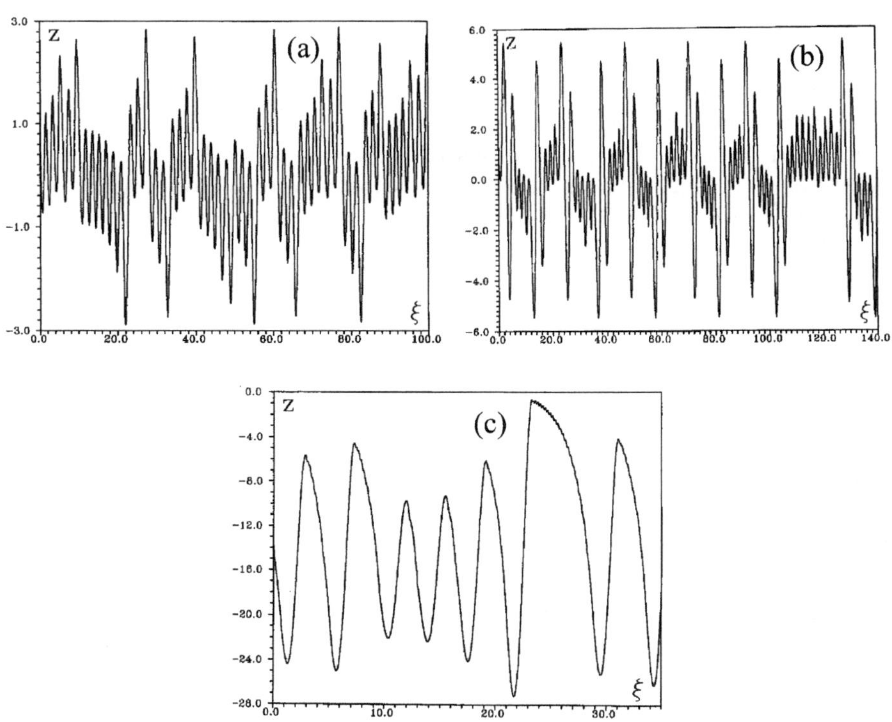

Fig. 4.54. The profiles $z(\xi)$ of chaotic waves obtained for small $\mu = 0.01$. (**a**) $\gamma = -0.4$, $k = 4.135$. (**b**) $\gamma = -0.4$, $k = 4.135$. (**c**) $\gamma = -0.6$, $k = 22$

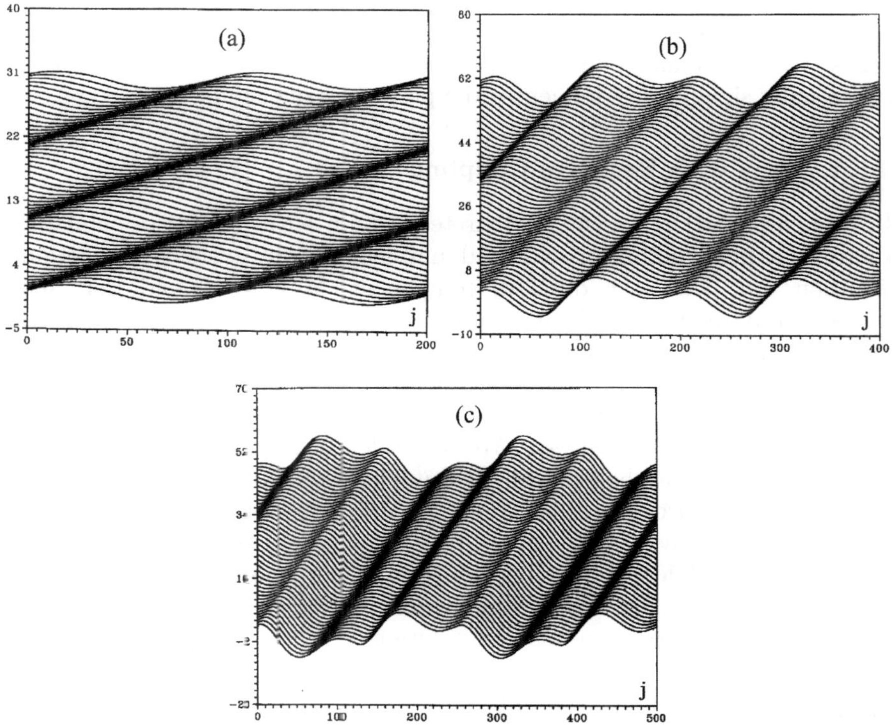

Fig. 4.55. Evolution of periodic waves in a circular array of coupled modified Chua's circuits for small coupling: $\mu = 0.01$. The y-coordinate is $z + st$, where s is a scaling coefficient (**a**) The simplest profile, corresponding to the orbit shown in the Fig. 4.41a: $\gamma = -0.6$, $k = 3.5$, $d = 0.01$, $s = 5$. (**b**) $\gamma = -0.7$, $k = 4.55$, $d = 0.01$, $s = 10$: the profile associated with the orbit of Fig. 4.53a. (**c**) $\gamma = -0.6$, $k = 5.25$, $d = 0.01$, $s = 10$: the profile associated with the orbit of Fig. 4.54b

observed 30 turns aroud the circular array). With increasing d, $d > 1$, different disturbances show up, and the waves stop to propagate. Figure 4.55 illustrates the evolution of three stable periodic waves corresponding to different types of stable periodic orbits of (4.87). The wave shown in Fig. 4.55a is associated *qualitatively* (i.e. the shape) with that of the stable periodic orbit shown in Fig. 4.40d. The wave in Fig. 4.55b corresponds to one of the orbits of Fig. 4.53c, and, correspondingly, the wave in Fig. 4.55c to the orbit shown in Fig. 4.53a.

4.5 Salient Features of Chua's Circuit in a Lattice

We have seen that an array of coupled Chua's circuits may be considered as a useful model to mimic nonequilibrium discrete media. The spatio-temporal dynamics of these media includes such typical nonlinear phenomena as structure (pattern) formation and the propagation of nonlinear waves of different types. We have found that the type of coupling (resistive or inductive) between units significantly influences the properties of the medium.

4.5.1 Array with Resistive Coupling

In this case, (4.4) represents a discrete medium with properties similar to those of a nonequilibrium extended multicomponent system of reaction–diffusion type with one diffusive component. We have found that such a medium allows pattern formation and, at variance with its corresponding extended analog, (4.4) shows high multistability. There is a continuum set of spatial structures of (4.4) with different profiles along the spatial coordinate j. These profiles can be homogeneous, periodic, chaotic, etc. Wave motions on (4.4) are also highly diverse. We have identified the conditions of front propagation and traveling pulses. By studying the heteroclinic orbits of the model, we have proved the existence of stable monotonic and oscillatory wave fronts. The investigation of the homoclinic orbits permitted us to prove the existence of different traveling pulses: from simple single-hump pulses to chaotic wave trains. By numerically integrating the equations, we were able to discriminate stable from unstable solutions.

4.5.2 Array with Inductive Coupling

We have shown that (4.34) represents a discrete medium whose properties are similar to those of a dissipation-modified BKdV equation (Chap. 2). We have shown that when a group of canonical Chua's circuits are inductively coupled to form a 1D array, solitary waves and periodic waves are possible for certain parameter values. These solitary waves may have a single hump or they may show a rather complicated form with two, three, or more humps. We have analyzed the dependence of their (phase) velocity on the coupling parameter. We have also shown that strictly speaking they are unstable. However, numerical integration showed that they can propagate with no appreciable change of profile for quite some time. Hence, a 1D chain of inductively coupled Chua's circuits may be considered as a nonequilibrium medium with properties similar to the properties of dissipative continuous media obeying the dissipation-modified BKdV equation.

We have also shown that in a circular array of inductively coupled Chua's circuits defined by (4.34) stationary periodic waves exist. These waves are stable (i.e. travel without an appreciable change in form) if a rather simple stability criterion is satisfied. This criterion contains a geometric condition limiting the "size" of the waves and conditions on the Lyapunov spectrum of the linear system (4.81). Numerically, we have been able to observe undeformed waves for more than 30 turns along the circular array.

We have also studied the bifurcation set, Π, corresponding to the appearance of homoclinic orbits associated with the fixed point, O, at the origin of the phase space of the circuit. In the parameter plane (β, γ), Π represents the bundle of curves originating from the same point, Π_0. We have shown that the saddle value of the saddle-focus O is positive, and hence in the neighborhood of Π, there exists a countable set of bifurcation curves corresponding to homoclinic orbits with any number of loops.

We have also studied the spatio-temporal dynamics of (4.88) of an inductively coupled *modified* Chua's circuits. We added to the Chua's circuits a voltage-controlled current source and obtained a circuit which has a number of interesting dynamic properties, such as becoming a generator of relaxation oscillations. The range of these oscillations is very wide: from regular to periodic to chaotic oscillations. We have shown that the dynamics of modified Chua's circuits can be modeled by a 2D system with hysteresis. This hysteretic system has chaotic attractors with different structures and characteristics. Two main scenarios for the appearance and evolution of these attractors have been analyzed:

(i) The first type of attractors is described by a *continuous* 1D point map. There are two intervals in which the map is unimodal. These intervals may be *independent*, and the map describes two independent identical chaotic attractors, or they may *overlap*, and the map describes a single, common chaotic attractor which may be interpreted as the result of the "interaction" between the two unimodal maps. The bifurcation mechanism leading to the appearance and evolution of these attractors is associated with the bifurcations of the unimodal map. In the phase space of the hysteretic system the independent attractors described by the map are symmetric with respect to the origin. By changing the control parameter, these attractors begin to interact and, eventually, coalesce into a common symmetric chaotic attractor.

(ii) The chaotic attractors of the second type are described by a *discontinuous* 1D map and characterized by the presence of a "river" of trajectories. The intervals of the map which correspond to the chaotic attractors have subintervals (or river sets) which can orient the map trajectories from one interval to the other thereby directing the interaction between the intervals, or "return" the trajectories back to the initial interval after a number of iterations by the map. We can change the map action on the river subintervals by changing the control parameter, thereby defining the different shapes of the chaotic attractors. In the phase space of the hysteretic system the attractors have a characteristic feature, namely, they contain *rivers of trajectories.* There can exist two identical chaotic attractors, or a symmetrically organized attractor, thanks to the interaction between attractors where a river of trajectories is created.

Finally we have studied traveling waves in an array of inductively coupled modified Chua's circuits. We have shown how waves can develop in the array with very different profiles, from periodic to chaotic. These profiles are determined by a 3D system describing the dynamics of the modified Chua's circuit. Numerical integration of the equations has permitted us to assess the stability of different types of periodic waves.

To conclude let us say that among the many publications dealing with Chua's circuit, the books by Madan [4.6] and by Chua [4.2], on the one hand, and the review of Shilnikov, on the other hand, are particularly relevant to the contents of this chapter.

5. Patterns, Spatial Disorder and Waves in a Dynamical Lattice of Bistable Units

5.1 Introduction and Motivation

The design of systems capable of storing and processing information is an important problem in modern science and technology. One of the directions intensively developing deals with neuro-inspired information processing systems. This implies designing systems using some operation principles of the nervous system of animals [5.1, 5.3, 5.4, 5.8, 5.21, 5.22]. The nervous system consists of a rich variety of neurons, which are huge in number with a still higher number of connections (synapses). Abstracting from details we may define three main states of a neuron. These are the base state, which can be at rest or in a so-called subthreshold oscillation mode which is a robust quasiharmonic oscillation; the excited state (or more than one as in the Inferior Olive); and the refractory state, where the neurons do not respond to inputs. By means of various connections (electrical or chemical synapses) the neurons form extremely complex spatially distributed neural networks [5.5–7,5.10,5.19,5.20]. It has been argued, and even experimentally established, that storing and processing information in the nervous system of animals is connected with the appearance of spatio-temporal structures of activity in the neural networks. Such structures are formed by the cooperative self-consistent action of neurons, hence self-organization of the network. From the viewpoint of nonlinear dynamics the neural network represents a spatially distributed discrete active system (or discrete active medium). In such a system an elementary unit of information is a spatial pattern, i.e. there is parallel information processing. Accordingly, neuro-inspired systems have a great advantage relative to traditional ways of sequential information processing. The simplest systems realizing such a neuro-inspired approach are chain and lattice systems with a local type of connection as illustrated in Fig. 1.4.

In this chapter we shall consider active lattice systems composed of elements possessing bistable properties. The bistability mimics in the simplest possible way the property of neurons of being either at rest or in one excited state. We study the spatio-temporal collective dynamics of such systems, and hence we shall describe how forms of self-organization arise there leading to dynamical structures which have the potential to store and process information. In a subsequent chapter we shall discuss how to replicate and copy dynamic patterns with a controllable degree of fidelity.

5.2 Spatial Disorder in a Linear Chain of Coupled Bistable Units

Let us consider the following model problem:

$$
\begin{aligned}
&\frac{\mathrm{d}x_j}{\mathrm{d}t} = y_j\,, \\
&\frac{\mathrm{d}y_j}{\mathrm{d}t} = -x_j - \mu[f(x_j)y_j - d(y_{j-1} - 2y_j + y_{j+1})]\,, \\
&j = 1, 2, 3, \ldots, N\,,
\end{aligned}
\tag{5.1}
$$

with boundary conditions

$$
y_0 = y_1\,,\ y_{N+1} = y_N\,, \tag{5.2}
$$

where $0 < \mu \ll 1$ and $f(x)$ are, respectively, the parameter and the function characterizing the dynamics of a unit in the chain and d is the intralattice coupling diffusion. Let $f(x) = ax^4 - ax^2 + 1$, where $a > 10$. In this case there is a fixed point at the origin and a stable limit cycle separated by an unstable limit cycle. The system of equations (5.1–2) belongs to a class of basic models of nonlinear dynamics [5.12, 5.15]. Let us focus on the spatial disorder in the chain (5.1).

5.2.1 Evolution of Amplitudes and Phases of the Oscillations

The system (5.1) is quasilinear, and therefore it can be studied by averaging methods. An averaged system for (5.1) is

$$
\begin{aligned}
\frac{\mathrm{d}r_j}{\mathrm{d}t} &= \frac{\mu}{2}\{-F(r_j) + d\,[r_{j-1}\cos(\varphi_j - \varphi_{j-1}) - 2r_j + r_{j+1}\cos(\varphi_{j+1} - \varphi_j)]\}\,, \\
\frac{d\varphi_j}{dt} &= \frac{\mu d}{2}\,[r_{j+1}/r_j \sin(\varphi_{j+1} - \varphi_j) - r_{j-1}/r_j \sin(\varphi_j - \varphi_{j-1})]\,,
\end{aligned}
\tag{5.3}
$$

where $j = 1, 2, \ldots, N$, and $F(r) = 2ar^5 - ar^3 + r$. From the boundary conditions (5.2), it follows that

$$
r_0 = r_1\,, \quad r_{N+1} = r_N\,, \quad \varphi_0 = \varphi_1\,, \quad \varphi_{N+1} = \varphi_N\,. \tag{5.4}
$$

The system (5.3–4) is gradient, and hence it has a 2π-periodic Lyapunov potential function U. Thus, the evolution of any initial condition is such that as time proceeds one of the steady states of (5.3) must be attained. Let us now find the coordinates of these states.

First note that only the states which satisfy $\varphi_{j-1} - \varphi_j = 0$ may be stable. This conclusion follows from the detailed analysis of the extrema of the potential U. The solutions $\varphi_j = \varphi^0$ satisfy this condition, where φ^0 is an arbitrary constant, i.e. an invariant line exists in the phase space of (5.3). The other coordinates of these steady states are solutions of

$$d(r_{j+1} - 2r_j + r_{j-1}) - F(r_j) = 0\,, \tag{5.5}$$

$$r_0 = r_1\,, \quad r_{N+1} = r_N\,. \tag{5.6}$$

We set $u_j = r_{j-1}$. Then (5.5–6) become

$$\begin{cases} u_{j+1} = r_j\,, \\ r_{j+1} = 2r_j - u_j + d^{-1}F(r_j)\,. \end{cases} \tag{5.7}$$

Smale's Horseshoe Map. Equation (5.7) will be interpreted as a dynamic system on the plane determined by the diffeomorphism S:

$$S\,:\,(u,r) \longrightarrow \left(r, 2r - u + d^{-1}F(r)\right)\,.$$

Then the portions of the discrete trajectories of the map $L : \{(u_2, r_2), \ldots, (u_N, r_N)\}$ on the (u, r) phase plane joining the curves

$$L_2 : \{r = u + d^{-1}F(u)\} \quad \text{and} \quad L_N : \{u = r + d^{-1}F(r)\}$$

define the solutions of (5.5–6) and, consequently, the steady states of (5.3). The map S is a Hénon-like map. We can isolate a domain in the parameter space such that the map acts like a "Smale's horseshoe" map (Sect. 7.3.3). There is the region D_{ch} (Fig. 5.1) determined by the following inequalities:

$$a > 10\,, \quad d < \min\left\{\frac{a-10}{20}, \frac{F(r_{max})}{2(r_3^0 - r_{max})}, \frac{-F(r_{min})}{2(r_3^0 + r_{min})}\right\}\,, \quad \Phi(u_{max}) > 2r_3^0\,,$$

with

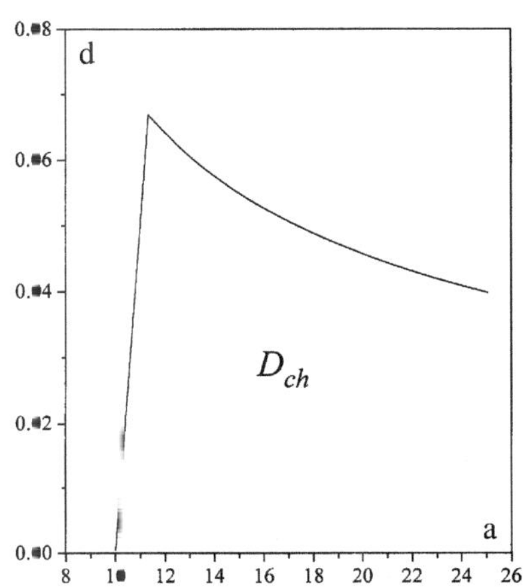

Fig. 5.1. Parameter range for the existence of spatial disorder

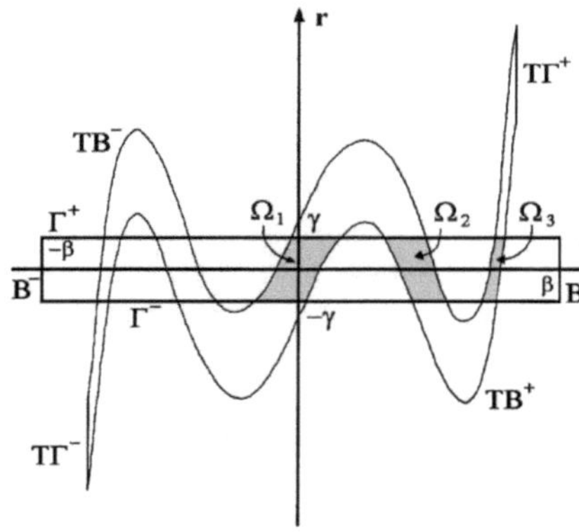

Fig. 5.2. The mapping of region Π due to the iteration of map S

$$r_{\substack{min \\ max}} = \sqrt{\frac{3a \pm \sqrt{9a^2 - 40a}}{20a}} \,, \quad r_3^0 = \sqrt{\frac{a + \sqrt{a^2 - 8a}}{4a}} \,,$$

$$u_{\substack{min \\ max}} = \sqrt{\frac{3a \pm \sqrt{9a^2 - 40a(1+2d)}}{20a}} \,, \quad \Phi(u) = 2u + d^{-1}F(u) \,,$$

where r_{min}, r_{max} and r_3^0 are, respectively, the abscissas of the extrema and of the zero of the function $F(r)$, and u_{max} is the abscissa of the maximum of the function $\Phi(u)$. Then, if the parameters of (5.7) belong to D_{ch}, there is a rectangle Π on the (u, r) phase plane such that S acts like a Smale's horseshoe map on Π (Fig. 5.2).

5.2.2 Spatial Distributions of Oscillation Amplitudes

Let us see how the initial curve L_2 is transformed by the map S. Note that $r_j \geq 0$, and therefore we only consider the portions of L_2 which are located in the first quadrant of the (u, r) plane. For $0 \leq u \leq \gamma$, L_2 lies between TB^+ and TB^- (Fig. 5.2). Consequently, the intersection of L_2 with the rectangle $\Pi^+ = \Pi \bigcap \{u \geq 0, r \geq 0\}$ has three components: $\Omega_i^+ = \Omega_i \bigcap \{u \geq 0, r \geq 0\}, i = 1, 2, 3$. These components are monotonically increasing curves joining the lines $r = 0$ and Γ^+ (Fig. 5.2). We label the curves located inside Ω_1^+ and Ω_3^+ as L_2^0 and L_2^1, respectively. The solutions of (5.5–6) corresponding to the trajectories of S passing through Ω_2^+ are unstable.

To study the stability of solutions we use the Gershgorin theorem (Appendix G). Due to the instability of the above-mentioned solutions, we disregard the portions of $S^k L_2, k = 0, 1, 2, \ldots, N$, belonging to Ω_2^+. As L_2^0 lies in Ω_1^+, the iterations of L_2^0 follow the "shoe" rules (on Π, S acts like a Smale's horseshoe map). Namely, L_2^0 is transformed to curve L_3, which has one component in both $\Omega_1^+$and Ω_3^+. We designate them as $L_3^{00}$and L_3^{01}. These components are also monotonically increasing curves, similar to L_2, and join the lines $r = 0$ and Γ^+. Note that with our notation, $L_k^{0m_1m_2\ldots m_{N-2}}$, the superscripts, taking the values 0 or 1, characterize the path traced by L_2^0 under iterations and the subscript k is associated with the number of iterations, which is equal to $k-2$. Proceeding as before, we can divide the curve L_3^{00} into two components, L_4^{000} and L_4^{001}, and L_3^{01} into L_4^{010} and L_4^{011}. Figure 5.3 illustrates the qualitative form of these components. Thus, S splits each curve located in Ω_1^+ or Ω_3^+ into two portions. This process is shown schematically in Fig. 5.4. Then after $(N-2)$ iterations, there appears a set consisting of 2^{N-2} monotonically increasing curves $L_N^{0m_1m_2\ldots m_{N-2}}$, $m_i \in \{0,1\}$, joining the lines $r=0$ and Γ^+ and located in the domains Ω_1^+ and Ω_3^+. A similar process occurs for the component L_2^1. It results in a set consisting of 2^{N-2} curves $L_N^{1m_1m_2\ldots m_{N-2}}$, $m_i \in \{0,1\}$, joining the lines $r=0$ and Γ^+. Thus, on the one hand, the set consisting of 2^{N-1} curves, $L_N^{0m_1m_2\ldots m_{N-2}}$ and $L_N^{1m_1m_2\ldots m_{N-2}}$, is formed as a result of $(N-2)$ iterations of L_2^0 and L_2^1. On the other hand, only the portions of the trajectories of S which start on L_2^0 or L_2^1 and get to L_N are of interest to us. Inside the rectangle Π^+, L_N has two monotonically increasing curves joining the lines $u=0$ and B^+. That is why L_N intersects both $L_N^{0m_1m_2\ldots m_{N-2}}$ and $L_N^{1m_1m_2\ldots m_{N-2}}$ at two points. As

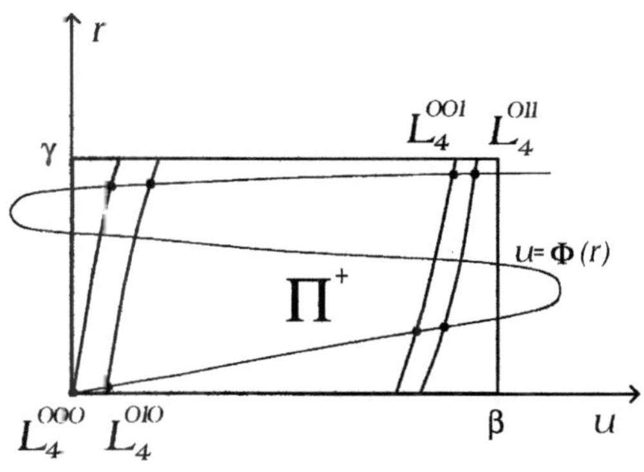

Fig. 5.3. The qualitative form of the components after second division L_2^0. The *dots* mark the points of intersection of the components with L_N

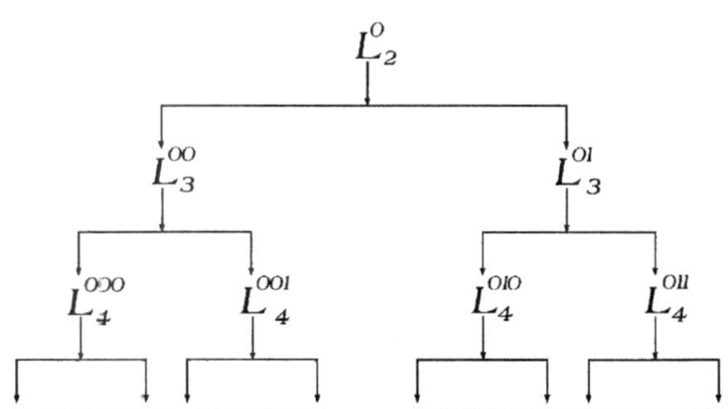

Fig. 5.4. Scheme of splitting the component L_2^0 by map S

a result of these intersections there exist 2^N points located in the increasing portions of the curve L_N. [The points lying in the decreasing portion of L_N are not considered because they correspond to the maxima of the function U and unstable solutions of (5.3).] Hence, 2^N portions of the trajectories of S joining L_2 and L_N exist in the first quadrant of the (u, r) phase plane. The coordinates of the points of these portions determine 2^N sets of solutions $r_j = r_j^*, j = 1, 2, \dots, N$, of (5.5–6). Note that there are two homogeneous solutions independent of $j : \{r_j = 0\}$ and $\{r_j = r^*\}$.

Thus (5.2) has 2^N steady states with coordinates $r_j = r_j^*, \varphi_j = \varphi^0$, where $j = 1, 2, \dots, N$. Consequently, the oscillations in the chain (5.1) become synchronized. When this occurs the distribution of oscillation amplitudes along the chain is described by the map (5.7) and can be very complex. Indeed, the amplitude distribution of oscillations along the spatial coordinate j is determined by the portions of the trajectories of S and can be described by a sequence of two symbols, $(0, m_2, m_3, \dots, m_{N-1}), (1, m_2, m_3, \dots, m_{N-1})$, with $m_i \in \{0, 1\}$. Moreover, the symbol "0" corresponds to a rather low oscillation amplitude, and the symbol "1" corresponds to the amplitude that is close to the amplitude of a limit cycle of an individual oscillator that is not affected by its (nearest) neighbors. There are $2^N - 1$ such motions, and hence (5.1) exhibits multistability. Since S has chaotic dynamics, the alternations of the symbols 0 and 1 will be very diverse, and for $N \longrightarrow \infty$, it can be described by a Bernoulli shift. Consequently, the oscillations in (5.1) are regular in time, while they are disordered in space. An example of the possible amplitude distribution "along" the spatial coordinate j is given in Fig. 5.5 for a chain composed of $N = 40$ units. The initial amplitude distribution was

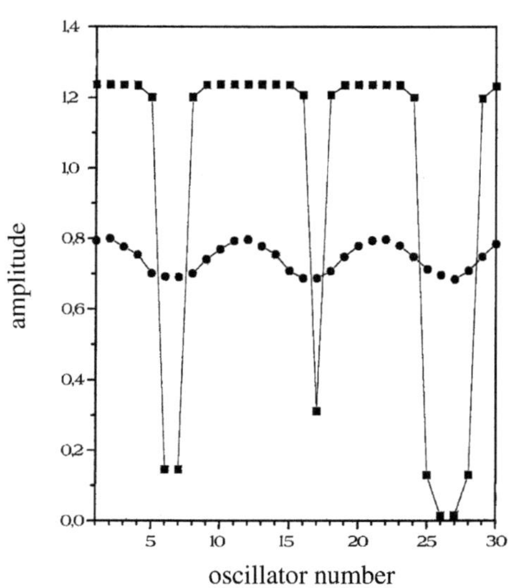

Fig. 5.5. Amplitude of oscillations vs oscillator number (*circles*: initial condition; *squares*: final state)

chosen to be sinusoidal. Our numerical experiments show that the system has high sensitivity to variations in the initial distributions of $x(t)$, a widespread property of complex spatially extended systems.

5.2.3 Phase Clusters in a Chain of Isochronous Oscillators

From the numerical study of (5.1) ($N = 10, a = 14, d = 0.07$ and $\mu = 0.01$) we infer that the onset of synchronization occurs in two stages having different time scales: (i) *Formation of clusters* — We define a cluster as a set of oscillators in which $\varphi_i \cong \varphi_j$ for i, j belonging to a single set. First, the oscillators aggregate in groups where the elements are phase-locked. Figure 5.6a illustrates the process of formation of two clusters. (ii) *Interaction of clusters* — The clusters just mentioned interact among themselves, and their phase differences changes. However this difference does not vanish and we rather observe a quasiperiodic solution of (5.1) in the numerical experiment (φ^0 is the first component of the asymptotic expansion of a phase of this solution). Following the interaction process, clusters move very slowly. The distribution of oscillator amplitudes drastically influences this process: phase singularities appear at the values of j corresponding to an inhomogeneity in the amplitude distribution. Figure 5.6b shows the phase distribution at different instants of time. One can say that here the formation of phase clusters is a transient process only. Furthermore, under certain conditions clusters are available at all times. Let us consider this phenomenon in detail.

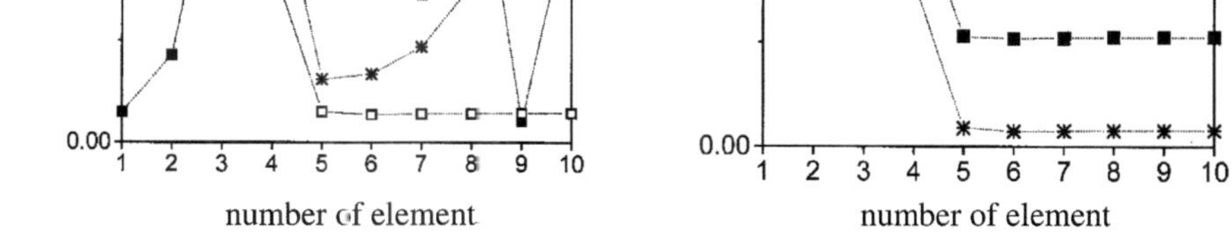

Fig. 5.6. Mean phase vs oscillator number at three different instants of time: **(a)** $t = 0$ (*stars*), $t = 8000$ (*solid squares*), $t = 45000$ (*open squares*); **(b)** $t = 90000$ (*stars*), $t = 220000$ (*solid squares*), $t = 300000$ (*open squares*)

5.3 Clustering and Phase Resetting in a Chain of Bistable Nonisochronous Oscillators

In this section we show that stable phase and frequency clusters can exist in a system with diffusive coupling, like (5.1), provided two basic conditions are fulfilled [5.14]. These conditions are:

(i) The amplitude distribution is chaotic along the chain;
(ii) Each unit of the chain is a *nonisochronous* oscillator, i.e. the oscillation frequency of each unit depends on its amplitude.

To illustrate the phenomena just described, let us consider an oscillator obeying the equations in polar coordinates:

$$\begin{cases} \dot{r} = rF(r)\,, \\ \dot{\varphi} = \omega(r)\,, \end{cases} \tag{5.8}$$

where r and φ are the oscillation amplitude and phase, respectively, $\omega(r) = \alpha r^2$ is the angular frequency of oscillations which depends on its amplitude, α, and $rF(r)$ is a nonlinear function having three zeros (Fig. 5.7). Equations (5.8) describe a bistable, nonisochronous oscillator that can be either at rest (stable steady state) or in the oscillatory state (stable limit cycle). Then for a chain consisting of N oscillators with diffusive nearest-neighbor coupling the equations are

$$\begin{aligned} \dot{z}_j &= z_j[F(|z_j|) + \mathrm{i}\omega(|z_j|)] + d(z_{j-1} - 2z_j + z_{j+1})\,, \\ j &= 1, 2, \dots, N\,, \quad z_0 = z_1\,, \quad z_{N+1} = z_N\,, \end{aligned} \tag{5.9}$$

where $z_j = r_j e^{\mathrm{i}\varphi_j}$ and d accounts for the *local* intralattice coupling between oscillators (d is real and positive). The index j stands for the variables of the j-th unit of the chain.

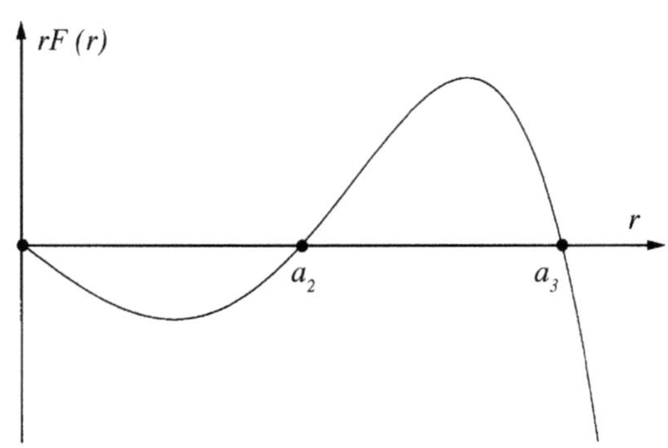

Fig. 5.7. The function characterizing the dynamics of a bistable unit or element in the chain

5.3.1 Amplitude Distribution along the Chain

To determine the amplitude distribution along the chain, let us construct invariant domains (we will come back to this procedure in Sect. 5.6).

Construction of Invariant Domains. Consider (5.9) using amplitudes and phase shifts ($\psi_j = \varphi_{j+1} - \varphi_j$):

$$\begin{aligned} \dot{r}_j &= r_j F(r_j) + d(r_{j-1}\cos\psi_{j-1} - 2r_j + r_{j+1}\cos\psi_j)\,, \\ r_k r_{k+1}\dot{\psi}_k &= r_k r_{k+1}[\omega(r_{k+1}) - \omega(r_k)] + d\,[r_k r_{k+2}\cdot\sin\psi_{k+1} \\ &\quad - (r_k^2 + r_{k+1}^2)\sin\psi_k + r_{k+1}r_{k-1}\sin\psi_{k-1}]\,, \\ & j = 1,2,\ldots,N\,, \quad k = 1,2,\ldots,N-1\,, \\ & r_0 = r_1\,, \quad r_{N+1} = r_N\,, \quad \psi_0 = 0\,, \quad \psi_N = 0\,. \end{aligned} \tag{5.10}$$

Location of the Bounded Solutions $\{r_j(t)\}$ *in the Phase Space.* Let us consider in the phase space of (5.10) the following regions:

$$V_B = \{\mathbf{r} : 0 \le r_j \le B\,,\, j = 1,2,\ldots,N\}\,,$$

where B is a parameter. From (5.10) and the form of $rF(r)$ (Fig. 5.7) it follows that at $B > a_3$ on the boundary of each region V_B the vector field of (5.10) points into V_B. Now consider the orientation of the vector field on the surfaces $\{r_j = 0\}$. As for $r_j = 0$ the value of the phase φ_j is left undetermined, we take φ_j such that

$$\dot{r}_j|_{r_j=0} = d(r_{j-1}\cos\psi_{j-1} + r_{j+1}\cos\psi_j) \ge 0\,. \tag{5.11}$$

Thus the vector field of (5.10) is brought inside V_B, and all bounded solutions $\{r_j(t)\}$ are located in the region V_B, where $B = a_3$.

Invariant Domains. As we shall further elaborate in Sect. 5.6, let us here simply say that we construct in the phase space of (5.10) a family of narrower invariant domains. For this purpose we consider the orientation of the vector field of (5.10) on the surfaces $\{r_k = c\}$, where $c = \text{const.} < a_3$. From (5.10) we obtain

$$\begin{aligned} \dot{r}_k|_{r_k=c} &= cF(c) + d(r_{k-1}\cos\psi_{k-1} - 2c \\ &\quad + r_{k+1}\cos\psi_k)\,. \end{aligned} \tag{5.12}$$

Let us demand that

$$\dot{r}_k|_{r_k=c} < 0\,, \quad \text{for} \quad r_j \in V_B, \quad \forall j \ne k. \tag{5.13}$$

This condition is satisfied for the values of the parameters taken in the region

$$D_{ch} = \left\{ d < \min\left(\frac{r_{\max}F(r_{\max})}{2(a_3 + r_{\max})}, \frac{|r_{\min}F(r_{\min})|}{2(a_3 - r_{\min})}\right)\right\}\,, \tag{5.14}$$

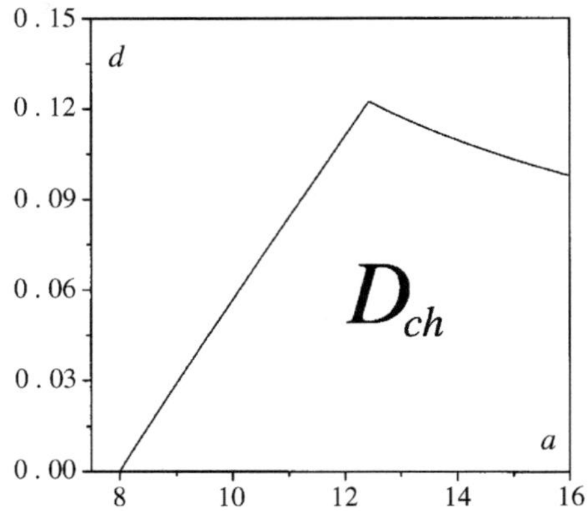

Fig. 5.8. Domain, D_{ch}, of amplitude disorder along the chain for $rF(r) = -2ar^5 + ar^3 - r$

where $r_{\min}$ and $r_{\max}$ are the coordinates of the minimum and maximum of $rF(r)$, respectively. In this case $c = b_0$, where b_0 is the smallest root of

$$cF(c) - 2d(c - a_3) = 0\,. \tag{5.15}$$

Similarly, for points of D_{ch} the inequality

$$\dot{r}_{k|r_k=c} > 0\,, \quad \text{for} \quad r_j \in V_B\,, \quad \forall j \neq k\,, \tag{5.16}$$

is satisfied when $c = b_1$, where b_1 is the largest root of

$$cF(c) - 2d(c + a_3) = 0\,. \tag{5.17}$$

Using the surfaces $\{r_k = b_0\}$ and $\{r_k = b_1\}$ and the boundary of the domain V_B, we define

$$\begin{aligned} V_j^0 &= \{0 \le r_j \le b_0\,, \quad 0 \le r_k \le B\,, \quad \forall k,\ k \neq j\}\,, \\ V_j^1 &= \{b_1 \le r_j \le B\,, \quad 0 \le r_k \le B\,, \quad \forall k,\ k \neq j\}\,, \end{aligned} \tag{5.18}$$

where b_0, b_1 and B are constants which depend on d and the specific nonlinear function $F(r)$. Then the vector field of (5.10) and, consequently, of (5.9) at the border of the regions V_j^0 and V_j^1 is oriented toward these regions. Thus for the points of D_{ch} there exist 2^N invariant domains which are intersections of the domains V_j^0 and V_j^1. Figure 5.8 illustrates the region, D_{ch}, corresponding to the choice $F(r) = -2ar^4 + ar^2 - 1$.

Distribution of Oscillation Amplitudes. If the initial amplitudes, $\{r_j(0)\}$, belong to the region V_j^0 or V_j^1, then $\{r_j(t)\}$ are located, and remain for ever, inside these regions and stay either small (V_j^0) or large (V_j^1). Thus for D_{ch} there exist at least 2^N different distributions of the amplitudes of oscillations along the chain (5.9). As each distribution can be "coded" by an arbitrary sequence of two symbols, the distribution of the amplitudes can be very diverse, including regular and disordered, spatially chaotic distributions.

5.3.2 Phase Clusters in a Chain of Nonisochronous Oscillators

Let us start first with an α that is small enough. To obtain a disordered amplitude distribution along the chain, we choose initial amplitudes, $\{r_j(0)\}$, from either V_j^0 or V_j^1. The initial phases are randomly distributed in the range $[-\pi, \pi]$. The numerical integration of (5.9) shows that the process of formation of stationary amplitude and phase distributions once more occurs in two stages with different time scales. First, there is a short stage at which the equilibrium amplitude distribution ($r_j = r_j^0$, where r_j^0 are constants) is formed (Fig. 5.9a). Then, slowly, phase clusters form along the chain (Fig. 5.9b,c). The number and spatial location of these clusters is determined by the former amplitude distribution. The chain splits into phase clusters at sites of gaps in the amplitude distribution (i.e. at the sites where oscillation amplitudes, r_j, are small). In Fig. 5.9b three phase clusters are shown: PC1 (oscillators at sites 1–10), PC2 (oscillators at sites 12–27), PC3 (oscillators at sites 28–36); one oscillator (at site 11) which does not belong to these clusters is also shown. Inside each cluster, the phase differences, $\psi_j = \varphi_{j+1} - \varphi_j \simeq 0$, are very small, but between clusters there are sharp jumps in phase, i.e. ψ_{10}, ψ_{11} and ψ_{27} are nonzero constants.

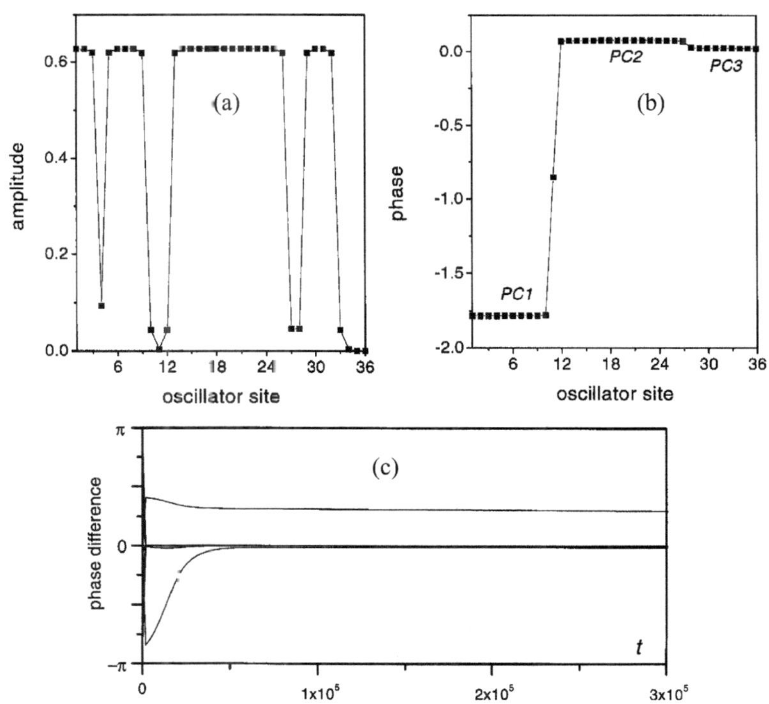

Fig. 5.9. Formation of phase clusters ($\alpha = 0.0018$, $d = 0.08$, $a = 12$). (**a**) Final amplitude distribution; (**b**) snapshot of the distribution of phases along the chain; (**c**) phase differences ($\psi_j = \varphi_{j+1} - \varphi_j$) vs time

As in the chain a phase cluster mode corresponds to the quantities r_j and ψ_j being constant, it corresponds to a steady state in the phase space of (5.10). From the equations for the phase differences, $\sin\psi_j$, we have a system of $(N-1)$ linear equations. Its solution is

$$\sin\psi_j = \frac{\alpha\left(\sum\limits_{i=1}^{j} r_i^2 \times \sum\limits_{i=j+1}^{N} r_i^4 - \sum\limits_{i=j+1}^{N} r_i^2 \times \sum\limits_{i=1}^{j} r_i^4\right)}{d r_{j+1} r_j \sum\limits_{i=1}^{N} r_i^2} . \tag{5.19}$$

From (5.19) it follows that the phase differences between neighboring oscillators are proportional to α/d and depend on the chosen distribution of amplitudes. As $\sin\psi_j \sim \alpha(d r_j r_{j+1})^{-1}$ a sharp phase jump between neighboring oscillators can be observed in positions where gaps in amplitude distribution appear, i.e. when r_j and r_{j+1} have low values and hence there is splitting of the chain into phase clusters. Figure 5.9a shows three such gaps where the amplitudes of neighboring oscillators, r_j and r_{j+1}, are small. They appear at sites 10–12, 27–28, and 33–36. In Fig. 5.9b we see that there is a small phase shift between clusters in oscillators at sites (27, 28). The largest phase shifts are observed between oscillators at sites 10, 11 and 11, 12 (Fig. 5.9b). This comes from the fact that the values $(r_{10}r_{11})$ and $(r_{11}r_{12})$ are much smaller than $(r_{27}r_{28})$. The lack of splitting into clusters in the gap at sites 33–36 is due to the compensation of the sums contained in the (5.19). Thus, although the distribution of amplitudes and phases along the chain is rather complex and the frequency of each unit, taken separately, depends on its amplitude, phase clustering appears.

5.3.3 Frequency Clusters and Phase Resetting

This section deals with the formation of frequency clusters and the complex behavior of phases of oscillations.

Clustering and Phase Resetting. Let us now consider an α that is large enough, e.g. so large that phase clusters cannot be maintained in the chain. As in the previous section, we choose initial conditions for amplitudes $\{r_j\}$ belonging to the region V_j^0 or V_j^1. Numerical integration of (5.1) shows that, after a transient stage, a certain distribution of amplitudes along the chain is established. In this case the amplitudes are not constants, but they are bounded and remain inside the regions V_j^0 and V_j^1. Figure 5.10a is a snapshot of the amplitude distribution along the chain. It is very similar to the stationary amplitude distribution shown in Fig. 5.9a.

Let us now consider the behavior of the phases and frequencies, $\omega_j = \dot{\varphi}_j$, of the oscillators. Figure 5.10b shows the behavior of the instantaneous frequencies of oscillation. Three groups appear. Each frequency group has a different

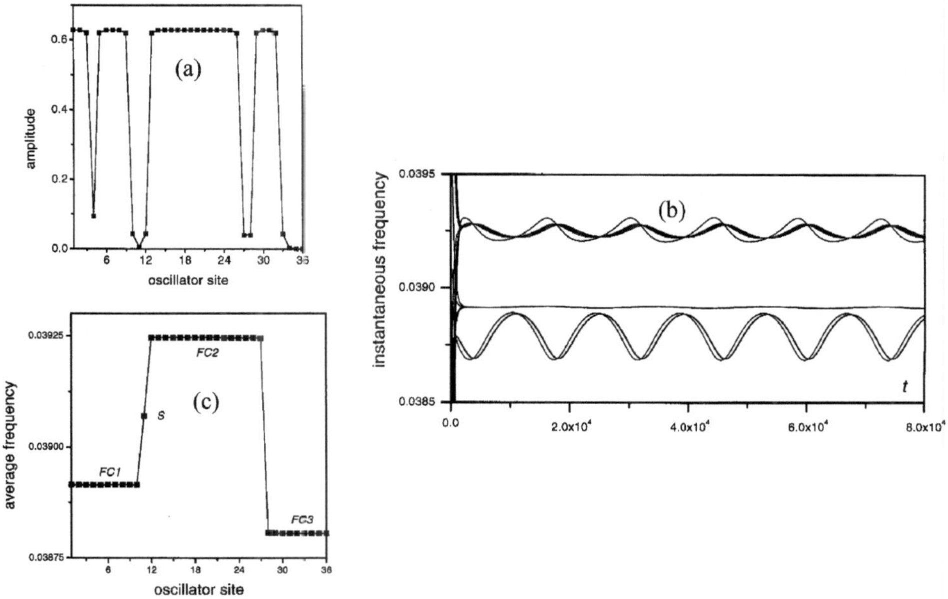

Fig. 5.10. Formation of frequency clusters ($\alpha = 0.1$, $d = 0.08$, $a = 12$). (**a**) Snapshot of the amplitude distribution; (**b**) instantaneous frequencies $[\omega_j(t) = \dot{\varphi}_j]$ vs time, (**c**) average frequency along the chain

number of elements. For each of these groups an average frequency can be defined. Figure 5.10c shows the distribution of average frequencies along the chain. We see that the chain splits into three frequency clusters: FC1 (oscillators at sites 1–10), FC2 (oscillators at sites 12–27), FC3 (oscillators at sites 28–36) and one isolated oscillator at site 11 (hereafter S). Let us denote the average frequencies of FC1, FC2 and FC3 as Ω_1, Ω_2 and Ω_3, respectively. Note that in each frequency cluster there are oscillators with low- and high-amplitude oscillations. Comparing Figs. 5.9b and 5.10c we see that the splitting into clusters takes place at the same points in both the phase and the frequency cases, namely where amplitude gaps exist (at sites 10–12 and 27–28). Thus, basically, cluster formation in the chain is a similar process for both phase clusters and frequency clusters.

Let us now investigate the behavior of S in further detail. The time evolution of amplitude $r_{11}(t)$, of phase $\varphi_{11}(t)$ and of frequency $\omega_{11}(t)$, are shown in Fig. 5.11a,b and c, respectively. From time to time the function $r_{11}(t)$ vanishes (Fig. 5.11a). At these instants of time we see very short peak singularities in the time dynamics of the instantaneous frequency ω_{11} (Fig. 5.11c). Simultaneously, jumps of π occur in the phase $\varphi_{11}(t)$ (arrow in Fig. 5.11b). Then there is *phase resetting* in the chain. Phase resetting is the result of the collective dynamics of the chain. S is affected by the cluster formation. FC1 and FC2 try to "enslave" this isolated element. In such a competition process

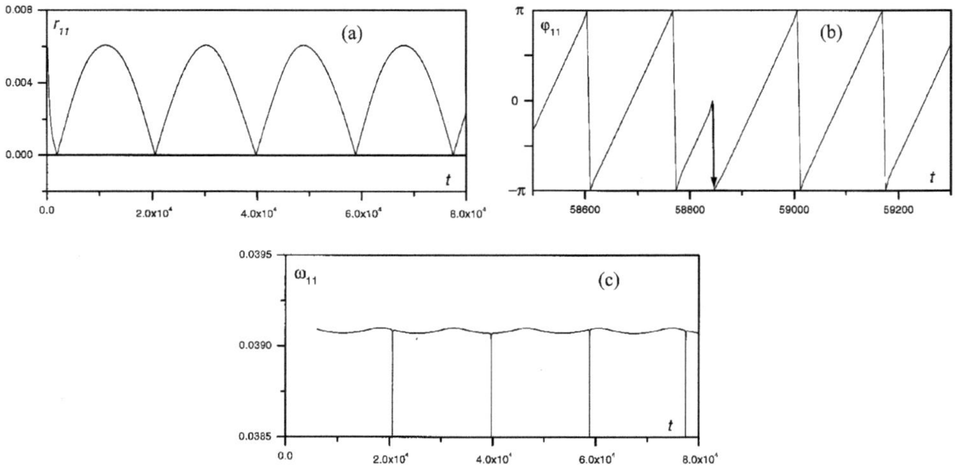

Fig. 5.11. Phase resetting. All conditions correspond to the formation of frequency clusters (Fig. 5.10). (**a**) Amplitude of oscillation at site 11 vs time; (**b**) phase resetting process for oscillator at site 11 (used different time scale); (**c**) instantaneous frequency of oscillations at site 11 vs time, singularities of frequency

none of the clusters is a winner. S keeps its average frequency. However, from time to time it is forced to change its phase by a value π, so that parity between clusters FC1 and FC2 is maintained.

Phenomenological Model of Phase Resetting. Let us consider oscillators belonging to FC1 as a single element which rotates at Ω_1, and the oscillators belonging to FC2 as another element with Ω_2. They are independent and act on S. Comparing Figs. 5.10b and 5.11c we see that the instantaneous frequency of S is between the frequencies of FC1 and FC2. Let the frequency of S be $\Omega_s = \Omega_2 - \Omega_1/2$, and its phase identical to the phase taken by both clusters. As the amplitude of oscillations of S is small, we expand the nonlinear function $rF(r)$ and omit all terms beyond the first order, hence $rF(r) \approx -r$. Then from (5.10) we obtain a single equation describing the time evolution of the amplitude of S:

$$\dot{r} = -(1-d)r + dR_1R_2 \cos \frac{\Omega_2 - \Omega_1}{2} t \,, \tag{5.20}$$

where R_1 and R_2 are constants characterizing the interaction between the corresponding cluster and S. The solution of (5.20) is

$$r(t) = Ce^{-\gamma t} + \frac{dR_1R_2}{(\Delta\Omega)^2 + \gamma^2} \left(\gamma \cos \Delta\Omega t + \Delta\Omega \sin \Delta\Omega t\right) \,, \tag{5.21}$$

with C an arbitrary constant, $\Delta\Omega = \Omega_2 - \Omega_1/2$ and $\gamma = 1 - d$. From (5.21) it follows that $r(t)$ is a function which almost periodically reaches zero and, eventually, reverses its sign. As $r(t)$ corresponds to the amplitude of oscillation of S, inversion of the sign corresponds to a π change in the phase of oscillations in S, hence *phase resetting*. The period, T, between successive zeros of the function $r(t)$ can be estimated from (5.21):

$$T = \frac{2\pi}{\Omega_2 - \Omega_1} . \tag{5.22}$$

Taking the values of Ω_1 and Ω_2 from Fig. 5.10c to (5.22) we get $T \cong 1.9 \times 10^4$. This value agrees satisfactorily with the period of frequency peaks observed in Fig. 5.11c.

5.4 Clusters in an Assembly of Globally Coupled Bistable Oscillators

Let us now study cluster formation when the bistable oscillators are globally coupled [5.16]. Thus, we consider an assembly of N *globally* coupled identical *bistable* oscillators evolving in time according to

$$\begin{aligned} \dot{W}_j = &-W_j \left\{ f(|W_j|) - \mathrm{i}\left[\omega + \alpha g(|W_j|)\right]\right\} \\ &+(\beta + \mathrm{i}\gamma)\left(\overline{W} - W_j\right) , \quad j = 1, 2, \dots , N , \end{aligned} \tag{5.23}$$

where W_j is a complex variable, $f(|W|) = 2a|W|^4 - a|W|^2 + 1$, $g(|W|) = |W|^2 - 2|W|^4$, and

$$\overline{W} = \frac{1}{N} \sum_{k=1}^{N} W_k . \tag{5.24}$$

The functions $f(|W|)$ and $g(|W|)$ and the parameters ω, a and α characterize each unit in the assembly. ω and a have positive values and $\alpha \leq 0$. When $\alpha = 0$, motions are isochronous, and when $\alpha < 0$ they are nonisochronous. For $a > 8$ the stable limit cycle and the stable steady state coexist with basins of attraction separated by an unstable limit cycle. The parameters β and γ characterize the strength of the coupling between the oscillators and have positive values. This interaction operates only through the mean field $\overline{W}$. We assume that the assembly size, N, is suitably large. An alternative description of (5.23) is obtained using "polar" coordinates, $z_j = W_j e^{i(\gamma-\omega)t}$. Then (5.23) becomes

$$\dot{z}_j = -z_j \left[h(|z_j|) - \mathrm{i}\alpha g(|z_j|)\right] + (\beta + \mathrm{i}\gamma)\bar{z} , \quad j = 1, 2, \dots , N , \tag{5.25}$$

with $h(|z|) \equiv f(|z|) + \beta$, $\bar{z} = \frac{1}{N} \sum_{k=1}^{N} z_k$.

5.4.1 Homogeneous Oscillations

Homogeneous oscillations of the assembly (5.23) correspond to the solutions of (5.25), independent of j. The system (5.25) has three homogeneous solutions:

$$z_j(t) = r^0 \exp(\mathrm{i}(g(r^0)t + \varphi^0)), \quad j = 1, 2, \ldots, N, \tag{5.26}$$

with

$$r^0 = \begin{cases} 0, \\ r^{(1)} \equiv \frac{1}{2}\sqrt{1 - \sqrt{1 - 8/a}}, \\ r^{(2)} \equiv \frac{1}{2}\sqrt{1 + \sqrt{1 - 8/a}}. \end{cases} \quad \varphi^0 = \text{const.},$$

Linearizing (5.25), around each solution (5.26), we obtain for perturbations, $\xi_j \in C$, the following equations for disturbances upon the trivial solution:

$$\dot{\xi}_j = -\xi_j + (\beta + \mathrm{i}\gamma)\left(\frac{1}{N}\sum_{k=1}^{N}\xi_k - \xi_j\right); \tag{5.27}$$

for disturbances around the other homogeneous oscillations we obtain

$$\begin{aligned} \dot{\xi}_j =& - H(r^0)\,(a + i\alpha)\,(\xi_j + \xi_j^*) \\ &+ (\beta + i\gamma)\left(\frac{1}{N}\sum_{k=1}^{N}\xi_k - \xi_j\right). \end{aligned} \tag{5.28}$$

Starred quantities denote complex conjugation, and $H(r) = r^2(4r^2 - 1)$. The matrices associated with (5.27–28) are circulant, and their eigenvalues, which are the Lyapunov exponents of the solutions of (5.26), can be easily found. To do this, the matrices are reduced to block-diagonal form. The trivial solution $z_j = 0$ has the Lyapunov exponents

$$\begin{gathered} \lambda_1 = \lambda_2 = -1, \quad \lambda_{2+s} = -(\beta + 1) \pm \mathrm{i}\gamma, \\ s = 1, 2, \ldots, N - 2, \end{gathered} \tag{5.29}$$

and the other solutions have $\lambda_1 = 0, \lambda_2 = -2aH(r^0)$ and $(N-1)$ pairs of exponents, which are the roots of

$$\lambda^2 + 2[aH(r^0) + \beta]\lambda + \beta^2 + \gamma^2 + 2H(r^0)\,(a\beta + \alpha\gamma) = 0. \tag{5.30}$$

The analysis of the distribution of eigenvalues in the complex plane shows that the trivial solution of the assembly is stable and that homogeneous oscillations with amplitudes $r^{(1)}$ are unstable while those with $r^{(2)}$ are stable in the region defined by

$$\beta^2 + \gamma^2 + 2H(r^{(2)})\,(a\beta + \alpha\gamma) > 0. \tag{5.31}$$

Hence for some initial conditions all the oscillators of the assembly (5.23) can be at rest, and the others may exhibit homogeneous periodic oscillations.

5.4.2 Amplitude Clusters

Let us consider the collective dynamics of (5.23), consisting of isochronous oscillators ($\alpha = 0$) globally coupled with $\beta(\gamma = 0)$ now taken real. For $\alpha = \gamma = 0$, (5.25) is *gradient* as

$$\frac{\mathrm{d}z_j}{\mathrm{d}t} = -\frac{\partial U}{\partial z_j^*}, \tag{5.32}$$

with

$$U = \tfrac{1}{2} \textstyle\sum_{j=1}^{N} \left[G\left(|z_j|^2\right) + \frac{\beta}{N} \sum_{k=1}^{N} |z_k - z_j|^2 \right],$$
$$G\left(|z|^2\right) = 2\, z|^2 \left(1 - a/2|z|^2 + 2a/3|z|^4\right).$$

Consequently, (5.25) has steady states only, and then for any initial conditions all trajectories tend to one of them. Amplitudes and phases of stable oscillations of (5.23) correspond to those stable fixed points of (5.25), of which the simplest correspond to homogeneous oscillations.

In-Phase Motions. It follows from (5.25) that in the phase space of the system there exists a manifold of *in-phase* motions $S = \big\{ \varphi_j = \varphi^0 = \text{const.}, j = 1, 2, \ldots, N \big\}$. On the manifold S the equations describing the evolution of amplitudes have the form

$$\dot{r}_j = -F(r_j) + \frac{\beta}{N} \sum_{k=1}^{N} (r_k - r_j), \tag{5.33}$$

with $z_j = r_j e^{i\varphi^0}$ and $F(r) = 2ar^5 - ar^3 + r$. The coordinates of the steady states of (5.33) determine the amplitudes of *in-phase* oscillations of (5.23).

Let us now find conditions of existence of inhomogeneous states of (5.33). We find from (5.33) that for each $i = 1, 2, \ldots, N-1$, the coordinates of the steady states obey the following relationship:

$$F(r_{i+1}) - F(r_i) = -\beta(r_{i+1} - r_i). \tag{5.34}$$

On the other hand, according to Lagrange's theorem there exist $\rho_i \in (r_i, r_{i+1})$ such that

$$F(r_{i+1}) - F(r_i) = F'(\rho_i)(r_{i+1} - r_i). \tag{5.35}$$

Comparing (5.34) and (5.35), we conclude that

$$F'(\rho_i) = -\beta. \tag{5.36}$$

Hence, taking into account the form of $F(r)$, we see that (5.36) is satisfied in the parameter region delineated by

$$\beta + 1 < 9a/40 \quad \text{and} \quad a > 8. \tag{5.37}$$

For the parameter values from this region (5.37) there are two points on the curve $F(r)$ for $r \geq 0$ where the angular coefficients of the tangents are equal to $-\beta$ (Fig. 5.12, lines L_1 and L_2). From (5.35–36) it follows that the coordinates of the steady states of (5.33) must be abscissas of the points of intersection of the curve $F(r)$ with the secant, which is parallel to the tangents L_1 and L_2 (Fig. 5.12). Consequently, the coordinates of the steady states form sets $\{r_j^0\}$, $j = 1, 2, \ldots, N$, whose elements have either two or three different digits. As shown below only in-phase oscillations, corresponding to the stationary points whose coordinates satisfy the condition $F'(r) + \beta > 0$, can be stable. Abscissas of the "middle" point in Fig. 5.12 obviously violate this condition. Thus we consider the steady states corresponding to the "extreme" points in Fig. 5.12. We look for sets, $\{r_j^0\}$, consisting of two different positive numbers, denoted by p_0 and q_0. Since (5.33) is symmetric under permutation of the N indices, one can assume, without loss of generality, that

$$r_j = \begin{cases} p_0 \,, \; j = 1, 2, \ldots, n \,, \\ q_0 \,, \; j = n+1, \ldots, N \,. \end{cases} \tag{5.38}$$

From (5.33) we obtain the following system of equations for p_0 and q_0:

$$\begin{aligned} q_0 &= p_0 + N/(N-n)\beta\, F(p_0) \,, \\ p_0 &= q_0 + N/n\, \beta\, F(q_0) \,. \end{aligned} \tag{5.39}$$

Note that outside the region defined by (5.37), i.e. above the curve K in Fig. 5.13, (5.39) can have only solutions that satisfy the condition $p_0 = q_0$ and correspond to homogeneous oscillations (as shown in Sect. 5.3.1). The largest number of real solutions of (5.39), such that $p_0 \neq q_0$, is six (Fig. 5.13). But since (5.39) is invariant under the transformations $\{n \to N-n,\ N-n \to n,\ q_0 \to p_0,\ p_0 \to q_0\}$, the solutions, satisfying the condition $p_0 > q_0$, coincide with the corresponding solutions with $p_0 < q_0$, which are numbered in a different way. Therefore, for example, we can restrict consideration to just three

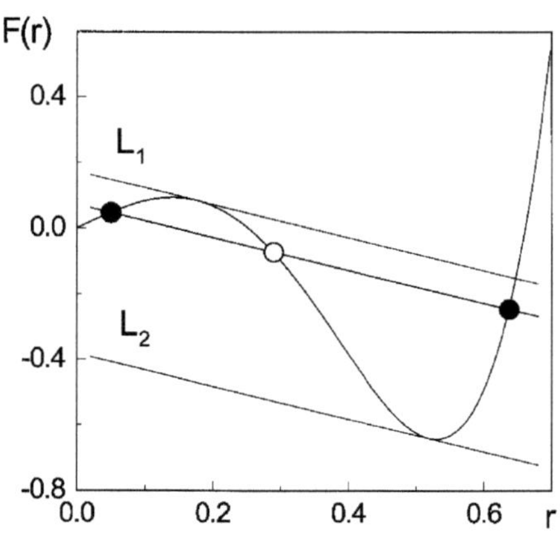

Fig. 5.12. Geometrical interpretation of conditions (5.34–35)

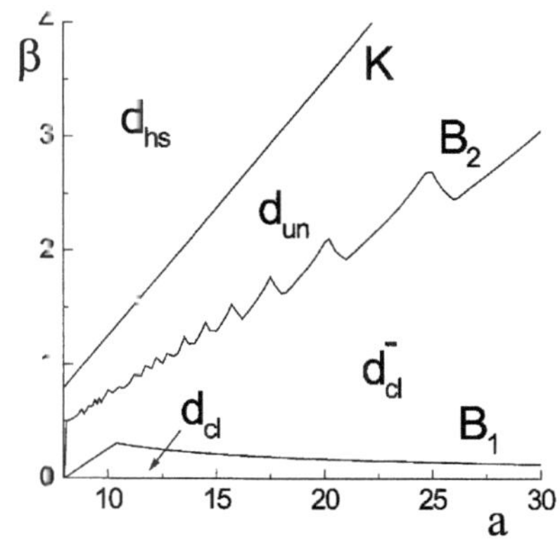

Fig. 5.13. Regions of existence and stability of homogeneous and inhomogeneous oscillations of the assembly ($N = 30$). The region d_{cl} defines the parameter values, for which $2^N - 2$ inhomogeneous states are stable. The region d^-_{cl} is where the amplitude-phase clusters become unstable. The line B_2 bounds the region of stability of the inhomogeneous states. Above the curve K only homogeneous states can exist

of them with $p_0 > q_0$. We show below that stable inhomogeneous in-phase oscillations have amplitude distributions, $\{r^0_j\}$, of values p_A and q_A only, that are the coordinates of the point A in Fig. 5.14. For $\beta \ll 1$, using regular perturbation theory, one can find that

$$
\begin{aligned}
p_A &= \beta \frac{(N-n)}{N} r^{(2)} + O(\beta^2)\,, \\
q_A &= r^{(2)} - \beta \frac{n}{2N r^{(2)} \sqrt{a(a-8)}} + O(\beta^2)\,,
\end{aligned}
\tag{5.40}
$$

where $r^{(2)}$ is the amplitude of one of the homogeneous solutions (5.26).

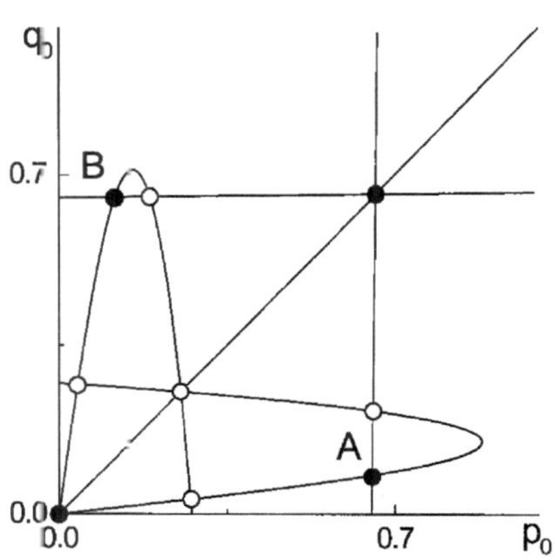

Fig. 5.14. Graphic solutions of (5.39)

In conclusion, in the assembly described by (5.23), we have shown that both homogeneous and inhomogeneous in-phase oscillations are possible.

Stability of the Inhomogeneous In-Phase Oscillations. Let $\{r_j = r_j^0,\ \varphi_j = \varphi_j^0\}$ be the solution of (5.25), providing inhomogeneous in-phase oscillations of the assembly. Linearization of (5.25) around this solution gives the following equations for perturbations, $\xi_j = r_j - r_j^0$ and $\eta_j = \varphi_j - \varphi^0$:

$$\begin{aligned} \dot{\xi}_j &= -\alpha_p \xi_j + d\sum_{k=1}^{N} \xi_k\,, \quad j = 1, 2, \ldots, n\,, \\ \dot{\xi}_j &= -\alpha_q \xi_j + d\sum_{k=1}^{N} \xi_k\,, \quad j = n+1, \ldots, N\,, \end{aligned} \tag{5.41}$$

and

$$\begin{aligned} \dot{\eta}_j &= -d\sigma_1 \eta_j + d\sum_{i=1}^{n} \eta_i + d\frac{q_0}{p_0} \sum_{i=n+1}^{N} \eta_i\,, \quad j = 1, 2, \ldots, n\,, \\ \dot{\eta}_j &= -d\sigma_2 \eta_j + d\frac{p_0}{q_0} \sum_{i=1}^{n} \eta_i + d \sum_{i=n+1}^{N} \eta_i\,, \quad j = n+1, \ldots, N\,, \end{aligned} \tag{5.42}$$

with

$$\begin{gathered} \alpha_p \equiv \beta + F'(p_0)\,, \quad \alpha_q \equiv \beta + F'(q_0)\,, \\ d \equiv \beta/N\,, \quad \sigma_1 \equiv n + q_0/p_0(N-n)\,, \quad \sigma_2 \equiv N - n + p_0/q_0 n\,. \end{gathered}$$

It follows from (5.41–42) that amplitude and phase disturbances evolve in an independent way.

Consider first (5.41), describing the evolution of amplitude perturbations. Let us introduce "difference" variables:

$$\begin{aligned} \xi_{i+1} - \xi_i &= u_i\,, \qquad i = 1, 2, \ldots, n-1\,, \\ \xi_{i+1} - \xi_i &= v_{i-n}\,, \quad i = n+1, \ldots, N-1\,. \end{aligned}$$

It follows from (5.41) that

$$\begin{aligned} \dot{u}_i &= -\alpha_p u_i\,, \quad i = 1, 2, \ldots, n-1\,, \\ \dot{v}_k &= -\alpha_q v_k\,, \quad k = 1, 2, \ldots, N-n-1\,. \end{aligned} \tag{5.43}$$

If $\alpha_p > 0$ and $\alpha_q > 0$, in the phase space of (5.41) there exists the stable manifold

$$\begin{aligned} M = \{\ &\xi_1 = \xi_2 = \ldots = \xi_n = u(t), \\ &\xi_{n+1} = \xi_{n+2} = \ldots = \xi_N = v(t)\}\,. \end{aligned}$$

On M the evolution obeys

$$\begin{aligned} \dot{u} &= (nd - \alpha_p)u + d(N-n)v\,, \\ \dot{v} &= dnu + \big(d(N-n) - \alpha_q\big)v\,. \end{aligned} \tag{5.44}$$

Consequently, if M and the trivial solution of (5.44) are stable, all perturbations decay, $\xi_j \to 0$. This is the case in the parameter region defined by

$$\begin{gathered} \beta + F'(p_0) > 0\,, \quad \beta + F'(q_0) > 0\,, \\ \beta + F'(p_0) + F'(q_0) > 0\,, \\ \beta\left\{\frac{n}{N}F'(p_0) + \frac{(N-n)}{n}F'(q_0)\right\} + F'(p_0)F'(q_0) > 0\,. \end{gathered} \tag{5.45}$$

One can show that one of the eigenvalues of (5.42) vanishes and all others are negative for all possible values of the parameters β and a.

Thus (5.45) delineate the region where inhomogeneous steady states of (5.25) and, consequently, inhomogeneous oscillations of the assembly are stable. Using (5.45), it can be shown that all inhomogeneous oscillations with amplitude distribution of values different from p_A and q_A are unstable, while those having these values can be either stable or unstable.

Formation of Amplitude Clusters. From (5.40) it follows that for $\beta \ll 1$, $F'(p_0) > 0$ and $F'(q_0) > 0$, which ensures fulfillment of the stability conditions (5.45) for all n. Thus in this case there are $2^N - 2$ stable inhomogeneous states of (5.25). Solving, numerically, (5.39), with (5.45), we obtain that all these solutions exist and are stable not only for $\beta \ll 1$, but also in a region d_{cl} (Fig. 5.14). Each of such steady states has its own set $\{r_j^0\}$, $j = 1, 2, \ldots, N$, of values, p_A and q_A, that determines the amplitudes of inhomogeneous *in-phase* oscillations of (5.23). But due to symmetry (5.25) only $N-2$ of these states of (5.23) have a genuinely distinct behavior. Here elements form two clusters: n of them have "high" amplitude p_A and $(N-n)$ have "low" amplitude q_A, where $n = 1, 2, \ldots, N-1$.

As an example, Fig. 5.15b shows amplitude clusters formed from the initial distribution shown in Fig. 5.15a. Note that this property of equal amplitude in the clusters in a system with global coupling is *exact*, and not approximate as in systems with diffusive coupling. The stability conditions of amplitude clusters depend not only on the parameters of (5.25), but also on the values of N and n (to be precise, on their ratio). Therefore, roughly speaking, solutions with different n (for fixed N and a) lose stability for different values of β. The stability boundary and regions of existence for amplitude clusters with different n for the assembly with $N = 30$ are shown in Fig. 5.16. Numerical exploration shows that lines of stability and existence are very close and, for each fixed a, have a maximum whose location depends on the value of a. Hence clusters having a marked predominance of elements with high or low amplitude lose stability first

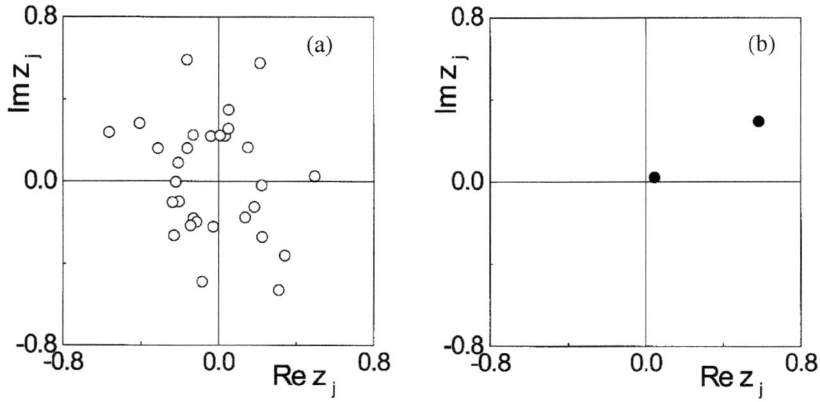

Fig. 5.15. One of the stable in-phase states ($N = 30, a = 18, \beta = 0.2$). (**a**) Initial state; (**b**) final distribution of complex amplitudes (14 elements have "low" amplitude and 16 "high" amplitude)

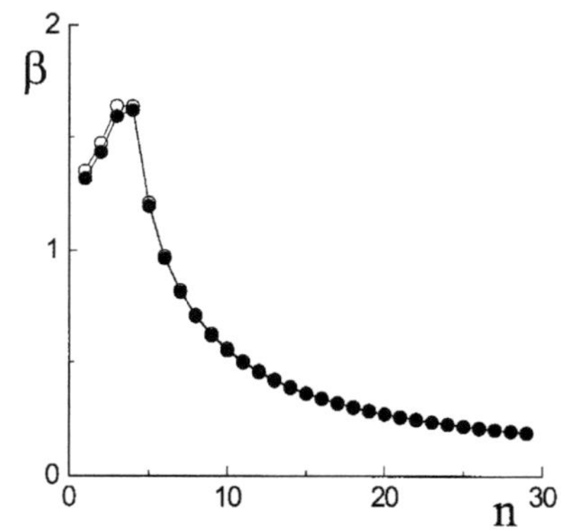

Fig. 5.16. Boundaries of existence (*open circles*) and stability (*solid circles*) of amplitude clusters with different n for $N = 30$ and $a = 18$

Thus when the parameters change within the region d_{cl}^{-} from curve B_1 to curve B_2 in Fig. 5.13, amplitude clusters lose stability one at a time. And when B_2 is reached, all of them become unstable. For the value of the parameters in the region above B_2 only homogeneous oscillations are stable.

5.4.3 Amplitude-Phase Clusters

Consider now the collective behavior of (5.23), when all oscillators are *nonisochronous* ($\alpha \neq 0$). We take the intralattice coupling to be complex ($\beta, \gamma \neq 0$). In this case from the equations for the phases, φ_j $\left(z_j = r_j e^{i\varphi_j}\right)$, it follows that in-phase motions are impossible for (5.25).

Let us look for the solution of (5.25) in the form

$$r_j(t) = \begin{cases} p(t)\,, & j = 1, 2, \ldots, n\,, \\ q(t)\,, & j = n+1, \ldots, N\,, \end{cases} \tag{5.46}$$

and

$$\varphi_j(t) = \begin{cases} \psi_1(t)\,, & j = 1, 2, \ldots, n\,, \\ \psi_2(t)\,, & j = n+1, \ldots, N\,. \end{cases} \tag{5.47}$$

From (5.25) the evolution equations for amplitudes, $p(t)$ and $q(t)$, and for the phase difference, $\psi = \psi_2 - \psi_1$, are as follows:

$$\begin{aligned} \dot{p} &= -Q(p) + q\frac{N-n}{N}(\beta\cos\psi - \gamma\sin\psi)\,, \\ \dot{q} &= -R(q) + p\frac{n}{N}(\beta\cos\psi + \gamma\sin\psi)\,, \\ \dot{\psi} &= \alpha[g(q) - g(p)] + \gamma\cos\psi\left(\frac{np}{Nq} - \frac{(N-n)q}{Np}\right) \\ &\quad - \beta\sin\psi\left(\frac{np}{Nq} + \frac{(N-n)q}{Np}\right) + \frac{N-2n}{N}\gamma\,, \end{aligned} \tag{5.48}$$

with

$$Q(p) \equiv p\left(f(p) - \beta\frac{N-n}{N}\right), \quad R(q) \equiv p\left(f(p) + \beta\frac{N-n}{N}\right).$$

Let us assume that β (5.48) satisfies

$$\beta < \begin{cases} \frac{(a-8)N}{8(N-n)}\,, & n \le N/2\,, \\ \frac{(a-8)N}{8n}\,, & n > N/2\,. \end{cases} \tag{5.49}$$

Then the functions $Q(p)$ and $R(q)$ satisfy the conditions:

$$\begin{array}{lll} Q(p_i) = 0\,, \quad R(q_i) = 0\,, & i = 0, 1, 2, & \\ Q(p) > 0\,, & \text{if} \quad p \in (0, p_1) & \text{and} \quad p > p_2\,, \\ Q(p) < 0\,, & \text{if} \quad p \in (p_1, p_2)\,, & \\ R(q) > 0\,, & \text{if} \quad q \in (0, q_1) & \text{and} \quad q > q_2\,, \\ R(q) < 0\,, & \text{if} \quad q \in (q_1, q_2)\,, & \end{array} \tag{5.50}$$

with

$$p_0 = 0\,, \quad p_{1,2} = \frac{1}{2}\sqrt{1 \mp \sqrt{1 - \frac{8}{a}\left(1 + \beta\frac{(N-n)}{N}\right)}}\,,$$

$$q_0 = 0\,, \quad q_{1,2} = \frac{1}{2}\sqrt{1 \mp \sqrt{1 - \frac{8}{a}\left(1 + \beta\frac{n}{N}\right)}}\,.$$

Let us show that there exists an invariant domain in the phase space of (5.48). Consider the following region:

$$\Omega = \{p, q : p_2 - A \le p \le p_2 + A,\ 0 \le q \le B\},$$

with parameters A and B that satisfy

$$0 < A < P_2 - p_1 \quad \text{and} \quad 0 < B < p_1 . \tag{5.51}$$

The boundary of Ω is a cylindrical surface. Consider the orientation of the vector field at this surface.

Let us take $q > 0$. In this case from (5.48) we obtain

$$\begin{aligned} \dot{p}_{|p=p_2+A} &\le -Q(p_2 + A) + q\tfrac{N-n}{N}\sqrt{\beta^2 + \gamma^2}\,, \\ \dot{p}_{|p=p_2-A} &\ge -Q(p_2 - A) - q\tfrac{N-n}{N}\sqrt{\beta^2 + \gamma^2}\,, \\ \dot{q}_{|q=B} &\le -R(B) + p\tfrac{n}{N}\sqrt{\beta^2 + \gamma^2}\,. \end{aligned} \tag{5.52}$$

Let us find the conditions such that for $q > 0$ the vector field of (5.48), at the boundary of Ω, is oriented into this region. This happens when A and B satisfy (5.51) and

$$A \le -p_2 + \frac{NR(B)}{n\sqrt{\beta^2 + \gamma^2}} \quad \text{and} \quad B \le -\frac{NQ(p_2 - A)}{(N - n)\sqrt{\beta^2 + \gamma^2}} . \tag{5.53}$$

If (5.53) has a solution, the existence of the region D in the (A, B) plane is ensured (Fig. 5.17). Note that not all parameter values allow a solution of (5.53). Let us mark by d_r the set of the parameter values of (5.48) for which D exists. For example, this region always exists if the following condition is fulfilled:

$$R(q_{max}) \ge \frac{n}{N}\sqrt{\beta^2 + \gamma^2}\,(2p_2 - p_{max})\,, \tag{5.54}$$

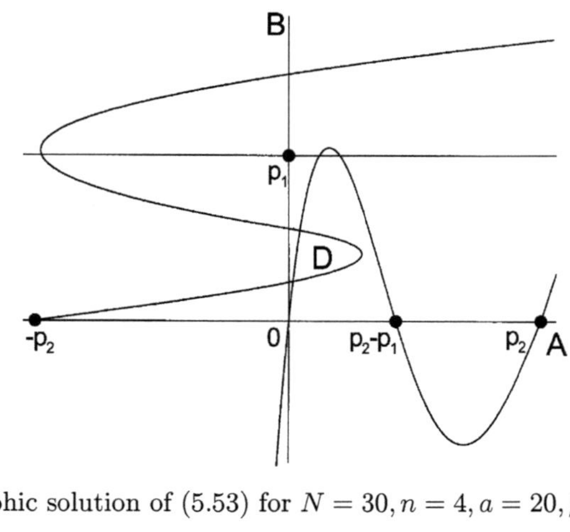

Fig. 5.17. Graphic solution of (5.53) for $N = 30, n = 4, a = 20, \beta = 1$ and $\gamma = 0.1$

with

$$q_{max} = \sqrt{\tfrac{3}{20} - \tfrac{1}{10}\sqrt{\tfrac{9}{4} - \tfrac{10}{a}\left(1 + \beta\tfrac{n}{N}\right)}}\,,$$
$$p_{max} = \sqrt{\tfrac{3}{20} + \tfrac{1}{10}\sqrt{\tfrac{9}{4} - \tfrac{10}{a}\left(1 + \beta\tfrac{N-n}{N}\right)}}\,;$$

(5.54) only gives an approximate location for the boundary of the region d_r.

Consider the orientation of the vector field at the surface $\{q = 0\}$. It follows from (5.48) that for $q = 0$ the phase ψ_2 can have an arbitrary value; however it does satisfy

$$\gamma\cos(\psi_2 - \psi_1) - \beta\sin(\psi_2 - \psi_1) = 0\,. \tag{5.55}$$

Substituting (5.55) into (5.48) we obtain that, in the plane $\{q = 0\}$, the vector field obeys

$$\dot{q}_{|q=0} = p\frac{n(\beta^2 + \gamma^2)}{N\beta}\cos(\psi_2 - \psi_1)\,. \tag{5.56}$$

We can determine the phase ψ_2 such that

$$\dot{q}_{|q=0} > 0\,. \tag{5.57}$$

Thus for the parameter values taken from the region d_r the vector field of (5.48) is oriented inwards to Ω. A qualitative sketch of the intersection of Ω with the plane $\{\psi = \text{const.}\}$ and the orientation of the vector field of (5.48) at the boundary of Ω are given in Fig. 5.18. Let us mark by Ω^+ the part of Ω between the planes $\{\psi = 0\}$ and $\{\psi = \pi - \arctan(\gamma/\beta)\}$. The orientation of the vector field on these planes can be found from (5.48):

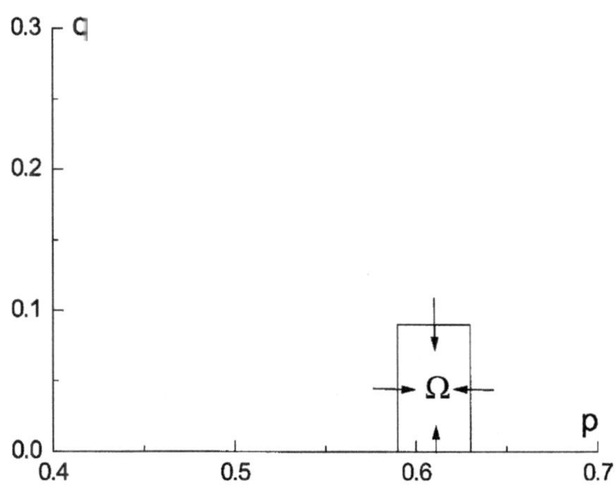

Fig. 5.18. Orientation of vector field at the boundary of the region Ω

$$\begin{aligned}\dot{\psi}_{|\psi=0} &= \alpha\left[g(q)-g(p)\right]+\tfrac{N-2n}{N}\gamma \\ &\quad +\gamma\left[\tfrac{np}{Nq}-\tfrac{(N-n)q}{Np}\right], \\ \dot{\psi}_{|\psi=\pi-\arctan\gamma/\beta} &= \alpha\left[g(q)-g(p)\right] \\ &\quad +N-2nN\gamma-\tfrac{2\beta\gamma}{\sqrt{\beta^2+\gamma^2}}\tfrac{np}{Nq}.\end{aligned} \tag{5.58}$$

The trajectories of (5.48) belonging to Ω satisfy

$$\begin{aligned}p_2-A_0<p(t)<p_2+A_0\,, \\ 0<q(t)<B_0\,.\end{aligned} \tag{5.59}$$

Using (5.59) we obtain from (5.58) that if the parameters of (5.48) belong to d_ψ, given by

$$\begin{aligned}&|\alpha|G_{max}+\tfrac{|N-2n|}{N}\gamma-\tfrac{n\beta\gamma}{\sqrt{\beta^2+\gamma^2}}\tfrac{p_2-A_0}{B_0}<0\,, \\ &|\alpha|G_{max}-\tfrac{|N-2n|}{N}\gamma \\ &\quad -\gamma\left(\tfrac{n(p_2-A_0)}{B_0}-\tfrac{(N-n)B_0}{p_2-A_0}\right)<0,\end{aligned} \tag{5.60}$$

with

$$G_{max}\equiv g(B_0)+\max\left\{g(p_2-A_0),-g(p_2+A_0)\right\}\,,$$

the vector field is oriented inwards to Ω^+. Thus, all along the surface of Ω^+ the vector field of (5.48) points inwards to this region; hence Ω^+ is the invariant domain.

Due to invariant property of Ω^+, in the phase space of the system there exists at least one trajectory, L, satisfying

$$\big(p(t),q(t),\psi(t)\big)\in\Omega^+\,. \tag{5.61}$$

It can be shown that the components $p(t)$ and $q(t)$ [the amplitudes of oscillations of (5.25)] satisfy (5.59), and the phase differences satisfy

$$0<\psi_2(t)-\psi_1(t)<\pi-\arctan\gamma/\beta\,.$$

Thus, the trajectory L determines the oscillatory behavior of (5.25) in the form of (5.46–47), i.e. in the form of amplitude-phase clusters. Such clusters exist for the values of the parameters from the region $\{d_r\bigcap d_\psi\}$.

Numerical integration of the evolution equations (5.25) permits verification of the conditions obtained above. The amplitude-phase clusters exist in a rather wide range of parameter values. Figure 5.19 shows two typical solutions in the form of the amplitude-phase clusters. The initial state is an almost in-phase state with a random distribution of amplitudes. Numerical integration of (5.25) shows not only the existence of the amplitude-phase clusters, but also their stability.

Note that the existence of the amplitude-phase states is impossible in assemblies composed of oscillators with a single attractor (e.g. limit-cycle oscillators).

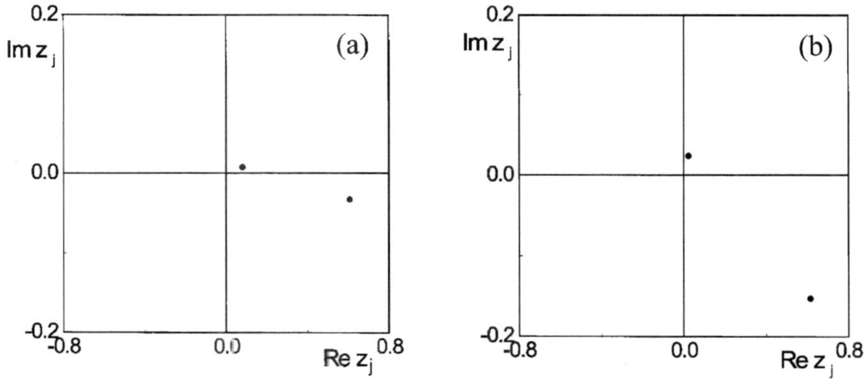

Fig. 5.19. Amplitude-phase clusters. Parameter values: $a = 20$; $N = 30$ (26 with "low" amplitude and 4 with "high" amplitude). (**a**) $\alpha = -0.1, \beta = 1, \gamma = 0.1$; (**b**) $\alpha = -10$, $\beta = 0.5, \gamma = 0.2$

5.4.4 "Splay-Phase" States

The collective state of an assembly when the amplitudes of oscillations are the same but the phases are different and $\varphi_j(t) = \Phi_j$, where Φ_j are constants, satisfying the condition $\sum_{j=1}^{N} e^{\mathrm{i}\Phi_k} = 0$, is called a "splay-phase" state. Let us show that (5.23) has stable "splay-phase" states in another albeit related way.

Taking into account the condition $\sum_{j=1}^{N} e^{\mathrm{i}\Phi_k} = 0$, we find from (5.25) the following solution:

$$z_j(t) = \begin{cases} r^{(3)} \exp\{\mathrm{i}\left[g(r^{(3)})t + \Phi_j\right]\}, & j = 1, 2, \ldots, n, \\ 0, & j = n+1, \ldots, N, \end{cases} \tag{5.62}$$

with

$$r = \frac{1}{2}\sqrt{1 + \sqrt{1 - 8(1+\beta)/a}} \quad \text{and} \quad \sum_{j=1}^{n} e^{i\Phi_j} = 0\,.$$

The solution (5.62) exists in the region of parameter values defined by $\beta < a/8 - 1$. Besides, for (5.25) there is a *splay-phase* state with amplitude $r = r^{(4)}$, where $r^{(4)} \equiv 1/2\sqrt{1 - \sqrt{1 - 8(1+\beta)/a}}$. But, as shown below, such a state is linearly (locally) unstable, and hence we disregard it. Note also that in (5.62) the index n is arbitrary and, in particular, can be equal to N. In the latter case, (5.62) defines a splay-phase state typical of assemblies of globally coupled limit-cycle oscillators.

Let us now consider the stability of splay-phase states (5.62). Linearizing (5.25) around the corresponding solution, we obtain for perturbations, $\xi_j \in C$, the following system:

$$\begin{cases} \dot{\xi}_j = \dfrac{\beta + \mathrm{i}\gamma}{N} \displaystyle\sum_{k=1}^{N} \xi_k e^{\mathrm{i}(\Phi_k - \Phi_j)} \\ \qquad - (a + i\alpha) H(r^{(3)})(\xi_j + \xi_j^*), \quad j = 1, 2, \dots, n, \\ \dot{\xi}_j = \dfrac{\beta + \mathrm{i}\gamma}{N} \displaystyle\sum_{k=1}^{N} \xi_k e^{\mathrm{i}(\Phi_k - \Phi_j)} - (\beta + 1)\xi_j, \quad j = n+1, \dots, N. \end{cases} \tag{5.63}$$

To (5.63) we associate a $2N \times 2N$ matrix whose eigenvalues are the Lyapunov exponents of (5.62). Upon transformation of the matrix to the block-diagonal form, it follows that all the eigenvalues split in two groups. The first group contains $N - n - 4$ negative eigenvalues equal to $-(\beta + 1)$ and $n - 4$ zero eigenvalues corresponding to the dimension of the manifold of locked fixed phases. The second group consists of eight eigenvalues, two of which are always negative and equal to $-(\beta + 1)$ and the rest of which are the roots of the characteristic equation

$$\begin{aligned} P(\lambda)P^*(\lambda) - (a^2 + \alpha^2)(\beta^2 + \gamma^2) H^2(r^{(3)}) \\ \times \Delta^2 (\lambda + \beta + 1)^2 = 0, \end{aligned} \tag{5.64}$$

with

$$\begin{aligned} P(\lambda) = {} & \lambda^3 + \lambda^2 (2aH(r^{(3)}) + 1 - i\gamma) \\ & + \lambda \Big\{ 2a(\beta + 1) H(r^{(3)}) - (\beta + i\gamma) \\ & \times \left[\left(2a - \tfrac{n(a + i\alpha)}{N} \right) H(r^{(3)}) + \tfrac{(\beta+1)n}{N} \right] \Big\} \\ & - \tfrac{(\beta+1)n}{N} [a\beta + \alpha\gamma + i(a\gamma - \alpha\beta)] H(r^{(3)}), \end{aligned}$$

$$\Delta = \left| \frac{1}{N} \sum_{k=1}^{N} e^{2i\Phi_k} \right| .$$

Accordingly, the stability conditions of (5.62) depend essentially on the parameter Δ, i.e. on the distribution of phase constants, Φ_j. For $\Delta = 0$ the stability boundary for splay-phase states of the form (5.62) is given in Fig. 5.20. For the parameter values from the region below the line, there can exist stable splay-phase states of two types. The first type is illustrated in Fig. 5.21a,b and the second in Fig. 5.21c,d. In the first case $(N - n)$ oscillators are at rest, and the other n are periodically oscillating. When the second type appears, all the oscillators are excited and indeed oscillate. Note that splay-phase states of the first type occur due to the *bistability* of the unit and do not exist in assemblies of oscillators with a single limit cycle. The solution of the second type is not related to the bistable properties of the units.

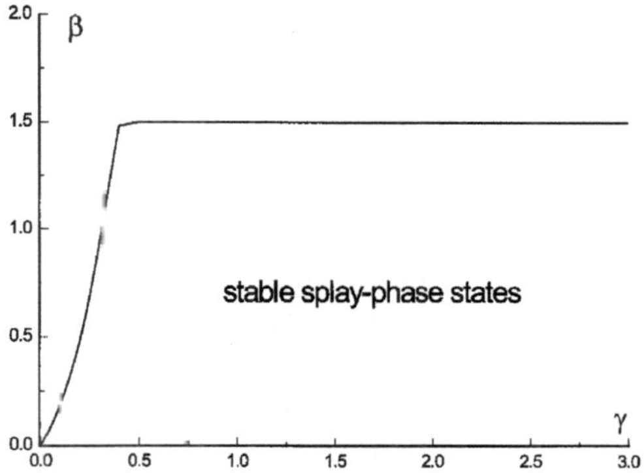

Fig. 5.20. Stability region of splay-phase states for $a = 20$ and $\alpha = -30$

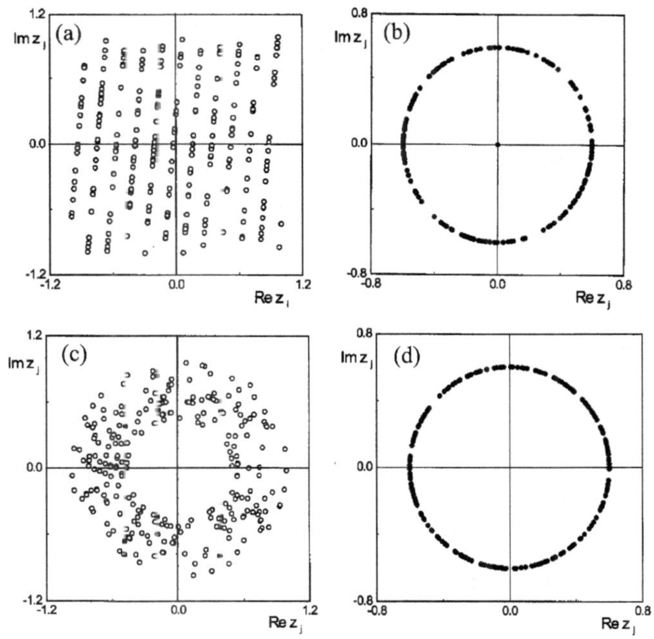

Fig. 5.21. Two types of splay-phase states for $a = 20$, $\alpha = -30$, $\beta = 1$, $\gamma = 2$ and $N = 256$: (**a**) initial state, (**b**) first type (39 elements have zero amplitude and 217 high amplitude); (**c**) initial state, (**d**) second type [final distribution for initial conditions (**c**)]

5.4.5 Collective Chaos

Numerical integration of (5.25) shows that for some parameter values the collective dynamics of (5.23) can be *chaotic* (there is a positive Lyapunov exponent). It can be shown that the salient dynamical properties of this regime are the same as for a "ρ-shaped type" regime in the case of globally coupled limit-cycle oscillators. The form of this ρ-shaped distribution of the oscillators of (5.23) is plotted in Fig 5.22a. Each oscillator in this regime is forced by the mean field, which itself is chaotic. The evolution of the order parameter, $M = |\bar{z}|$, is given in Fig. 5.22b. Individual oscillators move around the ρ-shaped loop with eventual jumps to the tail part. In (5.23) this regime has some new features. When the strength of the coupling becomes large enough (Fig. 5.22a), a number of oscillators separates from the tail part. These oscillators form a cluster (single dot in Fig. 5.22a) located at some distance from the ρ-shaped loop. Their motion is periodic and regular, while the dynamics of oscillators in the ρ-shaped loop is chaotic (Fig. 5.23).

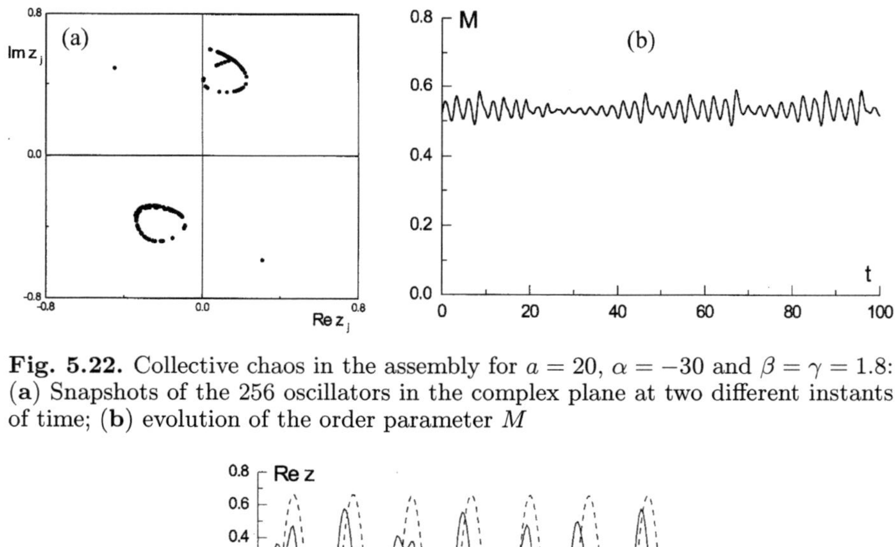

Fig. 5.22. Collective chaos in the assembly for $a = 20$, $\alpha = -30$ and $\beta = \gamma = 1.8$: (**a**) Snapshots of the 256 oscillators in the complex plane at two different instants of time; (**b**) evolution of the order parameter M

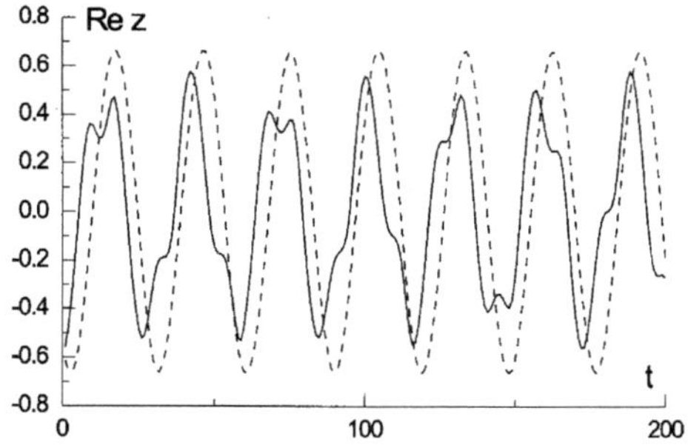

Fig. 5.23. Time evolution of two elements with regular and chaotic motion within the chaotic attractor for the parameter values of Fig. 5.22. The real parts of oscillation amplitudes are shown. The *broken line* shows the regular motion (element with $j = 1$), and the *solid line* shows the chaotic motion ($j = 5$)

It also appears that (5.23) possesses multistability. Depending on the initial conditions for the same parameter values one of the following regimes can be attained: (i) all the oscillators are at rest; (ii) the assembly exhibits collective chaos; and (iii) oscillators form a splay-phase state.

5.5 Spatial Disorder and Waves in a Circular Chain of Bistable Units

Let us consider (5.1) with periodic boundary conditions, and hence a circular chain,

$$y_{j+N} = y_j \,. \tag{5.65}$$

First let us see how the boundary conditions influence the spatial disorder [5.13]. Then we shall consider possible wave motions along the chain. Generally, to investigate waves in discrete systems a limiting transition from sets of coupled ordinary differential equations (ODE) to a partial differential equation (PDE) is carried out. One assumes that if the characteristic spatial scales of the wave regime considerably exceeds the spatial scale of the lattice this transition is meaningful.

Further we assume that the chain consists of an even number of oscillators, i.e. $N = 2k$. This restriction simplifies the analysis and is not essential for the results given below. We show below that in (5.1) and (5.65) oscillations having identical phase shift (in time) between neighboring "units" of the chain are possible. Such motions have wave-like form and are called phase waves.

5.5.1 Spatial Disorder

Let the parameters of (5.1) be in the region D_{ch} (Fig. 5.1). Then, the distribution of oscillation amplitudes, r_j, is determined by the chaotic dynamics of the map S. For the boundary conditions (5.65) each set of amplitudes is defined by a periodic trajectory (of period N) of S. For the points of D_{ch} the set of bounded trajectories of S is topologically conjugated to the shift-map of the topological Bernoulli scheme with two symbols. This suggests that S has $2^N - 1$ periodic trajectories of period N (the trivial orbit, $r_j = 0$, is omitted).

Thus, in (5.1) and (5.65) there are $2^N - 1$ phase-locked regimes with different oscillation amplitudes. For large N these distributions are very complex, due to the chaotic dynamics of S. In order to illustrate and substantiate the above-stated predictions (5.3) was numerically solved with (5.65), giving

$$\begin{aligned} r_{j+N}(t) &= r_j(t)\,, \\ \varphi_{j+N}(t) &= \varphi_j(t) + 2\pi m\,, \end{aligned} \tag{5.66}$$

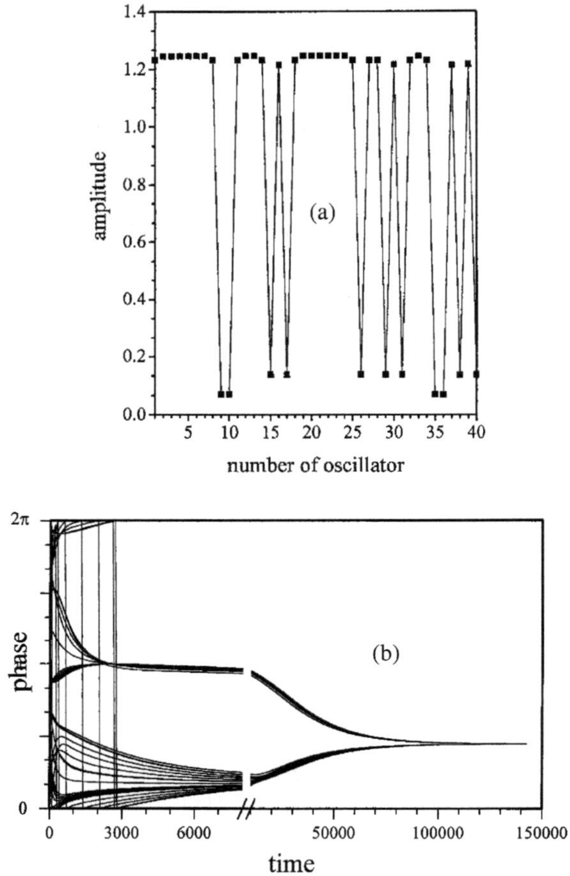

Fig. 5.24. (**a**) Final distribution of oscillation amplitudes; (**b**) evolution of phases of oscillations

with $m = 0, \pm 1, \pm 2, \ldots \pm N/2$. The results are shown in Fig. 5.24. We restrict consideration to the case $m > 0$, because the case of negative m is not different from positive m. The parameter values chosen are $N = 40$, $a = 11.5$ and $d = 0.06$. Initial phase and amplitude distributions were rather arbitrary. Figure 5.24a shows a typical final distribution of oscillation amplitudes. The evolution of phases displayed in Fig. 5.24b shows how the phase-locked oscillations appear with two separate sets of oscillators. In each set (or cluster) all oscillators are synchronized. The two clusters interact with one another until complete synchronization is achieved. The number of clusters and location of frontiers is directly associated to the heterogeneity in the amplitude distribution. Namely, breaks are generated where amplitude "gaps" appear, with the number of oscillators in the gap greater than one. In Fig. 5.24a there are two such gaps (oscillators 9/10 and 35/36) respectively.

5.5.2 Space-Homogeneous Phase Waves

Phase waves are determined by the solutions of (5.1), which correspond to the fixed points of the averaged system (5.3) with coordinates

$$r_j = \text{const.}, \quad \varphi_{j+1} - \varphi_j = \text{const.}, \quad j = 1, 2, \ldots, N.$$

From (5.3) and (5.66) we find that for all $m = 1, 2, \ldots, N/2$ and for the parameters in the domain H_m^{ex} separated by the inequalities

$$a > 8, \quad d < d^{ex},$$

where

$$d^{ex} \equiv \frac{a-8}{16(1-\cos M)}, \quad M = \frac{2\pi m}{N},$$

such fixed points exist. They can be subdivided into two families,

$$H^+ : \{r_j = r^+, \varphi_j = \varphi_j^*\} \quad \text{and} \quad H^- : \{r_j = r^-, \varphi_j = \varphi_j^*\},$$

where

$$\varphi_j^* \equiv \frac{2\pi m}{N} j + \varphi^0, \quad \varphi^0 = \text{const.},$$

$$r^\pm \equiv \sqrt{\frac{a \pm \sqrt{a^2 - 8a\,[1 + 2d(1-\cos M)]}}{4a}}.$$

To study the linear stability of the fixed points from $H^\pm$, we set $\xi_j = r_j - r^*$, $\eta_j = \varphi_j - \varphi_j^*$ for either H^+ or H^-. The linearization of (5.3) yields

$$\begin{aligned} 2\tfrac{d\xi_j}{dt} &= \mu\{d\cos M(\xi_{j-1} + \xi_{j+1}) \\ &\quad -[2d + F'(r^*)]\xi_j + dr^*\sin M(\eta_{j-1} - \eta_{j+1})\}, \\ 2\tfrac{d\eta_j}{dt} &= \mu[d\sin M/r^*(\xi_{j+1} - \xi_{j-1}) \\ &\quad - d\cos M(\eta_{j-1} - 2\eta_j + \eta_{j+1})]. \end{aligned} \tag{5.67}$$

Taking into account the periodic boundary conditions (5.66), ξ_j and η_j can be represented by the discrete Fourier series

$$\xi_j(t) = \sum_{s=1}^{N} \widetilde{\xi}_s(t)\exp(iSj), \quad \eta_j(t) = \sum_{s=1}^{N} \widetilde{\eta}_s(t)\exp(iSj), \tag{5.68}$$

with $S = 2\pi s/N$. After substitution of (5.68) into (5.67), we obtain

$$\begin{aligned} 2\tfrac{d\widetilde{\xi}_s}{dt} &= \mu\{[2d\cos M\cos S - 2d - F'(r^*)]\widetilde{\xi}_s - i2dr^*\sin M\sin S\widetilde{\eta}_s\}, \\ 2\tfrac{d\widetilde{\eta}_s}{dt} &= \mu\left[i2d\sin M\sin S/r^*\widetilde{\xi}_s - 2d\cos M(1-\cos S)\widetilde{\eta}_s\right]. \end{aligned} \tag{5.69}$$

The eigenvalues corresponding to (5.69) are given by

$$\begin{aligned}\lambda^2 - [4d\cos M\cos S - 2d(1+\cos M) - F'(r^*)]\lambda - 2d\cos M \\ \times(1-\cos S)[2d\cos M\cos S - 2d - F'(r^*)] \\ -4d^2\sin^2 M\sin^2 S = 0\,.\end{aligned} \tag{5.70}$$

One of the eigenvalues is zero. It is a consequence of the averaging method used. Thus, the stability of the fixed points from $H^{\pm}$ depends on the location in the complex plane of the remaining roots of (5.70). There are no stable fixed points from H^-, while from H^+ there are both stable and unstable fixed points. The condition $m < N/4$ is necessary for stability. A fixed point is (linearly) stable if the parameters a and d belong to the domain H_m^{st} separated for every $m = 1, 2, \ldots, \frac{N}{4} - 1$ by the inequalities

$$d < \min\left\{d^{ex}, d^{st}\right\}, \quad a > 8\,,$$

with

$$\begin{aligned}d^{st} \equiv \inf{}_{s\in[1,N]}\, d(s)\,, \\ d(s) \equiv \frac{\gamma(a-8)-2a(1-\cos M)+\sqrt{[\gamma(a-8)-2a(1-\cos M)]^2+8\gamma^2(a-8)}}{4\gamma^2}\,, \\ \gamma \equiv \frac{1+\cos S+4\cos M-6\cos^2 M}{\cos M}\,.\end{aligned} \tag{5.71}$$

The stable stationary points of (5.3) correspond to stable oscillations of (5.1) of the form

$$x_j = r^+\cos\left(t + \frac{2\pi m}{N}j + \varphi^0\right) + O(\mu). \tag{5.72}$$

This solution is a phase wave with wave number m. It propagates along the chain (5.1) with velocity $v = N/2\pi m$. Figure 5.25 illustrates with solid vertical lines in the (d, m) plane for $a = 11.5$ and $N = 40$, the domains of existence of stable phase waves. There are $N/4 - 1$ stable waves of the form (5.72). For any wave number m, increasing the intralattice coupling coefficient, d, leads to wave instability.

The conditions of linear stability are given by (5.71). Moreover space-homogeneous phase waves are evolutionary stable because (5.3) and (5.65) is a gradient system. To illustrate it and to investigate competition between space-homogeneous waves, we have numerically integrated (5.1) with (5.65), for $N = 40, a = 11.5, d = 0.6$ and $\mu = 0.01$. As shown above, waves with wave numbers $m = 1, 2, \ldots, 5$ are linearly stable for these parameter values (Fig. 5.25). The initial conditions are selected by varying values around a wave with $m = 3$, i.e.

$$\begin{aligned}r_j(t=0) = \delta_1\xi_j + 0.7\,, \\ \varphi_j(t=0) = 2\pi\cdot 3/40(j-1) + \delta_2\eta_j\,,\end{aligned}$$

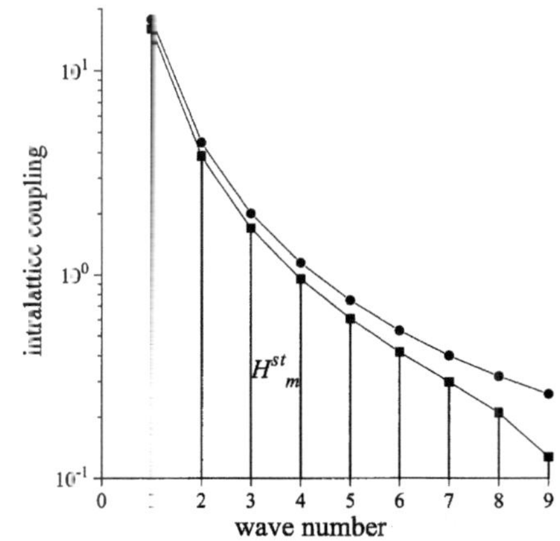

Fig. 5.25. Domain of existence (*circles*) and of stability (*squares*) of space-homogeneous phase waves

where ξ_j and η_j are random numbers taken in the range $[0, 1]$. For small enough δ_1 and δ_2 the wave with $m = 3$ reappears as time proceeds. However, for $\delta_1 = 0.8$ and $\delta_2 = 1.6$ appreciable modifications appear in the space–time evolution of the initial condition. System (5.1) is attracted to the wave with $m = 2$. For this case the evolution of the Lyapunov function U is depicted in Fig. 5.26. Based on the time behavior of U, the process of formation of a wave with $m = 2$ can be divided into three stages (see the sequence in Fig. 5.27):

(i) ($t \sim 0 - 500$): short transient process leading to a spatio-temporal structure close to a wave with $m = 3$ (Fig. 5.27a,b).

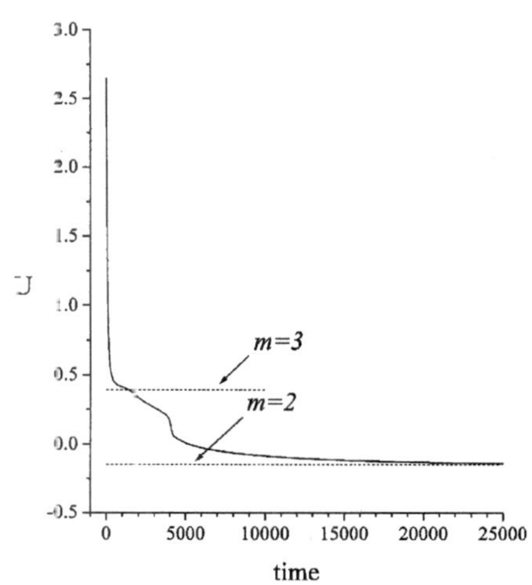

Fig. 5.26. Evolution of Lyapunov function (potential) U along a trajectory tending to the stationary state of H_2^{st}

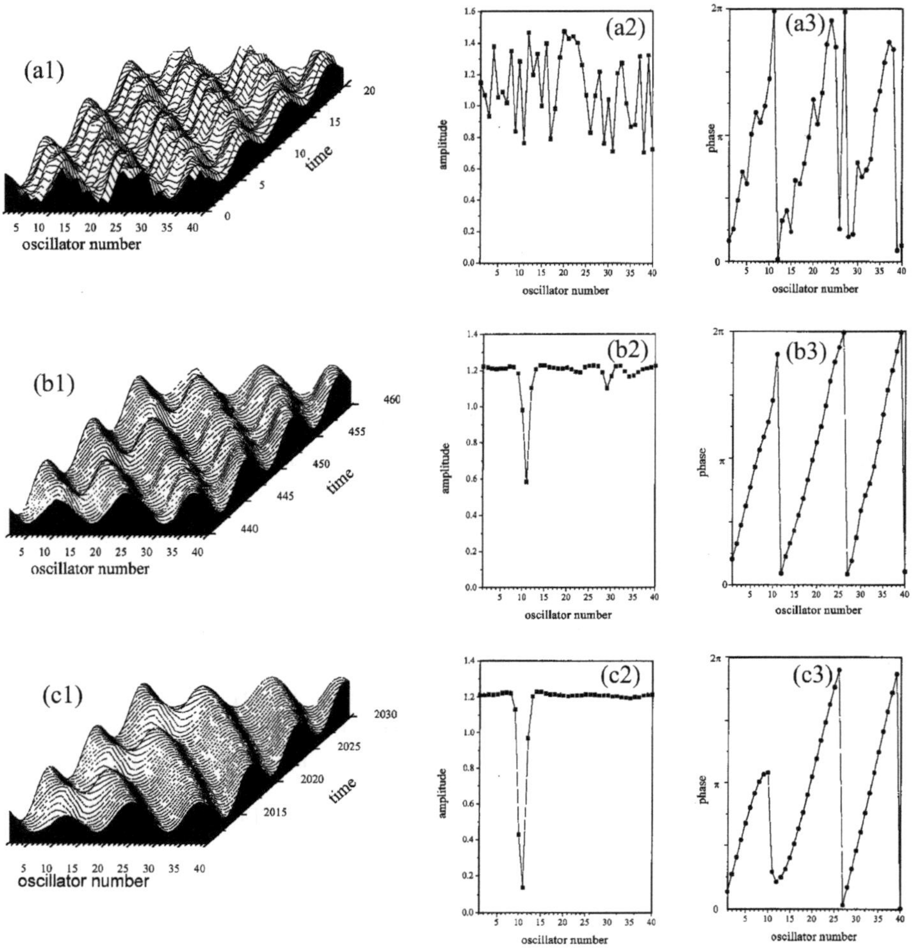

(ii) ($t \sim 1000 - 3000$): formation and growth of amplitude-phase inhomogeneity in oscillators 9–12 (Fig. 5.27c,d).

(iii) ($t \sim 4000 - 30000$): reorganization leading to the wave with $m = 2$ (Fig. 5.27e,f).

Thus, the transition from a wave with $m = 3$ leading to a wave with $m = 2$ occurs through the formation of a local breaking of the symmetry, i.e. it arises from a defect. When the defect disappears the wave with wave number $m = 2$ appears. Note that according to the initial conditions more than one defect can be formed. In this case the transition from a wave with m to another wave with $m - k$ (k is number of defects) occurs.

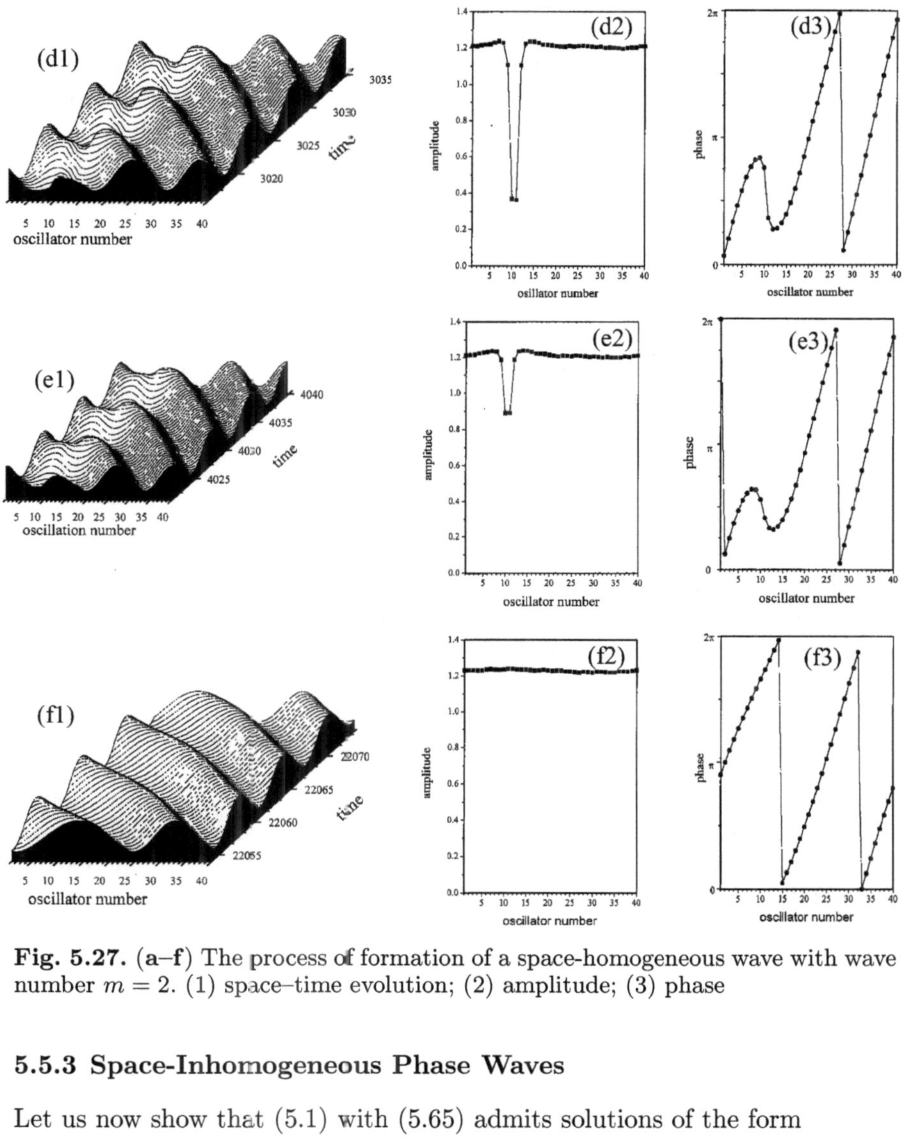

Fig. 5.27. (**a–f**) The process of formation of a space-homogeneous wave with wave number $m = 2$. (1) space–time evolution; (2) amplitude; (3) phase

5.5.3 Space-Inhomogeneous Phase Waves

Let us now show that (5.1) with (5.65) admits solutions of the form

$$x_j = r_j \cos(t + 2\pi m/Nj + \varphi^0) + O(\mu)\,, \quad m = 1, 2, \dots, N/2\,, \quad \varphi^0 = \text{const.} \tag{5.73}$$

These solutions differ from space-homogeneous waves by the fact that the amplitudes r_j vary along of the chain, and hence they are space-inhomogeneous phase waves. To prove their existence, we start again with the averaged system (5.3). For stationary points of (5.3), substituting $\varphi_j = 2\pi m/Nj + \varphi^0$ into the second equation of (5.3) we obtain

$$r_{j+1} - r_{j-1} = 0\,. \tag{5.74}$$

Consequently, r_j can be either independent of j or a 2-periodic function of j. We are interested in the latter case. Let ρ_1 and ρ_2 denote the quantities corresponded to a solution of (5.74). Using the first equation of (5.3) we obtain

$$\begin{aligned} -F(\rho_1) + 2d\,(\rho_2 \cos M - \rho_1) = 0\,, \\ -F(\rho_2) + 2d\,(\rho_1 \cos M - \rho_2) = 0\,. \end{aligned} \tag{5.75}$$

The number of real solutions of (5.75) varies according to the values of the parameters d, m and a. Figure 5.28 shows the relative location of curves (5.75) when there are nine (maximum) real solutions. Solutions with $\rho_1 = \rho_2$ have been discussed in the previous section; hence we omit them. To characterize these solutions we introduce quantities determining the slopes of the tangents to the curves at the point (ρ_1, ρ_2):

$$\begin{aligned} \sigma_1 = 2d + F^{'}(\rho_1)\,, \\ \sigma_2 = 2d + F^{'}(\rho_2)\,. \end{aligned}$$

There are only two solutions (solid circles in Fig. 5.28) with $\rho_1 \neq \rho_2$, $\sigma_{1,2} > 0$. Moreover, they are identical due to the imposed periodicity. Thus (5.3) admit a family of $N/2$ fixed points of the following form:

$$I_m : \left\{ \varphi_j = \frac{2\pi m}{N} j + \varphi^0,\ r_j = \begin{cases} r_1^*(m)\,,\ j \text{ is odd}\,, \\ r_2^*(m)\,,\ j \text{ is even}\,, \end{cases} \right\}$$

where $r_1^*(m) = \rho_1, r_2^*(m) = \rho_2$, $(\sigma_{1,2} > 0)$. As shown below, the other points (open circles in Fig. 5.28) correspond to unstable stationary points of (5.3). Figure 5.29 shows the domain I_m^{ex} of existence of the family I_m (circles in Fig. 5.29).

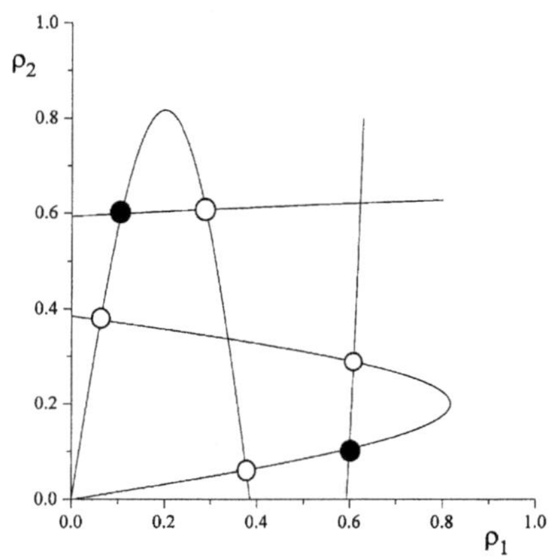

Fig. 5.28. Relative location of curves (5.75) for $N = 40, m = 2, a = 11.5$ and $d = 0.1$

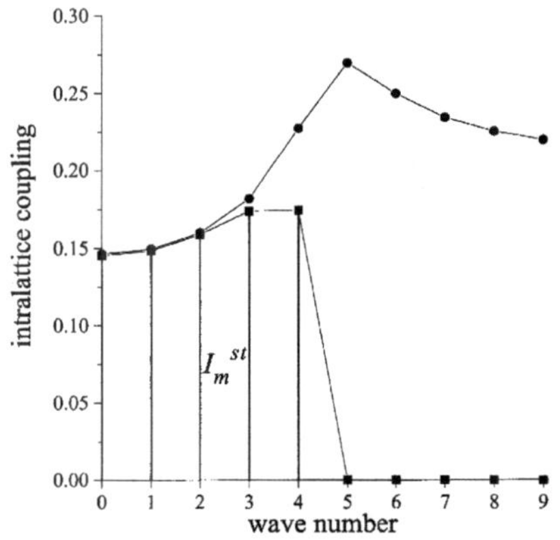

Fig. 5.29. Domain of existence (*circles*) and of stability (*squares*) of space-inhomogeneous phase waves

We now turn to the stability of the solutions from I_m. Linearizing (5.3) around one of the solutions we obtain the following matrix:

$$\vec{C} = \begin{pmatrix} C_1 & C_2 & 0 & \ldots\ldots & 0 & 0 & C_{N/2} \\ C_{N/2} & C_1 & C_2 & \ldots\ldots & 0 & 0 & 0 \\ \ldots\ldots & \ldots\ldots & \ldots\ldots & \ldots\ldots & \ldots\ldots & \ldots\ldots & \ldots\ldots \\ 0 & 0 & 0 & \ldots\ldots & C_{N/2} & C_1 & C_2 \\ C_2 & 0 & 0 & \ldots\ldots & 0 & C_{N/2} & C_1 \end{pmatrix},$$

with

$$\vec{C}_1 = \begin{pmatrix} -\sigma_1 & 0 & p & -qr_2^*(m) \\ 0 & -2pk & q/r_2^*(m) & pk \\ p & qr_1^*(m) & -\sigma_2 & 0 \\ -q/r_2^*(m) & p/k & 0 & -2p/k \end{pmatrix},$$

$$\vec{C}_2 = \begin{pmatrix} 0 & 0 & 0 & 0 \\ 0 & 0 & 0 & 0 \\ p & -qr_1^*(m) & 0 & 0 \\ q/r_2^*(m) & p/k & 0 & 0 \end{pmatrix}, \quad \ldots, \quad \vec{C}_{N/2} = \begin{pmatrix} 0 & 0 & p & qr_2^*(m) \\ 0 & 0 & -q/r_1^*(m) & pk \\ 0 & 0 & 0 & 0 \\ 0 & 0 & 0 & 0 \end{pmatrix},$$

$$p = d\cos M\,, \quad q = d\sin M\,, \quad k = r_2^*(m)/r_1^*(m)\,.$$

The matrix $\vec{C}$ is a $N/2 \times N/2$ block-circulant matrix, and hence its eigenvectors are

$$\vec{Y}_s = \begin{pmatrix} \vec{I} \\ \vec{Z}_s \\ \vec{Z}_s^2 \\ \vdots \\ \vec{Z}_s^{N/2-1} \end{pmatrix}, \quad \vec{Z}_s = \begin{pmatrix} z_s & 0 & 0 & 0 \\ 0 & z_s & 0 & 0 \\ 0 & 0 & z_s & 0 \\ 0 & 0 & 0 & z_s \end{pmatrix},$$

$$z_s = \exp(\mathrm{i}2S), \quad S = 2\pi s/N \quad s = 1, 2, \dots, N/2,$$

where $\vec{I}$ is the identity matrix. Then $(\vec{C} - \lambda\vec{I})$ can be diagonalized using the $N/2 \times N/2$ matrix $\vec{T}$ with columns consisting of the eigenvectors $\vec{Y}_s$, i.e.

$$\vec{T} = \left(\vec{Y}_1, \vec{Y}_2, \dots . \vec{Y}_{N/2}\right).$$

Indeed, by multiplying on the right $(\vec{C} - \lambda\vec{I})$ by $\vec{T}$ and on the left by the conjugate matrix $\vec{T}^*$, we obtain a block-diagonal $N/2 \times N/2$ matrix with diagonal units $\frac{N}{2}\vec{S}_s$, where

$$\vec{S}_s = \begin{pmatrix} -\sigma_1 - \lambda & 0 & p(1+e^{-\mathrm{i}2S}) & -qr_2^*(1-e^{-\mathrm{i}2S}) \\ 0 & -2pk - \lambda & q/r_1^*(1-e^{-\mathrm{i}2S}) & pk(1+e^{-\mathrm{i}2S}) \\ p(1+e^{\mathrm{i}2S}) & qr_1^*(1-e^{\mathrm{i}2S}) & -\sigma_2 - \lambda & 0 \\ -q/r_2^*(1-e^{\mathrm{i}2S}) & p/k(1+e^{\mathrm{i}2S}) & 0 & -2p/k - \lambda \end{pmatrix}.$$

Since the eigenvalues of the block-diagonal matrices coincide with eigenvalues of blocks, the eigenvalues of the matrix $\vec{C}$ are given by

$$\lambda^4 + a_1\lambda^3 + a_2\lambda^2 + a_3\lambda + a_4 = 0, \tag{5.76}$$

with

$$\begin{aligned} a_1 &= 2p(k+1/k) + \sigma_1 + \sigma_2, \\ a_2 &= 4p^2 + 2p(k+1/k)(\sigma_1+\sigma_2) + \sigma_1\sigma_2 \\ &\quad -4\left[p^2(1+\cos 2S) + q^2(1-\cos 2S)\right], \\ a_3 &= 2p\sigma_1\sigma_2(k+1/k) + 4p^2(\sigma_1+\sigma_2) \\ &\quad -2\left[\sigma_1+\sigma_2+2p(k+1/k)\right]\left[p^2(1+\cos 2S) + q^2(1-\cos 2S)\right], \\ a_4 &= 4p^2\sigma_1\sigma_2 - 4pq^2(1-\cos 2S)(k\sigma_2 + \sigma_1/k) \\ &\quad -2p^2(1+\cos 2S)(\sigma_1\sigma_2 \\ &\quad +4p^2) + 4\left[p^2(1+\cos 2S) - q^2(1-\cos 2S)\right]^2, \\ S &= 2\pi s/N, \quad s = 1, 2, \dots, N/2. \end{aligned}$$

One of the eigenvalues is always equal to zero ($a_4 = 0$, for $s = N/2$). The (linear) stability conditions of the stationary state from I_m are

$$\begin{aligned} &a_{1,2,3} > 0, \quad a_1a_2 - a_3 > 0, && \text{if } s = N/2, \\ &a_{1,2,3,4} > 0, \quad a_1a_2a_3 - a_3^2 - a_1^2a_4 > 0, && \text{if } s = 1, 2, \dots \tfrac{N}{2} - 1. \end{aligned} \tag{5.77}$$

It appears that the stationary points with $m \geq N/4$ are unstable for all values of the parameters. The other points can be either stable or unstable. Let the domain of parameters of stationary point with index m be I_m^{st}

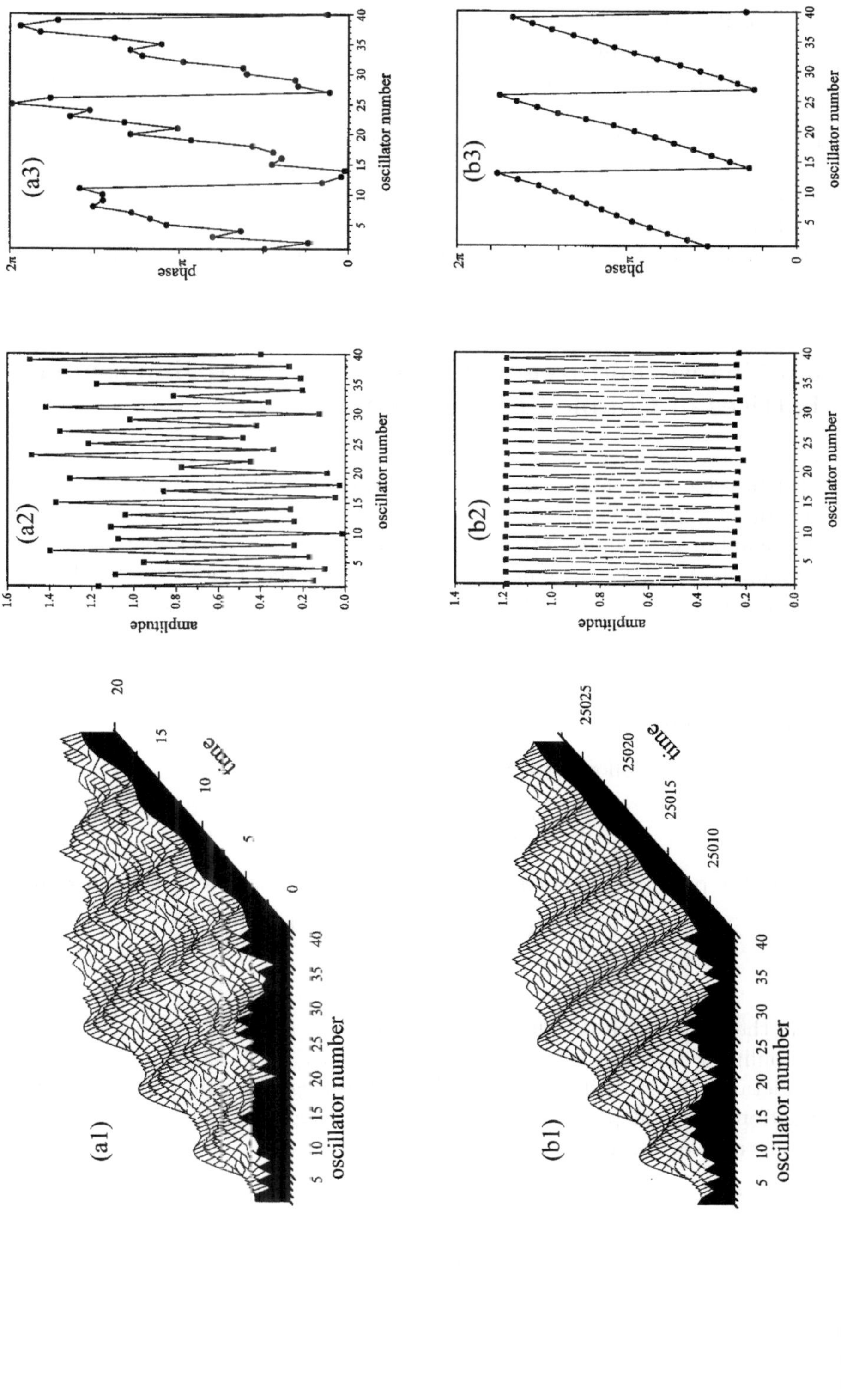

Fig. 5.30. The process of formation of a space-inhomogeneous wave with wave number $m = 3$

(Fig. 5.29, vertical lines). At variance with the space-homogeneous case, I_m^{st} sharply decreases after some m ($m = 4$ in Fig. 5.29). The stability analysis of the other fixed points of (5.3) can be performed along the same lines. However, in these cases (5.77) are not valid. Consequently, for (5.3) there is only one family of stable fixed points whose coordinates r_j vary along the chain. It is I_m^{st}. Every point of I_m^{st} corresponds to a stable space-inhomogeneous wave of the form of (5.73) for (5.1).

For completeness, the stability of space-inhomogeneous waves has been numerically checked. Figure 5.30 illustrates the time and space evolution of a space-inhomogeneous wave with $m = 3$.

5.6 Chaotic and Regular Patterns in Two-Dimensional Lattices of Coupled Bistable Units

Let us now consider the spatio-temporal dynamics of a 2D lattice of coupled bistable elements or units with continuous time, and hence a discrete version of a 2D reaction–diffusion equation (we shall come back to a similar problem but with discrete time in Chap. 7). The earlier given methodology is here applied to two particular cases: a lattice system representing a discrete version of a 2D FitzHugh–Nagumo–Schlögl (FNS) equation and a lattice of weakly coupled bistable oscillators, which is a generalization of (5.1) to the 2D case [5.15].

5.6.1 Methodology for a Lattice of Bistable Elements

We consider the following 2D lattice system:

$$\begin{aligned} \dot{u}_{j,k} &= f(u_{j,k}) + d(\Delta u)_{j,k}\,, \\ j,k &= 1,2,\,\dots\,,N\,, \end{aligned} \tag{5.78}$$

where the pair (j,k) denotes a site, $(\Delta u)_{j,k}$ is the discrete Laplace operator

$$\begin{aligned} (\Delta u)_{j,k} &= (\Delta u)_j + (\Delta u)_k - 4u_{j,k}\,, \\ (\Delta u)_j &= u_{j+1,k} + u_{j-1,k}\,, \\ (\Delta u)_k &= u_{j,k+1} + u_{j,k-1}\,, \end{aligned}$$

d is the intralattice coupling diffusion between units ($d > 0$) and $f(u)$ is a nonlinear function of cubic shape (Fig. 5.31). The choice of $f(u)$ here is dictated by its utility in physiology (FitzHugh–Nagumo model) and in reaction–diffusion problems (Schlögl model). The boundary conditions are taken to be of von Neumann type (zero flux):

$$u_{0,k} = u_{1,k}\,, \quad u_{j,0} = u_{j,1}\,, \quad u_{j,N+1} = u_{j,N}\,, \quad u_{N+1,k} = u_{N,k}\,. \tag{5.79}$$

When $d = 0$, in each point of the 2D lattice there is a one-dimensional element with two stable steady states whose basins of attractions are separated by an unstable steady state. Hence, for $d > 0$, (5.78) describes a 2D reaction–diffusion equation with nonlinear kinetics.

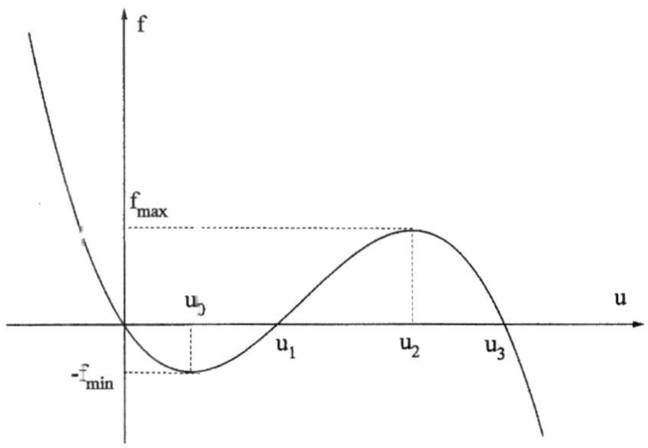

Fig. 5.31. Qualitative shape of the nonlinearity $f(u)$

Gradient Property of the System and Homogeneous Steady States. The system (5.78) is gradient, i.e. it can be rewritten in the form

$$\dot{u}_{j,k} = -\frac{\partial U}{\partial u_{j,k}}, \tag{5.80}$$

where

$$U = \sum_{j,k} \left(\frac{d}{2} (\nabla u_{j,k})^2 - \int_0^{u_{j,k}} f(\eta) d\eta \right),$$

$$(\nabla u_{j,k})^2 \equiv (u_{j+1,k} - u_{j,k})^2 + (u_{j,k+1} - u_{j,k})^2,$$

is the potential. Thus only steady states exist in the phase space $\mathbf{R}^{N^2}$. Accordingly, whatever the initial conditions, (5.78) tends to one of the stable steady states corresponding to a local minimum of U. In the 3D space $\{\mathbf{Z}^2, \mathbf{R}\}$ these minima represent stationary patterns.

The trivial steady states are spatially homogeneous states corresponding to the zeros of $f(u)$:

$$O_1: \ \mathbf{u} = \{0\}, \quad O_2: \ \mathbf{u} = \{u_1\}, \quad O_3: \ \mathbf{u} = \{u_3\},$$

with

$$\mathbf{u} = \{u_{j,k} : 1 \le j, k \le N\}.$$

Let us now turn to spatially inhomogeneous states.

Location of the Steady States in Phase Space. Let us consider in $\mathbf{R}^{N^2}$ of (5.78) the following regions:

$$\Omega_r = \{\mathbf{u} : |u_{j,k}| \le r\}, \tag{5.81}$$

where r is a parameter. From (5.78) and the form of f (Fig. 5.31) it follows that at $r > r^-$, where $r^- = u_3$ on the boundary of each region Ω_r, the vector field of (5.78) points into Ω_r. Hence, all steady states of (5.78) are located in the region Ω_{r^-}. Then the steady state O_3 is an "end" point of the boundary of Ω_{r^-}, and all other steady states are inside Ω_{r^-}.

Invariant Domains. Let us consider two auxiliary equations:

$$\dot{u} = f(u) - 4d(u - r)\,, \tag{5.82}$$

$$\dot{u} = f(u) - 4d(u + r)\,. \tag{5.83}$$

Then each of these equations, simultaneously, has three steady states if r satisfies $r < r^+$, where

$$r^+ = \min\left\{\frac{f_{min}}{4d} + u_0\ ,\ \frac{f_{max}}{4d} - u_2\right\}.$$

u_0 and u_2 denote, respectively, the minimum and the maximum of $f(u)$ (Fig. 5.14). The outer points are stable, while the inner one is unstable. We label the coordinates of the stable steady states with $u = u_1^+, u_3^+ \ (u_1^+ < u_3^+)$ for (5.82) and with $u = u_1^-, u_3^- \ (u_1^- < u_3^-)$ for (5.83), hence (Fig. 5.31)

$$u_1^- < 0 < u_1^+\,, \quad u_3^- < u_3 < u_3^+\,. \tag{5.84}$$

We require that $r^+ > r^-$. This condition will be satisfied for the parameter region D_{ch} selected by

$$d < \min\left\{\frac{f_{min}}{4(u_3 - u_0)}\ ,\ \frac{f_{max}}{4(u_3 + u_2)}\right\}. \tag{5.85}$$

For completeness, note that for a 1D lattice, i.e. for a linear chain, (5.85) becomes

$$d < \min\left\{\frac{f_{min}}{2(u_3 - u_0)}\ ,\ \frac{f_{max}}{2(u_3 + u_2)}\right\}.$$

Let the parameters of (5.78) belong to D_{ch}. We fix the parameter value $r = r_0 \in (r^-, r^+)$. All trajectories of (5.78), with the initial conditions outside the domain Ω_{r_0}, after some time enter Ω_{r_0} and do not leave it. Thus Ω_{r_0} is an invariant domain containing all steady states of (5.78). Therefore, we shall consider (5.78) inside Ω_{r_0}.

Let us construct in Ω_{r_0} still narrower invariant domains. For this purpose we consider in the phase space $\mathbf{R}^{N^2}$ the following regions:

$$\Omega_{j,k}^0 = \{u_1^- \le u_{j,k} \le u_1^+,\ |u_{j',k'}| \le r_0, \quad \forall j', k',\ (j - j')^2 + (k - k')^2 \ne 0\}\,,$$

$$\Omega_{j,k}^1 = \{u_3^- \le u_{j,k} \le u_3^+,\ |u_{j',k'}| \le r_0, \quad \forall j', k',\ (j - j')^2 + (k - k')^2 \ne 0\}\,.$$

Then the vector field of (5.78) at the border of the regions $\Omega_{j,k}^0, \Omega_{j,k}^1$ is oriented toward these regions. Thus for the points of D_{ch} there exist N^2 invariant domains $\Omega_{j,k}^0$, $\Omega_{j,k}^1$ in $\mathbf{R}^{N^2}$ of (5.78).

5.6.2 Stable Steady States

Existence of Steady States. First, let us show that, for parameter values taken in D_{ch}, (5.78) has 2^{N^2} stable steady states:

$$\mathbf{u}^* = \{u^*_{j,k}\}, \quad \mathbf{u}^* \in \bigcap_{j,k} \Omega^{a_{j,k}}_{j,k}. \tag{5.86}$$

Let us also show that an arbitrary $N \times N$ matrix (a_{jk}) consisting of the symbols 0 and 1 can be associated with one of them.

Let us consider an arbitrary matrix (a_{jk}) and the corresponding intersection $J = \cap \Omega^{a_{j,k}}_{j,k}$. The set J is convex as it is the direct product of the segments of the coordinate axes. Besides, J is a compact set. Since the field of (5.78) at the borders of $\Omega^0_{j,k}$ and $\Omega^1_{j,k}$ is oriented into these domains, $T_\tau(J) \subset J$, where T_τ is the time-shift operator determined by (5.78) and since $J \subset \Omega^{a_{jk}}_{jk}$, for any j, k

$$T_\tau(J) \subset T_\tau(\Omega^{a_{j,k}}_{j,k}) \subset \Omega^{a_{j,k}}_{j,k},$$

we have

$$T_\tau(J) \subset \bigcap_{j,k} \Omega^{a_{j,k}}_{j,k} = J.$$

Using Schauder–Tikhonov's theorem we find that the map T_τ has one fixed point in the set J. As (5.78) has no periodic trajectories, each fixed point of the map T_τ corresponds to a steady state of (5.78). Thus, (5.78) has 2^{N^2} steady states satisfying (5.86). This set of steady states also includes spatially homogeneous ones which correspond to the matrices $(a_{j,k})$ having only either 0 or 1.

Stability of the Steady States. Let us now find the conditions for stability of the steady states. Linearizing (5.78) in the vicinity of a steady state $\mathbf{u}^*$ we obtain

$$\dot{\xi}_{j,k} = d(\Delta\xi)_j - \sigma_{j,k}\xi_{j,k} + d(\Delta\xi)_k, \tag{5.87}$$

with $\sigma_{j,k} = -f'(u^*_{j,k}) + 4d$. Let us locate the complex eigenvalues of the linear operator L described by this system. First we consider the case of the spatially homogeneous steady states O_1, O_2 and O_3. In this case

$$\sigma_{j,k} = \sigma = \begin{cases} -f'(0) + 4d, & \mathbf{u} = \{0\}, \\ -f'(u_1) + 4d, & \mathbf{u} = \{u_1\}, \\ -f'(u_3) + 4d, & \mathbf{u} = \{u_3\}. \end{cases}$$

A solution of (5.87) can be taken in the form $\xi_{j,k} = \varphi_j \psi_k$. Substituting this expression into (5.87) and separating variables, we obtain equations for φ_j and ψ_k:

$$\dot{\varphi}_j = d\varphi_{j-1} - (\gamma + \sigma)\varphi_j + d\varphi_{j+1}\,, \tag{5.88}$$

$$\dot{\psi}_k = d\psi_{k-1} + \gamma\psi_k + d\psi_{k+1}\,, \tag{5.89}$$

where $j, k = 1, 2, \ldots, N$ and γ is an arbitrary real number. The spectrum of the linear operators determined by (5.88–89) is

$$\rho_s = -\gamma - \sigma - 2d\cos\frac{\pi s}{N+1}\,, \qquad \mu_l = \gamma - 2d\cos\frac{\pi l}{N+1}\,,$$

$$s, l = 1, 2, \ldots, N\,.$$

Hence, for spatially homogeneous steady states the spectrum of L is

$$\lambda_{sl} = -\sigma - 2d\left(\cos\frac{\pi s}{N+1} + \cos\frac{\pi l}{N+1}\right)\,, \tag{5.90}$$

$$s,\ l = 1, 2, \ldots, N\,.$$

Accordingly, the steady states O_1 and O_3 are locally stable, and O_2 is unstable.

For spatially inhomogeneous steady states the location of the spectrum of the operator L in the complex plane can be obtained using Gershgorin's criterion extended to a 2D lattice (see Chap. 7 and Appendix G). The operator L can be rewritten in coordinate form

$$\dot{\xi}_{jk} = \sum_{l=j-1}^{j+1} \sum_{s=k-1}^{k+1} a_{jkls}\xi_{ls}\,, \tag{5.91}$$

with

$$a_{jkjk} = -\sigma_{jk}\,, \quad a_{jkj-1k} = a_{jkjk-1} = a_{jkj+1k} = a_{jkjk+1} = d\,.$$

According to Gershgorin's criterion the spectrum of the operator L is located in the union of N^2 disks S_{jk} defined by

$$|f'(u^*_{jk}) - 4d - \lambda| \le 4d\,. \tag{5.92}$$

Since these steady states satisfy (5.86), then $f'(u^*_{jk}) < 0$, and it follows from (5.92) that all disks S_{jk} are on the left of the imaginary axis. Hence all spatially inhomogeneous steady states of (5.78) satisfying (5.86) are stable for the parameter values taken in D_{ch}.

Spatial Disorder and Patterns. With N^2 elements in the lattice there is a very large number, 2^{N^2}, of stable steady states in the phase space $\mathbf{R}^{N^2}$. In the space $\{\mathbf{Z}^2, \mathbf{R}\}$ each of these steady states corresponds to some spatial pattern. As each pattern can be "coded" by an arbitrary matrix (a_{jk}), consisting of 0 and 1 symbols, the structure of the patterns in $\{\mathbf{Z}^2, \mathbf{R}\}$ can be very diverse, including regular (spots, stripes, etc.) and spatially disordered patterns albeit all steady in time.

5.6.3 Spatial Disorder and Patterns in the FitzHugh–Nagumo–Schlögl Model

Consider (5.78–79) with the nonlinear function $f(u) = u(u - a)(1 - u)$, where the parameter a satisfies the condition $0 < a < 1$. In this case (5.78) can be interpreted as a discrete version of the 2D nonlinear diffusion equation. Applying the results of Sect. 5.3.1 we find the parameter region D_{ch} (Fig. 5.32) where the system has the 2^{N^2} stable steady states. To illustrate the theoretical results, the equations have been numerically integrated. Initial and final distributions of the variables $u_{j,k}$ on the plane $\mathbf{Z}^2$ are represented by a varying degree of grey color corresponding to the value of the variable $u_{j,k}$ at a site (j, k). Then the already relatively regular initial distribution shown in Fig. 5.33a evolves toward the regular pattern depicted in Fig. 5.33b.

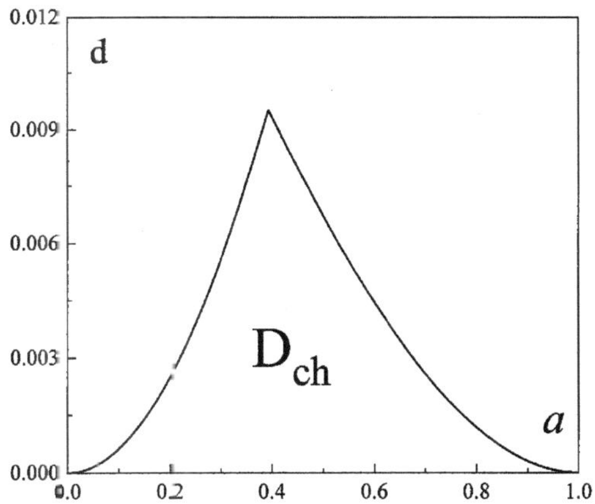

Fig. 5.32. The parameter region D_{ch} corresponding to the existence of spatial disorder in the 2D, diffusive FitzHugh–Nagumo–Schlögl (FNS) model

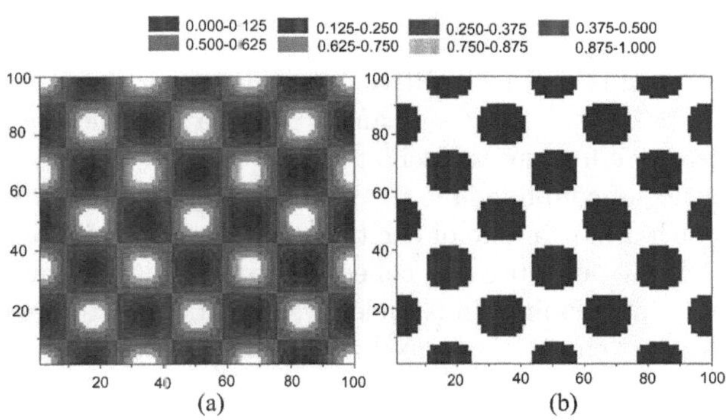

Fig. 5.33. Appearance of a pattern formed by "spots" in the lattice corresponding to the FNS model. Parameter values: $a = 0.4$, $d = 0.007$. (**a**) initial distribution; (**b**) final pattern

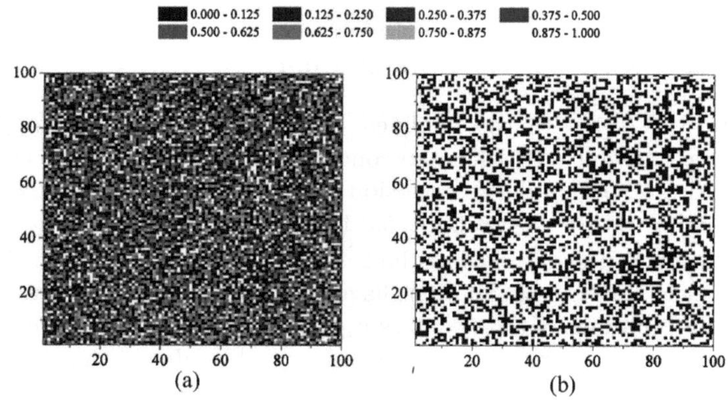

Fig. 5.34. Formation of an irregular pattern in the lattice corresponding to the FNS model. Parameter values: $a = 0.4$, $d = 0.007$. (**a**) initial distribution; (**b**) final pattern

Similarly, for the rather disordered initial distribution shown in Fig. 5.34a another irregular but stationary pattern appears in the lattice (Fig. 5.34b). The results depicted in Figs. 5.33 and 5.34 confirm the existence of very complex spatial configurations.

5.6.4 Spatial Disorder and Patterns in a Lattice of Bistable Oscillators

In this section we return to the unit considered in Sect. 5.1, and hence we consider the following lattice system:

$$\begin{cases} \dot{x}_{j,k} = y_{j,k}\,, \\ \dot{y}_{j,k} = -x_{j,k} - \mu\left[g(x_{j,k})y_{j,k} - d(\Delta y)_{j,k}\right]\,, \end{cases} \qquad (5.93)$$
$$j,k = 1, 2, \ldots, N\,,$$

where $0 < \mu \ll 1$, d is again a positive quantity proportional to the intralattice coupling, $g(x) = ax^4 - ax^2 + 1$, and $a > 8$ is a parameter describing the dynamics of a single unit as an oscillator.

In the absence of coupling, $d = 0$, at each site (j, k) of the lattice, there is an oscillator with a hard mode of excitation, as described in Sect. 5.1. Thus, when $d > 0$, (5.93) describes the collective dynamics of a square lattice of coupled bistable (fixed point, limit cycle) systems. Let us explore the behavior of (5.93) with von Neumann's (zero-flux) boundary conditions.

An Averaged System. A standard averaging procedure with (5.93) yields, as a first approximation using "amplitude-phase" variables,

$$\begin{aligned}
\dot r_{j,k} &= f(r_{j,k}) + d\,[r_{j-1,k}\cos(\varphi_{j-1,k}-\varphi_{j,k}) \\
&\quad + r_{j+1,k}\cos(\varphi_{j+1,k}-\varphi_{j,k}) + r_{j,k-1}\cos(\varphi_{j,k-1}-\varphi_{j,k}) \\
&\quad + r_{j,k+1}\cos(\varphi_{j,k+1}-\varphi_{j,k}) - 4r_{j,k}]\,, \\
r_{j,k}\dot\varphi_{j,k} &= d\,[r_{j-1,k}\sin(\varphi_{j-1,k}-\varphi_{j,k}) \\
&\quad + r_{j+1,k}\sin(\varphi_{j+1,k}-\varphi_{j,k}) + r_{j,k-1}\sin(\varphi_{j,k-1}-\varphi_{j,k}) \\
&\quad + r_{j,k+1}\sin(\varphi_{j,k+1}-\varphi_{j,k})]\,, \\
&\qquad j,k = 1,2,\ldots,N\,,
\end{aligned} \tag{5.94}$$

with $f(r) = -2ar^5 + ar^3 - r$. The quantities $r_{j,k}$ and $\varphi_{j,k}$ satisfy von Neumann's zero-flux boundary conditions. The system (5.94) is gradient as

$$\dot r_{j,k} = -\frac{\partial U}{\partial r_{j,k}}\,, \quad \dot\varphi_{j,k} = -\frac{\partial U}{r_{j,k}^2\partial\varphi_{j,k}}\,, \tag{5.95}$$

with the potential

$$\begin{aligned}
U = \sum_{j,k}\{G(r_{j,k}) + dr_{j,k}\,[2r_{j,k} - r_{j+1,k}\cos(\varphi_{j+1,k}-\varphi_{j,k}) \\
- r_{j,k+1}\cos(\varphi_{j,k+1}-\varphi_{j,k})]\}\,,
\end{aligned}$$

$$G(r) = \frac{a}{3}r^6 - \frac{a}{4}r^4 + \frac{1}{2}r^2\,.$$

Thus, at variance with the original system (5.93) now the attractors of (5.94) are *steady* states only.

Phase-Synchronization Modes. Let us restrict consideration to the steady states of (5.94) which for (5.93) correspond to phase-synchronization modes of the form

$$x_{j,k} = 2r_{j,k}\cos(t+\varphi^0) - O(\mu)\,, \quad y_{j,k} = -2r_{j,k}\sin(t+\varphi^0) + O(\mu)\,, \tag{5.96}$$

with $\varphi^0 = \text{const}$.

Existence of Phase-Synchronization Modes. From (5.94) it follows that the coordinates of steady states determining modes of phase synchronization satisfy the following equations:

$$\dot r_{j,k} = f(r_{j,k}) + d(\Delta r)_{j,k}\,, \tag{5.97}$$

$$\varphi_{j,k} = \varphi^0\,. \tag{5.98}$$

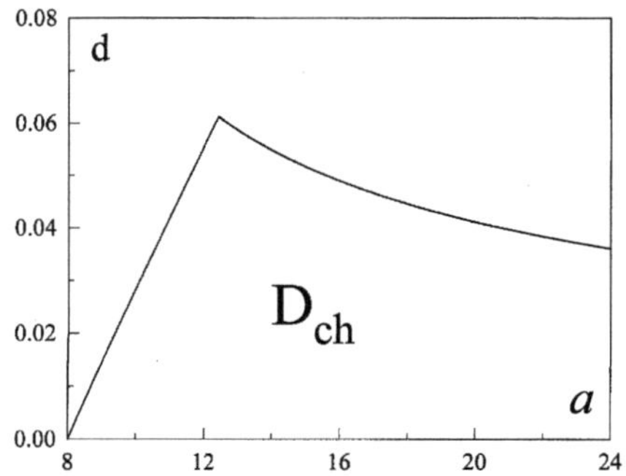

Fig. 5.35. The parameter region D_{ch} corresponding to the existence of spatial disorder in a lattice of coupled bistable oscillators

with $f(r)$ as illustrated in Fig. 5.14 in the domain $r > -r_{min}$ only, and

$$r_{min} = \sqrt{\frac{3a - \sqrt{9a^2 - 40a}}{20a}} .$$

However, as we are interested in the steady states of (5.97) with coordinates satisfying the inequality $r_{j,k} \geq 0$, the values of $f(r)$ for $r < -r_{min}$ are not relevant. Hence, the results proved in Sect. 5.3 can be applied to (5.97). Accordingly, 2^{N^2} steady states satisfying (5.86) exist for (5.97). As these steady states satisfy (5.97–98), they define phase-synchronization modes (5.96) of (5.93). Figure 5.35 shows the parameter domain D_{ch} where this occurs.

Stability of Phase-Synchronization Modes. An investigation of the stability of the spatially homogeneous and inhomogeneous steady states of (5.94) can be performed using Gershgorin's theorem once more. For all points in D_{ch}, stable phase-synchronization modes (5.96) correspond to steady states satisfying (5.86), and hence there are 2^{N^2} stable phase-synchronization modes.

In order to verify and make complete the theoretical predictions, a numerical integration of (5.94) has been performed. To consider phase-synchronization modes we take the initial amplitude distribution in the form shown in Fig. 5.36a. Initial phases $\varphi_{j,k}$ were arbitrary though close to $\varphi^0 = 0$ (Fig. 5.37a). After a short transient process, amplitudes form a stable pattern as shown in Fig. 5.36b. The onset of phase-synchronization takes quite a long time and follows two stages: (i) *Cluster formation* (here we define a cluster as a set of oscillators having $\varphi_{j,k} \approx \varphi^m$). The oscillators split into groups (clusters) where they are phase-locked (Fig. 5.37b,c). (ii) *Interaction of clusters.* The interaction among clusters leads to the phase-synchronization mode (Fig. 5.37d). These synchronous motions can display very complex amplitude distributions.

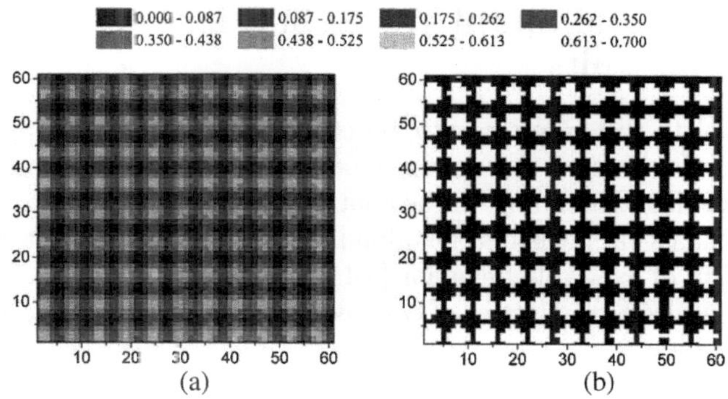

Fig. 5.36. Formation of a spatially inhomogeneous pattern of oscillation amplitudes, $a = 12.5, d = 0.05$. (**a**) Initial distribution; (**b**) final pattern

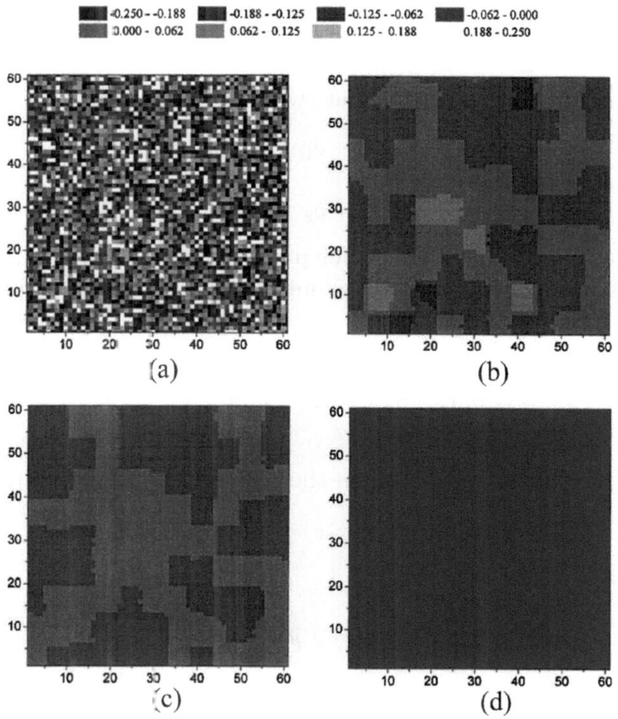

Fig. 5.37. Phase synchronization through *clustering*. $a = 12.5$, $d = 0.05$. (**a**) Initial distribution; (**b**) formation of clusters; (**c**) cluster interaction; (**d**) final distribution with phase-synchronization mode

5.7 Patterns and Spiral Waves in a Lattice of Excitable Units

Let us complete this chapter with the study of a lattice of coupled Chua's circuits (see also Sect. 4.1). As mentioned in Chap. 4, the evolution of such a unit is very rich, depending on the control parameters. Here we chose values such that Chua's circuit is in a bistable mode with two steady states and has the possibility of excitable behavior [5.11].

The dynamics of the lattice of coupled Chua's circuits is described by the following system:

$$\begin{cases} \dot{x}_{j,k} &= \alpha(y_{j,k} - x_{j,k} - f(x_{j,k})) + D(\Delta x)_{j,k}\,, \\ \dot{y}_{j,k} &= x_{j,k} - y_{j,k} + z_{j,k}\,, \\ \dot{z}_{j,k} &= -\beta y_{j,k} - \gamma z_{j,k}\,, \end{cases} \tag{5.99}$$

$$j,k = 1,2,\ \dots\ ,N\,.$$

where $f(x)$ is the three-segment, piecewise-linear function

$$f(x) = \begin{cases} b_1 x - a - b_1 & \text{if} \quad x \geq 1\,, \\ -a\ x & \text{if} \quad -1 < x < 1\,, \\ b_2 x + a + b_2 & \text{if} \quad x \leq -1\,, \end{cases} \tag{5.100}$$

with $a > 0$ and $b_1, b_2 > 0$. The other parameters of the system, α, β and γ, are also taken to be positive. We assume von Neumann (zero-flux) boundary conditions.

Let us take

$$\alpha = 2.5\,, \quad \beta = 0.5\,, \quad \gamma = 0.01\,, \quad a = 1.5\,, \quad b_1 = b_2 = b = 2\,. \tag{5.101}$$

Then there are three fixed points in the phase space of the unit:

$$O(0,0,0)\,, \quad P^{+}(x_0, y_0, z_0)\,, \quad P^{-}(-x_0, -y_0, -z_0)\,,$$

with

$$x_0 = \frac{(b+a)(\gamma+\beta)}{[\gamma b + \beta(b+1)]}\,, \quad y_0 = \frac{(b+a)\gamma}{[\gamma b + \beta(b+1)]}\,,$$

$$z_0 = -\frac{(b+a)\beta}{[\gamma b + \beta(b+1)]}\,.$$

For the chosen parameter values the "outer" points, P^+ and P^-, are stable foci, while the origin, O, is a saddle.

The system (5.100–101) can be considered as a discrete version of a spatially extended reaction–diffusion system. The "reaction kinetics" of such a medium is defined by the dynamics of the unit in the lattice. As (5.99) is a third-order ODE system, in a bistable mode for the chosen parameters, we have bistable "reaction kinetics".

5.7.1 Pattern Formation

To show the possibility of pattern formation in the lattice, we again follow the methodology developed in Sect. 5.3.

Existence of Steady Patterns. The steady patterns in the lattice correlate with the steady states of (5.99), which are the roots of the following system:

$$\begin{cases} -\dfrac{\alpha\beta}{\gamma+\beta}x_{j,k} - \alpha f(x_{j,k}) + D(\Delta x)_{j,k} = 0\,, \\ y_{j,k} = \dfrac{\gamma}{\gamma+\beta}x_{j,k}\,, \\ z_{j,k} = -\dfrac{\beta}{\gamma+\beta}x_{j,k}\,. \end{cases} \tag{5.102}$$

$$j = 1, 2, \ldots, N\,, \quad k = 1, 2, \ldots, N\,.$$

The coordinates $x_{j,k}$ of the roots of (5.102) also coincide with the coordinates of the steady states of the following auxiliary system:

$$\dot{x}_{j,k} = F(x_{j,k}) + D(\Delta x)_{j,k} = 0 \tag{5.103}$$

$$j = 1, 2, \ldots, N\,, \quad k = 1, 2, \ldots, N\,,$$

with

$$F(w) = -\frac{\alpha\beta}{\gamma+\beta}w - \alpha f(w)\,.$$

This system (5.103) is a discrete reaction–diffusion equation of FNS type (5.78). Thus (5.103) has 2^{N^2} stable steady states when the parameters of the system are located in the domain D_{ch} defined by

$$D < D^* = \min \begin{cases} \dfrac{\alpha[a(\gamma+\beta)+\beta][\gamma b+\beta(b+1)]}{4(\gamma+\beta)[\beta+(\gamma+\beta)(a+2b)]}\,, \\ \dfrac{\alpha[a(\gamma+\beta)-\beta][\gamma b+\beta(b+1)]}{4(\gamma+\beta)[\beta+(\gamma+\beta)(a+2b)]}\,. \end{cases} \tag{5.104}$$

For the parameter set (5.101) the critical value of the diffusion coefficient is $D^* \approx 0.15$. Each of these states defines a steady pattern in the $\{\mathbf{Z^2}, \mathbf{R}\}$ state space. Moreover, as earlier, every pattern can be coded by an $N \times N$ matrix of two symbols (for example, 0 and 1) and any arbitrary $N \times N$ matrix defines a possible spatial configuration of the pattern. Thus, again there exists a wealth of possible steady spatial patterns from simple homogeneous, periodic, regular to disordered and spatially chaotic. Low enough diffusion (5.104) implies the bistable character in the spatial distribution of the species (state variables) in the lattice. The coordinates $(x^*_{j,k}, y^*_{j,k}, z^*_{j,k})$ of the steady states are located in neighborhoods ("absorbing domains") of the fixed points $P^+(x_0, y_0, z_0)$ and $P^-(-x_0, -y_0, -z_0)$ of the unit at each site.

Needless to say, to be actually realized, these patterns must be at least locally asymptotically stable solutions of (5.99).

Stability of the Steady States. Let

$$\mathbf{x} = \{x^*_{j,k}\}, \quad \mathbf{y} = \{y^*_{j,k}\}, \quad \mathbf{z} = \{z^*_{j,k}\}$$

be one of the steady states of (5.99). Its coordinates satisfy the conditions $|x^*_{j,k}| \geq 1, \forall (j,k)$. Then the stability of this solution is defined by

$$\begin{cases} \dot{\xi}_{j,k} = \sigma_{j,k}\xi_{j,k} + D(x_{j-1,k} + x_{j+1,k} \\ \qquad\quad + x_{j,k-1} + x_{j,k+1}) + \alpha\eta_{j,k} , \\ \dot{\eta}_{j,k} = \xi_{j,k} - \eta_{j,k} + \nu_{j,k} , \\ \dot{\nu}_{j,k} = -\beta\eta_{j,k} - \gamma\nu_{j,k} , \end{cases} \tag{5.105}$$

with $\sigma_{j,k} = -\alpha[1 + f'(x^*_{j,k})] - 4D$. Due to symmetry, and to the piecewise linearity of f (5.100), $\sigma_{j,k} \equiv \sigma = -\alpha(1+b) - 4D$ for all steady states. Hence, there appears a drastic simplification in our study. The local stability of any

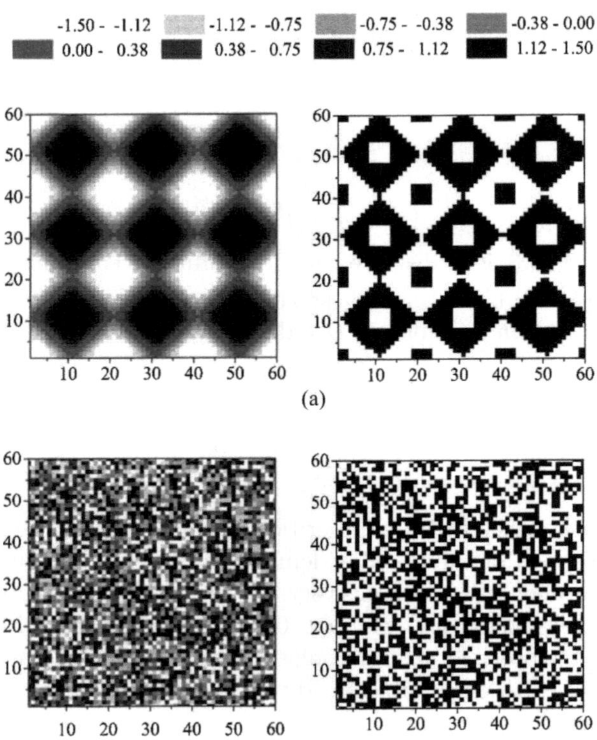

Fig. 5.38. Pattern formation in a lattice for a weak enough intralattice diffusion coefficient, $D = 0.06$. The grey scale represents the distribution of the state variable x in the lattice. Left: initial distribution of x; right: stable steady pattern. (**a**) Formation of a regular pattern; (**b**) formation of a spatially disordered (chaotic) steady pattern. The pictures on the left are the initial conditions

spatially inhomogeneous steady pattern is guaranteed by the stability of the homogeneous steady patterns corresponding to the fixed points P^+ and P^- of the unit. Solving the eigenvalue problem for the linear system (5.105), we find that for the chosen parameter set these homogeneous states are locally asymptotically stable.

Thus, for weak enough intralattice diffusion D the lattice (5.105), with the chosen "bistable kinetics", exhibits a huge number of stable steady patterns. The bistable character of the "species" distribution along the lattice yields any 2D black ("0") and white ("1") picture as a steady pattern. It is only necessary to choose an appropriate initial condition.

Figure 5.38a (right-hand panel) illustrates the regular, "spot" form of a pattern appearing in the lattice for the initial conditions (left-hand panel):

$$\begin{aligned} x_{j,k} &= x_0\left(\cos\tfrac{6\pi j}{N} + \cos\tfrac{6\pi k}{N}\right), \\ y_{j,k} &= \tfrac{\gamma}{\gamma+\beta}x_{j,k}, \\ z_{j,k} &= -\tfrac{\beta}{\gamma+\beta}x_{j,k}, \end{aligned} \tag{5.106}$$

$$j,k = 1,2,\ \dots\ ,N = 60\,.$$

An example of disordered or chaotic pattern is given in Fig. 5.38b. The initial conditions have been chosen with a random distribution (associated with a pseudorandom sequence obtained with the computer) of x (Fig. 5.38b, left-hand panel), while y and z are calculated from x similarly to (5.106).

5.7.2 Spiral Wave Patterns

Let us now consider wave solutions of (5.99) with, however, a restriction to spiral waves, for which, due to a high enough intralattice diffusion D, there is now pattern formation in the lattice.

"Dark" and "Bright" Spiral Waves and Excitability. An array (chain) of Chua's circuits in the described bistable mode supports a variety of stationary pulses and pulse trains including complex or chaotic profiles. Very much as in optical fibers there are "dark" and "bright" pulses propagating along the "background" homogeneous states P^+ and P^-, respectively. The solution $\{x_j(t), y_j(t), z_j(t)\}$ corresponding to a pulse in the chain defines a plane-wave solution $\{x_{j,k} \equiv x_j(t), y_{j,k} \equiv y_j(t), z_{j,k} \equiv z_j(t)\}$ in the 2D lattice.

Let us take the plane wave corresponding to a single "bright" pulse of the 1D chain and break this front at some instant of time (Fig. 5.39a). The edge of the front starts to twist and after some time forms a stationary spiral wave of bright type (see sequence in Fig. 5.39). Needless to say, there can be dark spirals having evolved from the dark plane pulse. Figure 5.40 illustrates a fully developed spiral rotating around its core. As in other excitable media the core consists of unexcited cells, while the other cells exhibit time-periodic, phase-shifted pulses as shown in Fig. 5.40b,c. Note that the lattice supports spiral

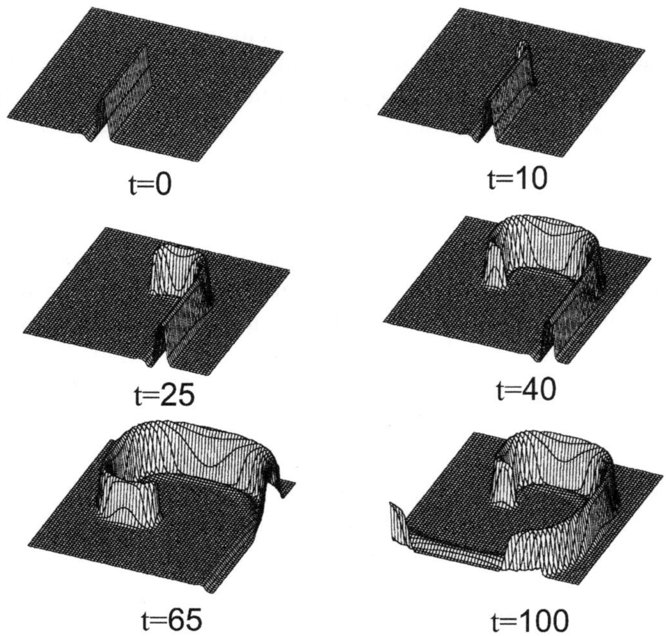

Fig. 5.39. Evolution of a "bright" excitable spiral wave following the rupture of the plane wave front. Intralattice diffusion $D = 0.8$

waves only when the diffusion D is strong enough. Decreasing the diffusion the spiral waves yield to a steady pattern as earlier shown (5.104).

Thus, for the parameter set (5.101), which provides the bistable mode of Chua's circuit, and for strong enough intralattice diffusion, the lattice is indeed a discrete reaction–diffusion medium with two excitable states. It follows from the symmetry of f (5.101) that the two states have equivalent properties allowing in the lattice dark and bright excitable spirals for the same parameter values. Note that we have considered only the simplest, basic types of spiral waves. More complex wave patterns can also be excited, including those obtained from the multihump 1D pulse trains and "multi-armed" spiral wave solutions.

Oscillatory Spiral Waves. When a unit exhibits a limit cycle, the lattice behaves like an oscillatory medium. Periodic trains in the 1D case (chain) and target patterns and oscillatory spirals in a 2D system are examples of the typical processes in such a medium. The spirals in an oscillatory medium look rather different to those found in excitable systems. In an oscillatory medium at each space point (lattice site) the system oscillates around the same limit cycle of the reaction kinetics, and the diffusion provides a definite global phase coherence of the local oscillations, hence the spiral wave. Our "reaction kinetics" does not have a limit cycle, as it only operates with two stable fixed points. However, let us show that the lattice (5.99) with (5.100) and (5.101)

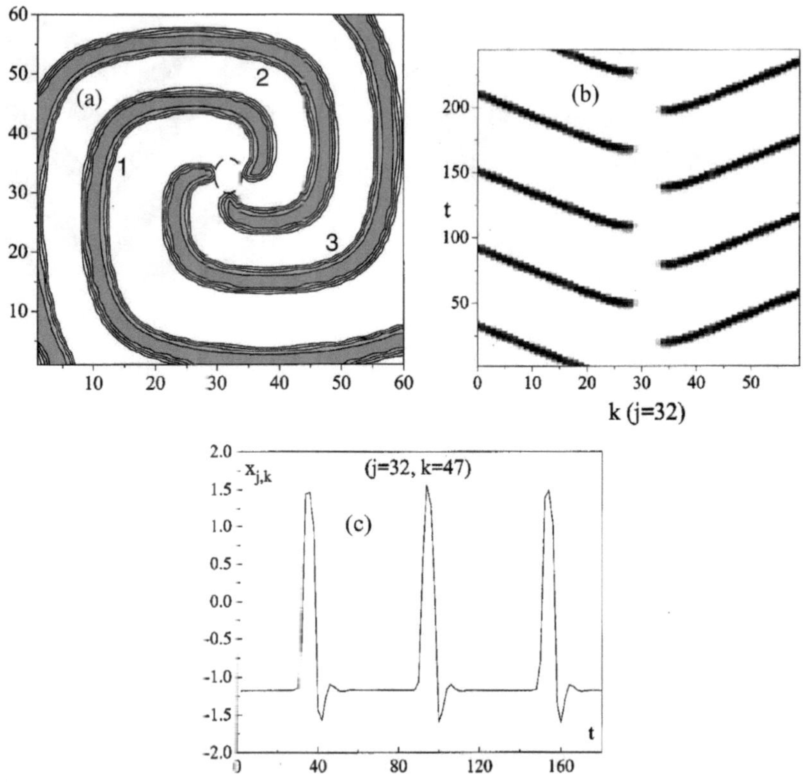

Fig. 5.40. Fully developed excitable spiral wave for $D = 0.8$. (**a**) 1, 2 and 3 are snapshots (x variable) of the wave rotating around its core. (**b**) Space–time diagram of the section of the lattice along $j = 32$. (**c**) Oscillations of a unit far away from the spiral core

supports spiral waves which behave like the waves of an oscillatory medium. Figure 5.41 illustrates the fully developed "oscillatory" spiral obtained in the lattice for the same "kinetics" and diffusion coefficient as for the excitable waves described in the previous subsection. The behavior of the spiral core and phase-shift relations between the other cells are illustrated in Fig. 5.41b,c. A cell near the core (Fig. 5.41c) oscillates (dashed curve) with the same period and a slightly smaller amplitude than the other cells (solid curve). This type of spiral wave can be obtained, for example, when there is wave re-entry in the two-layer system, as we show in Chap. 6.

What mechanism underlies the "oscillatory" spiral wave in a medium which has nonoscillatory reaction kinetics? Let us go deeper into the dynamics of the third-order Chua's excitable unit.

Metastable Periodic Behavior of the Excitable Unit as a Possible Origin of the Oscillatory Properties of the Medium. Possible oscillatory motions in the third-order phase space of the unit can be described by

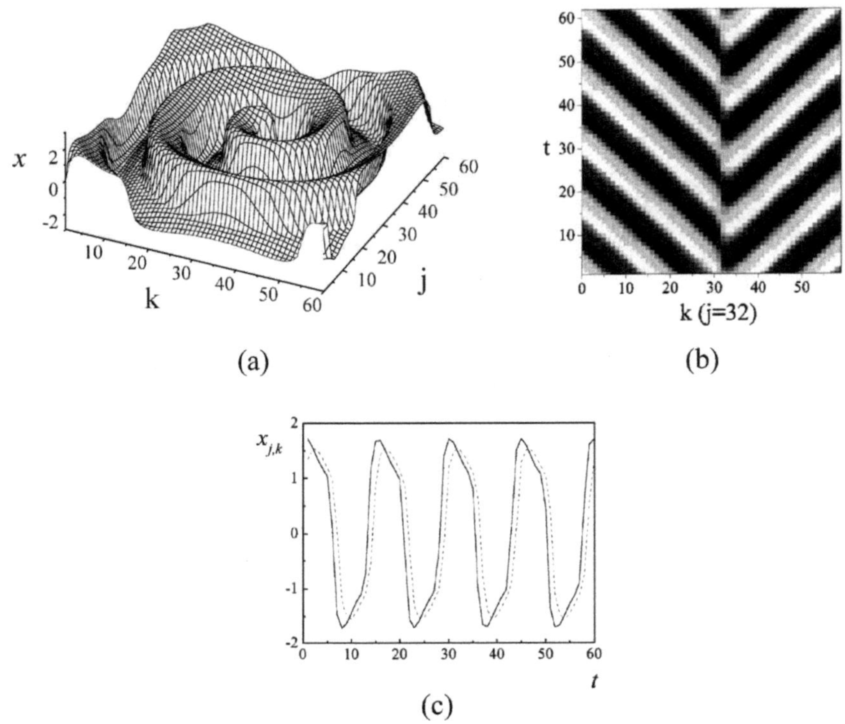

Fig. 5.41. Fully developed "oscillatory" spiral wave for $D = 0.8$. (**a**) A snapshot (x variable) of the wave. (**b**) Space–time diagram of the section of the lattice along $j = 32$. (**c**) Oscillations of a unit near the spiral core (*dashed curve*) relative to the oscillations of the other units (*solid curve*)

a 1D Poincaré return map. The behavior of the map for different values of b_2, taken as a control parameter from (5.101), is shown in Fig. 5.42a. A saddle-node bifurcation occurs in the system for a certain value of b_2. The map in case 1 has a stable fixed point, hence the unit exhibits limit-cycle behavior. Figure 5.42b shows its time evolution. Then, the lattice in this case actually represents an oscillatory medium. Further increasing the values of b_2, the fixed point disappears and the map is described by curve 3 (Fig. 5.42a). Then the trajectories of the system can stay for a rather long time near the site of the limit cycle earlier referred to and then decay to one of the two stable steady points (Fig. 5.42c). Thus, the intralattice diffusion acting as a "rigid" mechanism profits from such a long-lasting (metastable) local quasi-oscillatory stage to provide "global coherence" to the units, hence the self-sustained oscillatory spiral wave. How long need the metastable oscillations last (how far can the map stand from the bifurcation point; Fig. 5.42a) is defined by the ratio between the characteristic time scale of the local oscillations, τ_L (Fig. 5.42c), and that of the diffusion, τ_D. τ_D can be estimated, for example, from the velocity, c, of plane wave fronts propagating for a given diffusion

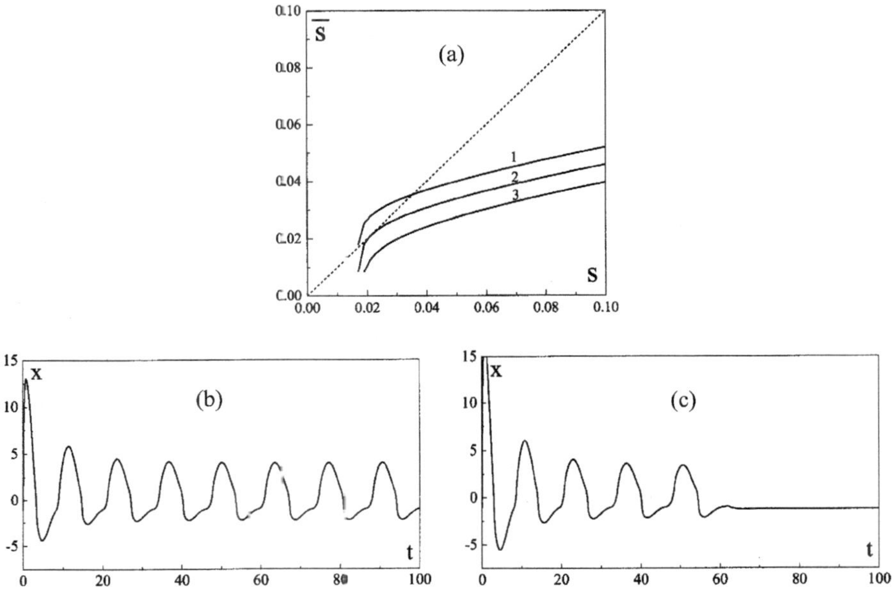

Fig. 5.42. (**a**) Poincaré (1D) return map characterizing oscillatory behavior of the unit near the saddle-node bifurcation. Curve 1: $b_2 = 0.1$; curve 2: $b_2 = 0.148$; curve 3: $b_2 = 0.2$. (**b**) Time evolution of the limit cycle corresponding to the fixed point of map 1. (**c**) Typical periodic-like time evolution for the map 3 displaying "long"-lasting metastable periodic-like behavior

coefficient D. For the parameter values of Fig. 5.41, we have $\tau_D \sim 1$. Then, as $\tau_D \ll \tau_L$, diffusion can create the oscillatory spiral wave as a kind of coherent (pseudo) oscillation of the units without them being in a true limit cycle.

Note that for the chosen reaction kinetics (5.101) ($b_2 = b = 2$) the map stands very far from the bifurcation point, and hence the oscillations of the unit decay rather quickly. However, since the characteristic duration of the periodic-like oscillations is rather long ($T \sim 15$ from Fig. 5.42b,c) even a few periods suffice for $\tau_D \ll \tau_L$, thus helping to create and sustain an oscillatory spiral. We have also observed this behavior with numerical integration (Fig. 5.41).

5.8 Salient Features of Networks of Bistable Units

The study of collective behavior, pattern and wave dynamics, in networks of coupled oscillators is of interest both from the nonlinear dynamics view point and for its potential for applications in arrays of coupled lasers, neural systems, arrays of Josephson junctions, reaction–diffusion systems, synchronization networks, networks of electronic oscillators, etc. Frequently, such networks possess regular structure and, in the simplest case, can be described by a p-dimensional lattice ($p = 1, 2, 3 \ldots$) characterized by coordinates $j \in \mathbf{Z}^p$. The elements or units of the network are located in the nodes or sites of a lattice. If the number of oscillators is large enough, then the coordinates j can be considered as points in space and the network can be interpreted as a nonequilibrium discrete medium.

Multistability (i.e. coexistence of two or more stable states) is a salient property of nonlinear systems. It is clear that when such systems are combined into a single system the property of multistability influences essentially not only its time evolution but also its spatial behavior.

1D chains with ring and open chain geometry can be treated as a nonequilibrium medium, since they possess the typical properties of such media, such as multistability, spatial disorder, space-homogeneous waves, etc. However, we would like to stress the fact that the chain has specific properties, such as clustering and space-inhomogeneous waves. The existence of these new properties comes, on the one hand, from the discreteness of the space coordinate, and on the other hand, from the bistability of the oscillators. We note that these properties are lost in the continuum analog of (5.1), i.e. in the PDE obtained by the corresponding limiting procedure.

The role of bistability is to allow competition between synchronous yet different motions. In general competition between one and another of those regimes is possible. To investigate this phenomenon we have made some numerical computations, as reported in Sects. 5.4.2–3. For instance, we have chosen initial conditions close to a space-homogeneous wave with wave number $m = 9$. After a transient process, we quasisteadily increased the coupling coefficient, d. The wave with $m = 9$ survives until $d \in H_9^{st}$. It agrees with the result of the linear stability analysis (Fig. 5.25). Outside H_9^{st} the wave loses stability through defect formation, as described in Sect. 5.4.2. It results in a wave with $m = 8$. Similar processes occur for the other waves. Finally for $d > H_1^{st}$ the space-homogeneous regime with $m = 0$ is realized.

We have investigated the processes of formation of phase and frequency clusters which may be relevant for understanding basic features of synchronization of oscillations in neurodynamics. Coexistence of chaos and order is fundamental in the dynamics of a chain of many locally, diffusively coupled bistable nonisochronous oscillators. Indeed, if the amplitude distribution of oscillations along the spatial coordinate j is chaotic, then such a distribution can be the *background* for clustering. Even though spatially they have chaotic amplitudes, the oscillators take on regular temporal patterns: *phase* and *frequency clusters*. Suitably choosing the initial conditions for amplitudes of oscillations from certain domains, V_j^0 or V_j^1 (5.18), we can prescribe a given final distribution of r_j, which in turn determines spatial areas of synchronous oscillations.

In Addition, we have found that the competition between frequency clusters is the basis for the phenomenon of *phase resetting*. It appears when the following two conditions are fulfilled:

(i) There are two (or more) frequency clusters in the chain.
(ii) A single oscillator appears isolated between them. Cluster competition yields jumps of π in the phase of the isolated oscillator.

The model studied in Sect. 5.3 is of potential interest in understanding the dynamics of large neural oscillatory networks or lattices. Such systems are characterized by quite complex intralattice connections. Our global coupling theory may prove valuable in mimicking real behavior. Furthermore, the bistability of units captures an ingredient of the bistability of neurons with coexistence of the state of rest or subthreshold oscillations and excited states. We have shown that this property of units leads to interesting new features in the collective behavior of the globally coupled network:

(i) Amplitude-phase clusters can form in the system (5.23). Then all neurons (here mimicked by bistable oscillators) break into two groups. The first group consists of "strongly" excited neurons having rather "high" oscillation amplitude, hence spiking. The second group is composed of "weakly" excited neurons with a "low" oscillation amplitude, hence subthreshold oscillators. Furthermore, neurons taken from different groups oscillate with a constant phase shift.
(ii) The system (5.23) can operate in a mode such that some of the neurons exhibit chaotic oscillations, while the others oscillate regularly, hence regular and chaotic oscillations form a "linked" state.
(iii) The system (5.23) can have two types of multistability. In the first case, collective chaos and regular dynamics (splay-phase states and the trivial state) "compete" with each other. In the second case, the competition occurs between the amplitude-phase clusters, splay-phase states and the trivial state. Note that, although a "winner" in this competition is one of the large number of such splay-phase states, there is a difference in the behavior of the assembly (5.23) in the two cases. In the first case, some of the neurons (bistable oscillators) are excited (splay-phase state) and the others part are at rest. In the second case, *all* the neurons are either excited or at rest.

We have studied the spatio-temporal dynamics of a rather general lattice system composed of N^2 coupled bistable elements. Such a system is the discrete version of a 2D reaction–diffusion equation. The analysis is based on the construction of the invariant domains in its phase space. Conditions for multistability in the form of an explicit inequality for the intralattice coupling coefficient, d, have been obtained. We have shown that, depending on parameter values and on the initial conditions, the system can display very diverse patterns determined by a matrix of two symbols given beforehand. The existence of a large number, 2^{N^2}, of attractors, among which there are the states chaotically distributed in space, provides evidence for spatial disorder in a medium modeled by a discrete 2D reaction–diffusion equation.

To illustrate the general results found we have considered three lattice systems: one with the FitzHugh–Nagumo–Schlögl model as unit, a network of weakly coupled bistable oscillators and a network of coupled Chua's circuits (all in the bistable-excitable mode). In the case of a lattice of oscillators we have given evidence for the formation of phase clusters with phase-locking among elements.

Finally, we have shown that a 2D lattice of coupled Chua's circuits with a rather strong intralattice diffusion coefficient may support "excitable" spiral waves of "dark" and "bright" types. They naturally originate from the excitable single pulses. Although the units in the lattice do not possess limit-cycle behavior, as they have long-lasting, metastable periodic-like stages the lattice can also support "oscillating" spiral waves like spiral waves in true oscillatory media.

To conclude let us recall the studies of the dynamics of chain systems composed of bistable oscillators by Defontaines, Pomeau and Rostand [5.2], by McNell [5.9] and by MacKay and Sepulchre [5.8] which are related to the contents of our chapter. To the readers interested in the phenomena of frequency cluster formation, i.e. frequency and phase synchronization, we recommend the review by Pikovsky, Rosenblum and Kurths. They discuss the problem of phase-synchronization of chaotic systems. A wealth of material on the dynamics of systems with global coupling can be found in the book by Tass [5.21]. It also contains an extensive bibliography on the dynamics of networks of oscillators with various types of coupling. For some basic notions and additional information, we recommend the books by Winfree [5.23], Zhabotinsky [5.24] and Zykov [5.25].

6. Mutual Synchronization, Control and Replication of Patterns and Waves in Coupled Lattices Composed of Bistable Units

6.1 Introduction and Motivation

The content of this chapter is closely linked with the theme treated in Chap. 5. Here we also discuss active lattice systems composed of elements with bistable properties. However, the geometrical architecture of the systems to be used is more sophisticated. We consider here a multilayer lattice architecture, with interacting lattices or layers and hence a 3D geometry or anatomy. Many systems from various areas of science and technology have such multilayer 3D structure. Take, for instance Josephson superlattices consisting of many stacked tunnel junctions, artificial reaction–diffusion dynamical systems using molecular electronics technology, neural networks, layered porous media, Cellular Neural Networks and so on [6.1–5,6.10–16,6.22–24,6.26,6.27]. In spite of significant differences between all these cases we may expect some common hence universal properties associated with the multilayer geometry. The purpose of this chapter is to identify and describe such properties.

For a lattice system, i.e. a structure composed of two or more interacting 2D lattice layers, we shall use the following model equations:

$$\dot{\mathbf{A}}_{\mathbf{p}}^{(l)} = \mathbf{F}(\mathbf{A}_{\mathbf{p}}^{(l)}) + \mathbf{D}^{(l)}(\Delta\mathbf{A})_{\mathbf{p}}^{(l)} + \mathbf{H}^{(l)}\left(\mathbf{A}_{\mathbf{p}}^{(l-1)} - 2\mathbf{A}_{\mathbf{p}}^{(l)} + \mathbf{A}_{\mathbf{p}}^{(l+1)}\right), \quad (6.1)$$

where $\mathbf{A}_{\mathbf{p}}^{(l)}$ is the state vector (real or complex) in the lth layer, $\mathbf{p}$ is a vector fixing the position or site occupied by a unit in each layer, l is the index of the layer ($l = 1, 2, \ldots, M$), $\mathbf{D}^{(l)}$ is a constant square matrix with positive elements characterizing the intralattice diffusion, $\mathbf{H}^{(l)}$ is a constant square matrix characterizing the strength of interlattice couplings, $\mathbf{F}$ is a nonlinear vector function describing the nonlinear dynamics of the units, e.g. $\mathbf{F}$ provides the bistability (the dimensions of $\mathbf{D}^{(l)}, \mathbf{H}^{(l)}$ and $\mathbf{F}$ coincide with the dimension of the vector $\mathbf{A}_{\mathbf{p}}^{(l)}$). For each layer, $(\Delta\mathbf{A})_{\mathbf{p}}^{(l)}$ describes a discrete Laplace operator which for $\mathbf{p} = \{j = 1, 2, \ldots, N\}$ (1D lattice or chain) and for $\mathbf{p} = \{j, k = 1, 2, \ldots, N\}$ (2D square lattice) are, respectively,

$$(\Delta\mathbf{A})_j = \mathbf{A}_{j+1} + \mathbf{A}_{j-1} - 2\mathbf{A}_j ,$$
$$(\Delta\mathbf{A})_{j,k} = \mathbf{A}_{j+1,k} + \mathbf{A}_{j-1,k} + \mathbf{A}_{j,k+1} + \mathbf{A}_{j,k-1} - 4\mathbf{A}_{j,k} .$$

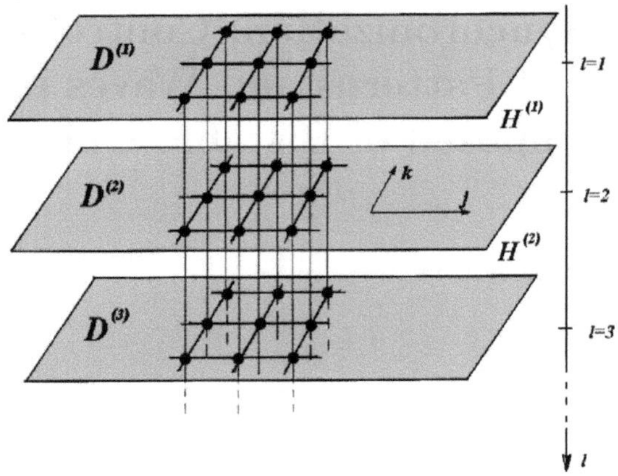

Fig. 6.1. Schematic of a multilayered architecture

We take von Neumann boundary conditions both for the boundary elements of a layer ($j, k = 1, N$) and for the elements of the outer layers ($l = 1, M$). Figure 6.1 illustrates the architecture of such layered lattice system.

6.2 Layered Lattice System and Mutual Synchronization of Two Lattices

6.2.1 Bistable Elements or Units

Consider a two-layer system composed of 2D square lattices [6.18–20]. Let us take a unit with two stable fixed points (the basins of attraction of these points are separated by an unstable fixed point). Thus we have units with 1D phase space. In (6.1) we put $l = 1, 2;\;\; p = \{j, k = 1, 2, \dots, N\}$, $A_p^{(l)}$ is a scalar variable and F is a scalar function that, for illustration, we take to be of the following form $F(x) = x(1-x)(x-a)$, $0 < a < 1$. Consequently, (6.1) reduces to

$$\begin{cases} \dot{u}_{j,k} = f(u_{j,k}) + d(\Delta u)_{j,k} - h_{j,k}(u_{j,k} - v_{j,k})\,, \\ \dot{v}_{j,k} = f(v_{j,k}) + d(\Delta v)_{j,k} - h_{j,k}(v_{j,k} - u_{j,k})\,, \end{cases} \tag{6.2}$$

with

$$f(w) = w(w-1)(a-w)\,, \quad 0 < a < 1\,;$$

$h_{j,k}$ and d are the corresponding coupling coefficients. The pair (j, k) defines the coordinates of a site in the lattice ($j, k = 1, 2, \dots, N$), and $(\Delta w)_{j,k}$ is the discrete Laplace operator (see Sect. 5.3.1). The boundary conditions are

$$
\begin{aligned}
&w_{0,k} = w_{1,k}\,, \quad w_{j,0} = w_{j,1}\,, \\
&w_{j,N+1} = w_{j,N}\,, \quad w_{N+1,k} = w_{N,k}\,.
\end{aligned}
\tag{6.3}
$$

If $h_{j,k} = 0$, (6.2) splits into two independent identical lattice systems. We denote each of these subsystems or layers by L_0. L_0 describes a network of interacting bistable units with two stable steady states which are the lowest (low state) and the highest (upper state) values of w in $f(w) = 0$. As shown in Chap. 5, L_0 is gradient for the parameters taken in the region D_{ch} (5.47). Moreover, it has 2^{N^2} attractors (steady states). As also shown, this set of attractors can be coded by arbitrary $N \times N$ matrices consisting of two symbols (for instance, "0" and "1"). Thus L_0 may display rather arbitrary spatially regular or chaotic states, albeit steady in time (Fig. 5.21). Let us now see the results of switching on the interlattice coupling, $h_{j,k}$.

Gradient Property of the System. As (6.2) can be rewritten as

$$
\dot{u}_{j,k} = -\frac{\partial U}{\partial u_{j,k}}\,, \quad \dot{v}_{j,k} = -\frac{\partial U}{\partial v_{j,k}}\,,
$$

it derives from the potential

$$
\begin{aligned}
U = \sum_{j,k=1}^{N} \Bigg(&\frac{d}{2}\left[(\nabla u_{j,k})^2 + (\nabla v_{j,k})^2\right] - \int_0^{u_{j,k}} f(\eta)\,\mathrm{d}\eta \\
&- \int_0^{v_{j,k}} f(\eta)\,\mathrm{d}\eta + \frac{h_{j,k}}{2}(u_{j,k} - v_{j,k})^2 \Bigg)\,,
\end{aligned}
$$

with

$$
(\nabla w_{j,k})^2 \equiv (w_{j+1,k} - w_{j,k})^2 + (w_{j,k+1} - w_{j,k})^2\,.
$$

Thus only steady states exist in the phase space, and accordingly, any initial conditions tend to one of the stable steady states corresponding to a local minimum of U. Generally, we restrict consideration to positive values of $h_{j,k}$.

Stable Manifold of Synchronous States. For convenience, let us change variables:

$$
x_{j,k} = u_{j,k} - v_{j,k}\,, \quad y_{j,k} = u_{j,k} + v_{j,k}\,.
$$

Then, (6.2) becomes

$$
\begin{aligned}
\dot{x}_{j,k} &= d(\Delta x)_j - (4d + 2h_{j,k} + a)x_{j,k} + d(\Delta x)_k \\
&\quad - \frac{x_{j,k}}{4}(x_{j,k}^2 + 3y_{j,k}^2) + (1+a)x_{j,k}y_{j,k}\,,
\end{aligned}
\tag{6.4}
$$

$$
\begin{aligned}
\dot{y}_{j,k} &= d(\Delta y)_j - (4d + a)y_{j,k} + d(\Delta y)_k \\
&\quad - \frac{y_{j,k}}{4}(3x_{j,k}^2 + y_{j,k}^2) + \frac{(1+a)}{2}(x_{j,k}^2 + y_{j,k}^2)\,.
\end{aligned}
\tag{6.5}
$$

It follows from (6.4) that a synchronization manifold defined by $M = \{x_{j,k} = 0,\ j,k = 1,2,\ \dots\ ,N\}$ exists in the phase space of (6.4–5). Let us show that the manifold M is globally asymptotically stable. To do this we consider the auxiliary Lyapunov function

$$V = \sum_{j,k=1}^{N} \frac{x_{j,k}^2}{2},$$

whose derivative with respect to (6.4–5) is

$$\dot{V} = -\sum_{j,k=1}^{N} (P_{j,k} + Q_{j,k}), \tag{6.6}$$

with

$$\begin{aligned} P_{j,k} \equiv &- d(x_{j,k}\,x_{j-1,k} + x_{j,k}\,x_{j,k-1}) + p_{j,k}x_{j,k}^2 \\ &- d(x_{j,k}\,x_{j+1,k} + x_{j,k}\,x_{j,k+1}), \\ Q_{j,k} \equiv x_{j,k}^2 &\left(\frac{1}{4}(x_{j,k}^2 + 3y_{j,k}^2) - (1+a)y_{j,k} + \frac{(1+a)^2}{3}\right), \end{aligned}$$

where

$$p_{j,k} \equiv 4d + 2h_{j,k} - \frac{(1-a+a^2)}{3}.$$

Then all $Q_{j,k}$ are positive definite. Let us prove that the function $P = \sum_{j,k=1}^{N} P_{j,k}$ is also positive definite. Consider the vector $\mathbf{z} = (z_1, z_2, \dots, z_{N^2})$, where $z_1 = x_{11}, z_2 = x_{12}, \dots, z_{N^2} = x_{NN}$. Using the components of $\mathbf{z}$, the function P takes the quadratic form

$$P = \sum_{i,j=1}^{N^2} a_{ij} z_i z_j \equiv \mathbf{z}^T A\mathbf{z},$$

where $a_{ij} = a_{ji}$, the superscript T denotes the transpose, and $A = \|a_{ij}\|$ is a square symmetric, $N^2 \times N^2$ matrix. The quadratic form P will be positive definite if the eigenvalues of the symmetric matrix A are positive. By applying Gershgorin's theorem to A, we find that if $p_{j,k} > 4d$ then the union of the Gershgorin disks, corresponding to A, is located to the right of the imaginary axis (see Appendix G). Therefore, the inequality

$$h_{j,k} > \frac{1-a+a^2}{6}, \quad \text{for all } j,k, \tag{6.7}$$

ensures the positiveness of all eigenvalues of A, and, consequently, P has a positive definite quadratic form. Then, using (6.6) we find that outside the manifold M the inequality $\dot{V} < 0$ is satisfied, while $\dot{V} = 0$ on the manifold; hence the synchronization manifold M is globally asymptotically stable. It follows from (6.5) that motions on M are governed by L_0, i.e. by only one of the two lattices.

Thus, when (6.7) are satisfied, (6.2) consisting of two interacting lattice systems shows mutual synchronization for all stationary states. Since for the parameter values taken from the region D_{ch} the set of stationary states contains spatially chaotic ones, i.e. steady states with an irregular structure along the spatial coordinates (Chap. 5), then there is also synchronization of spatially chaotic attractors for (6.5). We have the synchronization of two coupled spatially disordered, though time-independent, systems. To be sure that the terminal patterns are actually synchronized, we follow the evolution of the quantity

$$\mathrm{dist}\big(\mathbf{u}(t), \mathbf{v}(t)\big) = \frac{\sqrt{\sum_{j,k=1}^{N} (u_{j,k} - v_{j,k})^2}}{N} \tag{6.8}$$

which characterizes the distance between two patterns in the N^2 state space. When (6.8) vanishes we have the synchronization of the patterns.

Examples of Mutual Synchronization of Patterns. Let us see a few examples of the mutual synchronization just described. They have been obtained by numerical integration of (6.2) for different initial conditions. We consider (i) two competing different steady in time but spatially irregular patterns; (ii) steady patterns with different spatial regular structure; (iii) regular and irregular steady patterns; and (iv) interaction of regular patterns when some of the interlattice bonds, $h_{j,k}$, are broken or ill-functioning, hence not satisfying (6.7). Needless to say, the initial patterns are stable steady states of each L_0 taken separately. A fourth-order Runge–Kutta integration routine was used.

(i) The interaction of lattices with two steady in time but spatially irregular initial distributions shown in Fig. 6.2a,b leads to the formation of two synchronized patterns which are also steady in time and irregular in space (Fig. 6.2c,d). The initial condition is a random space distribution of lower and upper steady states. Then we switch on the interlattice interaction, which competes with the intralattice diffusion, and, as time proceeds, we end up with two seemingly identical copies of the same spatially chaotic state, like two clones. The action of the interlattice coupling appears dominant relative to the intralattice diffusion process. When (6.7) is fulfilled, in a first stage the two lattices move toward the synchronization manifold, then the intralattice diffusion controls the subsequent stage of the time evolution until the terminal state is reached. There is a spectrum of grey colors between black (lower state) and white (upper state) according to the values of $u_{j,k}$ and $v_{j,k}$ in the lattice sites. But there is no need to make this scale explicit.

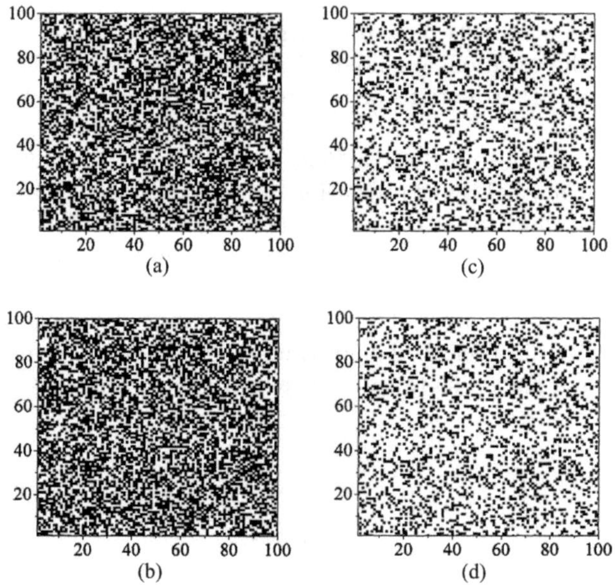

Fig. 6.2. Synchronization of steady irregular (spatially chaotic) patterns in two interacting lattices with 100×100 elements. Parameter values: $a = 0.4$, $d = 0.006$, $h = 0.2$. (**a–b**) Initial distributions, both steady in time but spatially random and different. (**c–d**) Synchronized patterns, both again steady in time, spatially random and identical. Although the initial conditions are steady states of each lattice, taken separately, evolution in time with intralattice diffusion ($d \neq 0$) occurs when we switch on the interlattice coupling ($h \neq 0$). The numerical simulation proceeds until we again reach steady states simultaneously in both lattices which can be considered as identical terminal patterns

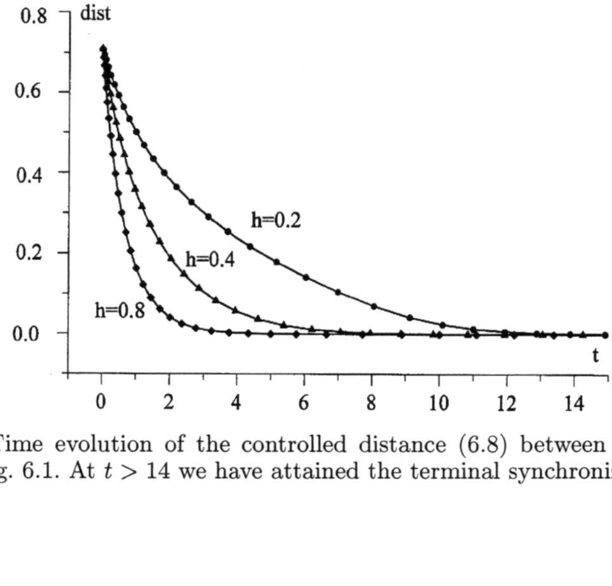

Fig. 6.3. Time evolution of the controlled distance (6.8) between the patterns shown in Fig. 6.1. At $t > 14$ we have attained the terminal synchronized states

Figure 6.3 illustrates the time evolution of the function for the patterns shown in Fig. 6.2 for different values of h; here we have considered all $h_{j,k} = h$. The distance (6.8) is a monotonically decreasing function in time and vanishes for long times ($t \to \infty$).

(ii) Another example is the synchronization originating in the competition of two different patterns of "stripe" form. Starting from the initial distributions shown in Fig. 6.4a,b, the two lattices synchronize and display identical patterns, as shown in Fig. 6.4c,d. The terminal states have features from both initial patterns, although we have no "natural selection", i.e. there is no winning pattern. Here again we have taken all $h_{j,k} = h$.

(iii) Let us illustrate the interaction of a regular and a spatially chaotic pattern, as shown in Fig. 6.5a,b. When all $h_{j,k} = h$, for h satisfying the inequality (6.7) the lattices synchronize and produce the identical patterns shown in Fig. 6.5c,d. The terminal patterns can be interpreted as a kind of "modulation" given by the regular pattern of Fig. 6.5a on the chaotic one of Fig. 6.5b. Note that the patterns contain a number of domains which are "filled" with spatial disorder. These domains alternate with clusters inherited from the initially regular pattern. [Here a cluster is a set of units of the lattice with $u_{j,k} \approx u_{s,l}$ ($v_{j,k} \approx v_{s,l}$) for j, k and s, l belonging to one set.] The evolution and shape of (6.8) for this and the previous case are similar to the curves given in the Fig. 6.3.

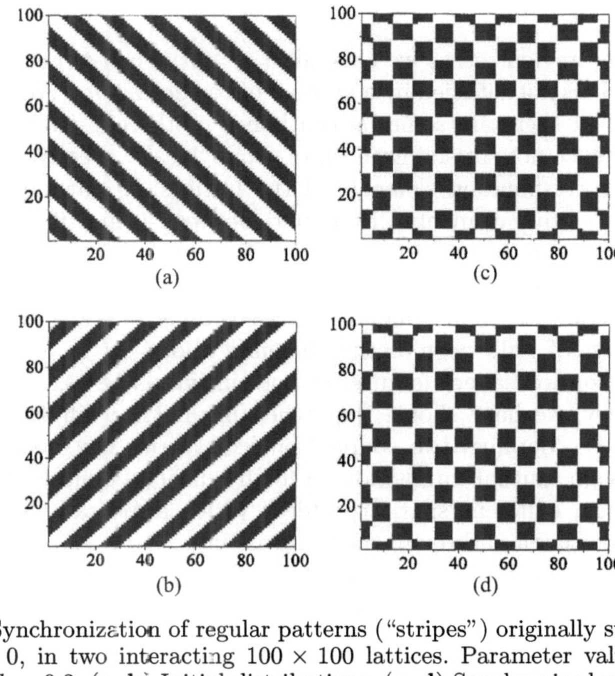

Fig. 6.4. Synchronization of regular patterns ("stripes") originally steady in time, when $h = 0$, in two interacting 100×100 lattices. Parameter values: $a = 0.5$, $d = 0.006$, $h = 0.2$. (**a–b**) Initial distributions. (**c–d**) Synchronized patterns

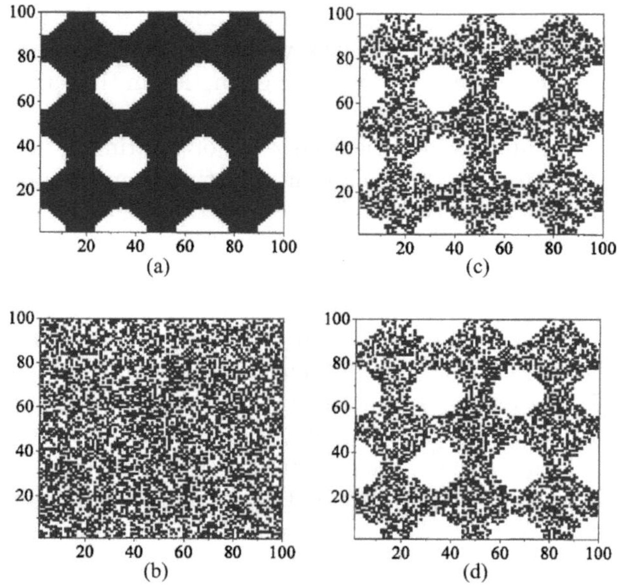

Fig. 6.5. Synchronization of regular ("spots") and spatially irregular patterns originally steady in time, when $h = 0$, in two interacting 100×100 lattices. Parameter values: $a = 0.4$, $d = 0.006$, $h = 0.2$. (**a–b**) Initial distributions. (**c–d**) Synchronized patterns

(iv) In the previous cases we have all $h_{j,k} = h = \text{const}$. Let us see what happens if (6.7) is not satisfied by a number of coefficients $h_{j,k}$. Such situation may occur when there are some broken bonds or there is "ablation" or ill-functioning, hence not satisfying (6.8), of some "local" interlattice couplings between the corresponding elements of the lattices. In particular, we take the matrix $h_{j,k}$ with elements distributed between 0 and 0.15 with equal probability (random space distribution). Figure 6.6 illustrates the results of the interaction of the regular initial patterns of Fig. 6.4a,b in this case. Figure 6.6c,d shows that, although there is no *complete* synchronization, the synchronization of only part of the elements allows us to identify in the terminal distribution salient features of both initial patterns.

In summary, we have shown that when intralattice diffusion competes with interlattice coupling, two coupled lattice systems can manage to synchronize through mutual interaction to finally display identical behavior in time and space. Different spatial patterns which are steady structures of the independent lattices through their mutual interaction force the two-layer lattice system to evolve, ending with two identical steady patterns whose structures can be very diverse and depend on the initial conditions and system parameters.

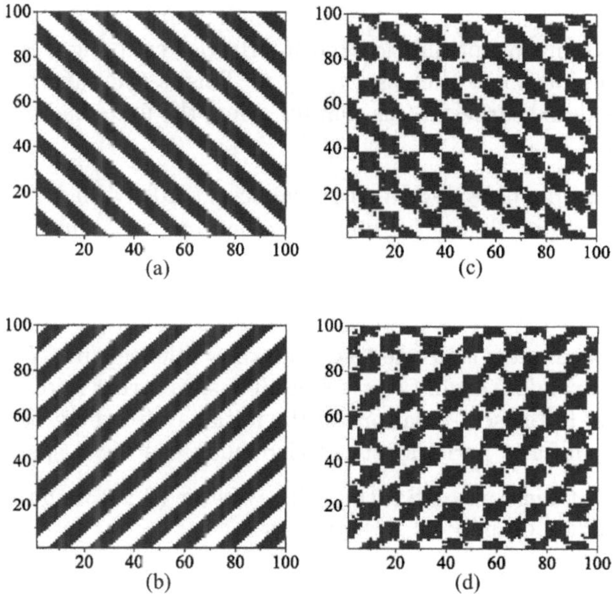

Fig. 6.6. Interaction of "stripe" patterns (Fig. 6.4) for an arbitrary distribution of constant but different values of the interlattice couplings $[h_{j,k} \in (0, 0.15)]$. The condition (6.7) is not satisfied by a number of elements, $h_{j,k}$, hence we have a case with broken bonds or ill-functioning interlattice couplings. Parameter values: $a = 0.5$, $d = 0.006$. (**a–b**) Initial distributions. (**c–d**) Terminal patterns

6.2.2 Bistable Oscillators

Let us consider a two-layer architecture of 2D square lattices in which units operate as Van der Pol oscillators with subcritical bifurcation [6.17]. Two different attractors co-exist in the phase plane of such oscillators. These are a stable fixed point and a stable limit cycle. For simplicity, we assume that the period of the oscillation does not depend on amplitude. In this case, one can describe the dynamics of such a unit by

$$\dot{A}_{jk} = -A_{jk} F(|A_{jk}|^2) \,, \tag{6.9}$$

where A_{jk} is the complex amplitude of the oscillation at the (j, k) site of the lattice. $F(|A|^2) = 2a|A|^4 - a|A|^2 + 1$, and $a > 8$, allows co-existence of the state of rest and the periodic oscillation in the phase space of (6.9). Then, (6.1) in this case can be rewritten in the form

$$\begin{aligned} \dot{u}_{jk} &= -u_{jk} F(|u_{jk}|^2) + d(\Delta u)_{jk} + h_{jk}(v_{jk} - u_{jk}) \,, \\ \dot{v}_{jk} &= -v_{jk} F(|v_{jk}|^2) + d(\Delta v)_{jk} + h_{jk}(u_{jk} - v_{jk}) \,, \end{aligned} \tag{6.10}$$

$$j, k = 1, 2, \ldots, N \,,$$

where $u_{jk} \equiv A_{jk}^{(1)}$ and $v_{jk} \equiv A_{jk}^{(2)}$ are complex amplitudes and, as earlier, d and h_{jk} account for intralattice and interlattice coupling, respectively. We restrict consideration to the case $h_{j,k} \geq 0$. This is the natural extension of the case considered earlier in this chapter where the units had only two stable steady states.

In-Phase Oscillations. For a single lattice, (6.10), with $h_{jk} = 0$, has 2^{N^2} stable steady states that set in-phase oscillations in the initial system and may co-exist in L_0 at sufficiently weak values of the intralattice, diffusive coupling. These oscillations exist for the values of the parameters in the region D_{ch} (Fig. 5.18). The amplitude distribution of in-phase modes "along" the spatial coordinates is in one-to-one correspondence with $N \times N$ matrices consisting of a random set of two symbols. Thus, as earlier said, the amplitude distribution of in-phase oscillations in the initial system may be very diverse: from simple (uniform, periodic, etc.) to irregular, chaotic in space (Fig. 5.19).

Gradient Features of the System. The system (6.10) with the interlattice coupling switched on may be rewritten as

$$\dot{u}_{j,k} = -\frac{\partial U}{\partial u_{j,k}^*}, \quad \dot{v}_{j,k} = -\frac{\partial U}{\partial v_{j,k}^*}, \tag{6.11}$$

where the potential is

$$\begin{aligned} U = \textstyle\sum_{j,k} \Big[& G(|u_{j,k}|^2) + G(|v_{j,k}|^2) + d\,\big(|u_{j+1,k} - u_{j,k}|^2 \\ & + |u_{j,k+1} - u_{j,k}|^2 + |v_{j+1,k} - v_{j,k}|^2 \\ & + |v_{j,k+1} - v_{j,k}|^2 \;\big) + h_{j,k}|u_{j,k} - v_{j,k}|^2 \Big] \,, \end{aligned}$$

$$G(|u|^2) = |u|^2 - \frac{a}{2}|u|^4 + \frac{2a}{3}|u|^6 \geq 0\,.$$

Thus, (6.11) is a gradient system, and under arbitrary initial conditions all its trajectories tend to one of the "equilibria" corresponding to a minima of U.

Manifold of In-Phase Oscillations. The existence of a manifold of in-phase oscillations follows immediately from (6.10):

$$S = \{\varphi_{j,k} = \psi_{j,k} = \theta_0 = \text{const.}\,,\; j,k = 1,2,\,\dots\,,N\}\,,$$

where the phases $\varphi_{j,k}$ and $\psi_{j,k}$ are related to the variables of (6.10):

$$u_{j,k} = r_{j,k}e^{i\varphi_{j,k}}\,, \quad v_{j,k} = \rho_{j,k}e^{i\psi_{j,k}}\,.$$

By linearizing (6.10) in the vicinity of S and using the Gershgorin theorem for the matrix of the linearized system, it can be shown that S is *locally* stable.

Mutual Synchronization of Oscillations. In a similar way as done for (6.2) it can be shown, as in Sect. 6.2.1, that due to the interlattice coupling the oscillations in either lattice are mutually synchronized on S if

$$h > h_1^*, \quad h_1^* = \frac{7a - 20}{40}, \tag{6.12}$$

is fulfilled, i.e. when the interlattice couplings exceed some critical values. Thus the different amplitude distributions of in-phase motions in two coupled lattices evolve to the amplitude patterns of identical spatial structure. This phenomenon may be interpreted as the *mutual* synchronization of oscillations in the two coupled lattices.

Since the qualitative shape of the amplitude patterns of (6.10) does not have any major differences with the steady patterns from (6.2) (see, for example, Figs. 6.2, 6.4 and 6.6), we do not present such patterns here.

6.2.3 System of Two Coupled Fibers

Consider a two-layer system composed of 1D-lattice (i.e. chains) diffusively coupled units that for illustration we take as Chua's circuits. Since each chain represents a spatially extended system and it is capable of sustaining different waves, making possible the transfer of an encoded information, it can be considered to be an *active* (electronic) *fiber*. Thus a fiber is a discrete version of a *multispecies reaction–diffusion medium*. By choosing an appropriate mode of the cell "reaction kinetics", we can mimic different types of such media.

Let a fiber be a chain or 1D lattice of N coupled electronic units as schematically shown in Fig. 6.7. Let us also assume that the interfiber interaction is provided by "point-by-point" nearest-neighbor connections with linear resistors between the corresponding elements of the two chains. Further, let us consider two such axon-like structures coupled together. The collective

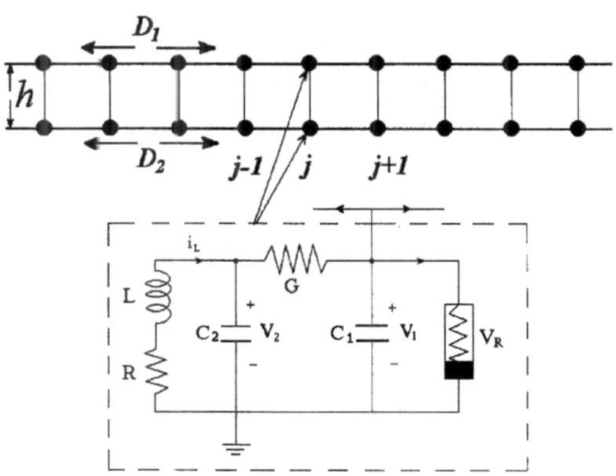

Fig. 6.7. Schematic of the discrete two-fiber system and details of the active electronic unit

dynamics of such a system is described by the following dimensionless coupled equations

$$\begin{cases} \dot{x}_j^1 = \alpha\left[y_j^1 - x_j^1 - f(x_j^1)\right] + D_1 \Delta x_j^1 \\ \qquad +h_j(x_j^2 - x_j^1)\,, \\ \dot{y}_j^1 = x_j^1 - y_j^1 + z_j^1\,, \\ \dot{z}_j^1 = -\beta y_j^1 - \gamma z_j^1\,, \\ \dot{x}_j^2 = \alpha\left[y_j^2 - x_j^2 - f(x_j^2)\right] + D_2 \Delta x_j^2 \\ \qquad +h_j(x_j^1 - x_j^2)\,, \\ \dot{y}_j^2 = x_j^2 - y_j^2 + z_j^2\,, \\ \dot{z}_j^2 = -\beta y_j^2 - \gamma z_j^2\,, \end{cases} \tag{6.13}$$

$$j = 1, 2, \ldots, N\,,$$

where superscripts $i = 1, 2$ denote the variables of the first and the second fiber, respectively; $\Delta x_j = x_{j-1} - 2x_j + x_{j+1}$ is the discrete Laplace operator; D_1 and D_2 are coefficients accounting for the strengths of the corresponding intrafiber diffusions; h_j characterizes interfiber coupling (which is also of diffusive type); and $f(x) = x(x-a)(x+b)$ with $a, b > 0$. Finally, von Neumann (zero-flux) boundary conditions are imposed on (6.13).

Motions along a Fiber and their Synchronization in Two Coupled Fibers. Let us introduce the new (difference) variables

$$u_j = x_j^1 - x_j^2,\ v_j = y_j^1 - y_j^2,\ w_j = z_j^1 - z_j^2\,,$$
$$s_j = x_j^1 + x_j^2,\ p_j = y_j^1 + y_j^2,\ q_j = z_j^1 + z_j^2\,.$$

Then, the two-fiber system (6.13) becomes

$$\begin{cases} \dot{u}_j &= \alpha\left[v_j - u_j - H(u_j, s_j)\right] + \frac{D_1+D_2}{2}\Delta u_j \\ &\quad +\frac{D_1-D_2}{2}\Delta s_j - 2h_j u_j\,, \\ \dot{v}_j &= u_j - v_j + w_j\,, \\ \dot{w}_j &= -\beta v_j - \gamma w_j\,, \\ \dot{s}_j &= \alpha\left[p_j - s_j - \Phi(u_j, s_j)\right] + \frac{D_1-D_2}{2}\Delta u_j \\ &\quad +\frac{D_1+D_2}{2}\Delta s_j\,, \\ \dot{v}_j &= s_j - p_j + q_j\,, \\ \dot{w}_j &= -\beta p_j - \gamma q_j\,, \end{cases} \tag{6.14}$$

$$j = 1, 2, \ldots, N\,,$$

with

$$H(u\ s) \equiv \frac{u^2 + 3s^2}{4} + (a - b)s - ab\,,$$

$$\Phi(u, s) \equiv \frac{s}{4}(3u^2 + s^2) + \frac{a-b}{2}(u^2 + s^2) - abs\,.$$

Synchronization of Two Identical Fibers with Interfiber Interaction. Let us first consider the case $D_1 = D_2 = D$. Then (6.14) has the trivial solution $u_j = v_j = w_j = 0$. In the phase space of (6.14) it defines the manifold of synchronized motions

$$M : \{u_j = 0,\ v_j = 0,\ w_j = 0\}\,, \quad j = 1, 2, \ldots, N\,,$$

which for suitable parameter values is globally asymptotically stable. Indeed, the quantity

$$V = \sum_{j=1}^{N} \left(\frac{u_j^2}{2} + \frac{\alpha v_j^2}{2} + \frac{\alpha}{\beta}\frac{w_j^2}{2} \right)$$

is a Lyapunov function. Its derivative with respect to (6.14) is

$$\dot{V} = -\sum_{j=1}^{N} (P_j + Q_j + R_j)$$

with

$$\begin{aligned} P_j &\equiv -Du_{j-1}u_j - [2D + 2h_j - \alpha(a^2 + ab + b^2)/3]u_j^2 \\ &\quad -Du_{j-1}u_j\,, \\ Q_j &\equiv \tfrac{\alpha u_j^2}{2}[u_j^2 + 3s_j^2 + 4(a-b)s_j + 4(a-b)^2/3]\,, \\ R_j &\equiv \alpha(u_j^2 - 2u_jv_j + v_j^2 + \gamma w_j^2/\beta)\,. \end{aligned}$$

Then, all Q_j and R_j are positive definite. The function P_j is a quadratic form. It is also positive definite if

$$h_j > h_2^* \equiv \frac{\alpha(a^2 + ab + b^2)}{6}\,, \quad \forall j = 1 \ldots N \tag{6.15}$$

It follows that outside M the inequality $\dot{V} < 0$ is satisfied and $\dot{V} = 0$ on the manifold. Thus, the synchronization manifold M is *globally asymptotically stable.* Therefore, any initial conditions in the identical coupled fibers tend to M, where the already *synchronized* motions are governed by the system describing the dynamics of a single fiber ($h_j = 0$).

Synchronization of Fibers with Different Diffusion Coefficients and Homogeneous Strong Interfiber Interaction. Let us now take $D_1 \neq D_2$ with very strong interfiber interaction coefficients $h_j = h \gg 1,\ \forall j = 1 \dots N$. Then, (6.14) has a smallness parameter, $\mu = 1/h \ll 1$, which affects the derivative $\dot{u}_j$; (6.14) has both *fast* and *slow* motions. In the phase space there exists a stable surface of *slow* motions defined by $\{u_j = 0,\ j = 1 \dots N\}$ and all trajectories of (6.14) after some time are trapped within thin layers (whose thickness is of the order of μ) near this surface. When $\mu = 0$ the *slow* motions are located exactly at the surface and are governed by the system

$$\begin{aligned} \dot{v}_j &= -v_j + w_j\,, \\ \dot{w}_j &= -\beta v_j - \gamma w_j\,, \end{aligned} \tag{6.16}$$

$$\begin{aligned} \dot{s}_j &= \alpha[p_j - s_j - \Phi(0, s_j)] + \tfrac{D_1 + D_2}{2} \nabla^2 s_j\,, \\ \dot{p}_j &= s_j - p_j + q_j\,, \\ \dot{q}_j &= -\beta p_j - \gamma q_j\,; \end{aligned} \tag{6.17}$$

(6.16) and (6.17) describe independent parts. The part (6.16) is linear and all trajectories asymptotically tend to the surface $\{v_j = w_j = 0\, j = 1 \dots N\}$. Thus, the slow motions occur on the surface that *coincides* with the M. The evolution on this surface is given by (6.17), describing a single chain with diffusion coefficient $D_s = (D_1 + D_2)/2$. Hence, for $h \gg 1$ evolution proceeds in two stages. Any initial condition in the first (fast) stage quickly comes to the thin layer which is very close to the stable surface of slow motions (M). In the second (slow) stage, motions are governed, approximately, by (6.17) for a single chain, with $D = D_s$, and tend to an attractor of this system (for example, a steady state or a wave pattern).

Effects of Interfiber Interaction for Varying "reaction kinetics". In the preceding section we analytically found a sufficient condition for the synchronization of all possible motions in the two coupled fibers. Let us now see the concrete effects of the interfiber ineraction. We have numerically integrated (6.13) for different modes (i.e. "reaction kinetics") of the unit and for varying intrafiber diffusion. For simplicity, only the homogeneous case $h_j \equiv h$ has been considered.

Wave Fronts in Coupled Interacting Fibers. Let us choose the parameters of the electronic circuit such that it possesses two stable steady states and take $\gamma \gg 1$. Then each fiber can be, approximately, described by a gradient system (Chap. 4). In particular, the dynamics of each unit is given by

$$\begin{cases} \dot{x} = -\,\alpha \dfrac{\partial U(x,y)}{\partial x}\,, \\[2ex] \dot{y} = -\,\dfrac{\partial U(x,y)}{\partial y}\,, \end{cases} \tag{6.18}$$

with the potential

$$U(x,y) = \frac{x^4}{4} - \frac{(a-b)x^3}{3} - \frac{(ab-1)x^2}{2} + \frac{y^2}{2} - xy\,. \qquad (6.19)$$

The two stable steady states of the unit correspond to the two minima of the potential. A single fiber represents a discrete, *bistable*, reaction–diffusion system of the FitzHugh–Nagumo–Schlögl (FNS) type. Increasing the diffusion coefficient, D, above some critical value, D_{fr}, such a fiber exhibits wave front propagation. A front is formed as a consequence of the transition of the units to the state with a lower value of the potential function U. Thus, for fixed D the direction of front propagation, and its velocity, are uniquely defined by the difference in the potential levels between the two minima of U. Let us show that when the fibers are coupled, $h \neq 0$, there is the possibility of dramatically changing the properties of the propagating fronts.

In the numerical calculations we fix the parameters of the cell to $a = 1.4, b = 1.8, \beta = 0.5, \alpha = 1$ and $\gamma = 10$, and hence for all practical purposes $\gamma \gg 1$. This set of parameter values provides the required properties for the cell to be a bistable unit.

Synchronization of Traveling Wave Fronts. Let us consider $D_1 = D_2 > D_{fr}$, and let the initial conditions be two fronts propagating in the same direction along parallel fibers with a finite time delay (Fig. 6.8, $t < t_0$). The level of grey color corresponds to the value of x in the junctions of the chains, according to the key given in the figure. Sudden switch-on of the interfiber interaction at $t = t_0$ makes the delay negligible after a transient process (Fig. 6.8, $t_0 < t < t_1$). Then the fronts become synchronized and propagate together (Fig. 6.8, $t > t_1$). When the fronts hit the boundary, a steady homogeneous distribution appears in the fibers, and hence each unit comes to the lower minimum of the potential U (marked white). The transient process in Fig. 6.8 occurs in the following way: The front in the first fiber, arriving with delay, significantly increases its velocity, while the first arriving front in the second fiber reverses motion and travels backward.

To describe these effects (front acceleration and front reversal) let us consider the gradient model (6.18). Let us take the front in the first fiber propagating relative to one of the two accessible homogeneous states in the second fiber. The interfiber interaction slightly affects the homogeneous state of the second fiber. Hence, approximately, the influence of the second fiber can be estimated as a constant in the difference term of (6.13), i.e.

$$h(x_j^2 - x_j^1) \approx h(x_0 - x_j^1)\,,$$

where x_0 is an x-coordinate of the steady homogeneous state. It takes the values $x_0 = a$ and $x_0 = -b$ for the two corresponding states. Then, the units of the first fiber are described by (6.18) with the *effective* potential function

$$U_h(x,y) = U(x,y) + \frac{h}{2\alpha}(x_0 - x)^2\,.$$

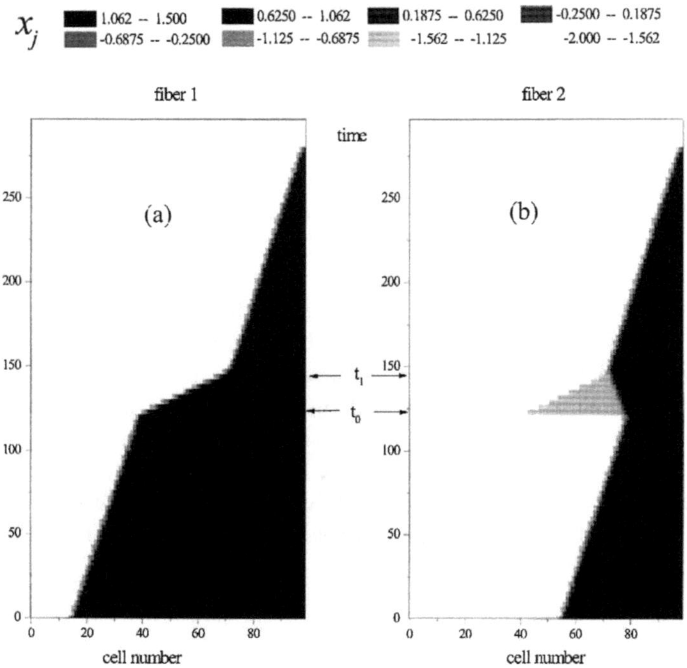

Fig. 6.8. Wave fronts in two identical bistable "fibers" with $D = 2$. (**a**) Synchronization of two wave fronts traveling with finite time delay. For $t < t_0 = 120$ the chains are independent, $h = 0$. At $t = t_0$, the interaction is switched on, $t > t_0$, $h = 0.5$. The time interval $t_0 < t < t_1 \approx 140$ corresponds to a transient process. The vertical axis accounts for time in arbitrary units

Figure 6.9a,b show cross-sections of the surface $U_h(x, y)$ by the plane $y = x$ for the values of x_0 corresponding to the steady states with lower (at $x_0 = -b$) and higher (at $x_0 = a$) potential levels, respectively. The dashed curve corresponds to the potential function, $U(x, y)$, for the units of the single fiber. In Fig. 6.9a the difference in the potential levels between the two minima increases (the right minimum practically disappears), and hence the front should increase its velocity. It is what we have seen, in Fig. 6.8, in the first fiber. Figure 6.9b shows that when x_0 is associated with the state of higher potential ($x_0 = a$), the left minimum of U_h increases. Then, the backward transition of the cell from the left minimum to the right becomes energetically preferable. Accordingly, there is a reversal of the wave fronts in the coupled fibers (the second fiber in Fig. 6.8).

Overcoming Propagation Failure in Coupled Fibers. Another example of the effect of interfiber action, allowing the behavior of the coupled fibers to change qualitatively as a whole, is the possibility of overcoming front propagation failure.

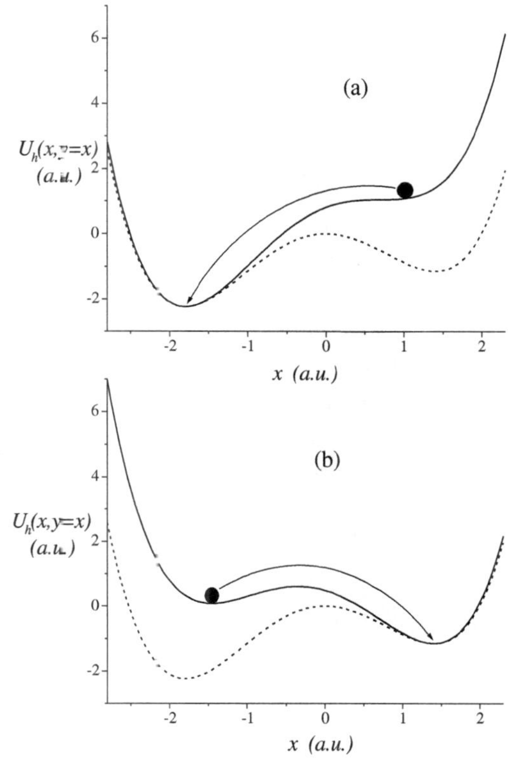

Fig. 6.9. Effective potential functions, U_h, for the coupled "fibers" describing (**a**) front acceleration and (**b**) front reversal propagation. Units are arbitrary (a.u.)

The existence of a huge number of stable steady patterns in discrete chains with bistable kinetics, when D is small enough, leads to wave front propagation failure. The perturbation of a front is always "pulled" by one of the stable steady patterns; hence it fails to propagate. This leads to the failure in the behavior of the fiber as a transfer system.

Let us take one chain with $D_1 > D_{fr}$ such that it allows front propagation, while the other chain with $D_2 < D^*$ is the ill-functioning fiber where waves fail to propagate. D^* is some given value not needed explicitely here. As shown in Sect. 6.2.3, with a strong enough interfiber interaction, $h \gg 1$, all motions in the "fibers" come very close to being synchronized. They are restricted within a thin layer near M, where both fibers behave, approximately, as a single one with the mean diffusion coefficient $D_s = \frac{D_1+D_2}{2}$. Hence, if we choose D_1 to be large enough, $D_s > D_{fr}$, the propagation failure in the ill-functioning fiber is overcome. This effect is illustrated in Fig. 6.10. When the fibers are independent ($t < t_0$) there is a wave front in the first one (well-functioning fiber) and a stable steady pattern in the second one (ill-functioning fiber). The strong interaction between the fibers is switched on

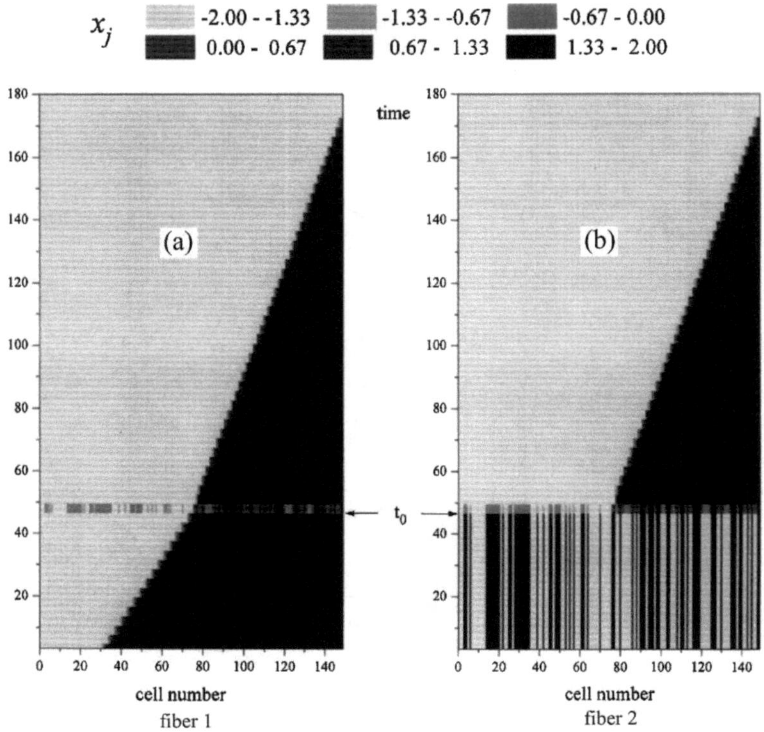

Fig. 6.10. Propagation failure overcome in coupled *bistable* "fibers". Diffusion coefficients: $D_1 = 3, D_2 = 0.2$. At $t_0 = 48$, the interfiber interaction is switched on, $h = 10$. The vertical axis accounts for time in arbitrary units. (**a**) Fiber 1; (**b**) fiber 2

instantaneously at t_0. After a very short transient process (fast motion stage) the steady pattern disappears in the ill-functioning fiber and a wave front appears almost identical to the front propagating in the first fiber ($t > t_0$). Note that the velocity of the synchronized fronts is lower than the velocity of the original front. This is due to the inequality $D_s < D_1$ and to the fact that the typical dependence of front velocity on diffusion strength in a reaction–diffusion system is generally an increasing function (Fig. 4.7).

Synchronization in Coupled Excitable Fibers. Another possible "reaction kinetics" of the unit allows a fiber mode with *excitable* properties (Chap. 4). Such is the case for a chain [(6.13), $h_j = 0$], when $a = b = 1.4$, $\gamma = 0.01$ and $\beta = 0.5$. We take α as the control parameter. Then, the chain sustains the propagation of stable solitary pulses or complex pulse trains.

The dynamics of excitations in systems of fibered or layered spatial architecture is of great importance in biology. In particular, excitation reentry in cardiac tissue is known to be responsible for several types of heart arrhythmia. Models of two coupled fibers of FNS type have been considered in the literature. Phenomena of different types of re-entry of propagating, action-potential-like pulses have been studied.

Let us study the dynamics of pulse-like excitations. We focus attention on the reentry phenomena as a result of interfiber synchronization. It appears that pulse-like excitations as well as the reentry typical for the classical FNS type models and quite different effects of initiations of pulse-bound states or traveling "bursts" are possible.

Simple Pulse Reentry and Pulse Failure. Let a single pulse be propagating in one fiber as an excitation of the rest state, while the second fiber is in the rest state. When the interfiber interaction occurs at a given instant of time, the fibers tend to be synchronized. Indeed, a pulse completely identical to the initial one can be excited in the second fiber, and the two pulses synchronously travel down to the bottom of the fibers. We have the simple "entry" of the excitation fiber into the fiber originally in its rest state. The situation is similar to that observed in fibers of FitzHugh–Nagumo–Schlögl type when the excitation threshold in the second fiber is low enough. Pulses may also *fail* to propagate in the coupled fibers as a result of the interfiber synchronization. In this case the pulse in the first fiber disappears and both chains come to the rest state. Apparently, the pulse is "pulled" by the second fiber at rest, which demands a higher threshold to be excited. At variance with the failure in chains, with simple *bistable* "kinetics", where failure is caused by ill-functioning behavior, here pulse propagation failure in the coupled fibers comes from the destruction of a given original excitation, while both fibers are functioning well.

Initiation of Trains of Pulses. A single fiber allows a variety of complex waves to steadily translate along it (see Chap. 4). In particular, one can easily excite pulse trains or *bound* states composed of an arbitrary number of humps or "spikes". Let us show how such bound states appear as a result of pulse reentry in coupled fibers.

The appearance of synchronized trains of pulses (bound states) traveling in opposite directions can be observed as follows: After the interaction is switched on there appears a "long-lasting, transient source of pulses" which may generate two or more pulses (multihump pulse) identical for both fibers (Fig. 6.11). Note that in this case the coupling coefficient h should be taken to be smaller than the predicted synchronization threshold (6.15). Thus the interfiber interaction looks more "elastic" but still enough to synchronize the chains. Further decreasing the coefficient h leads to the appearance of a long-lasting pulse source.

To illustrate how the two-fiber system exhibits such pulse trains, we introduce the quantity characterizing how the system reaches the synchronization manifold,

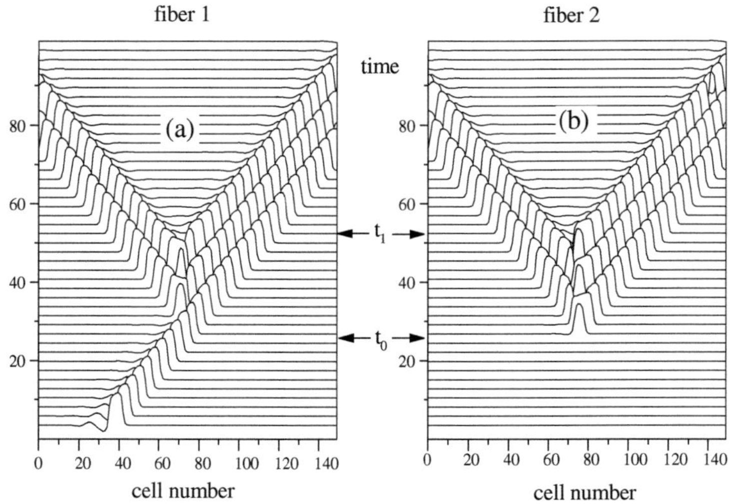

Fig. 6.11. Long-lasting pulse source generating trains of pulses. Parameter values: $\alpha = 4.5, D = 2, h = 1.05, t_0 = 25, t_1 \approx 52$. The vertical axis accounts for time in arbitrary units

$$\mathrm{dist}(t) = \frac{\sqrt{\sum_{j=1}^{N} \left[u_j^2(t) + v_j^2(t) + w_j^2(t)\right]}}{3N}, \tag{6.20}$$

which is, indeed, the distance between the vectors of the two chains in a $3N$-dimensional state space. Its vanishing value corresponds to the complete synchronization of the fibers. Figure 6.12 shows the time evolution of (6.20)

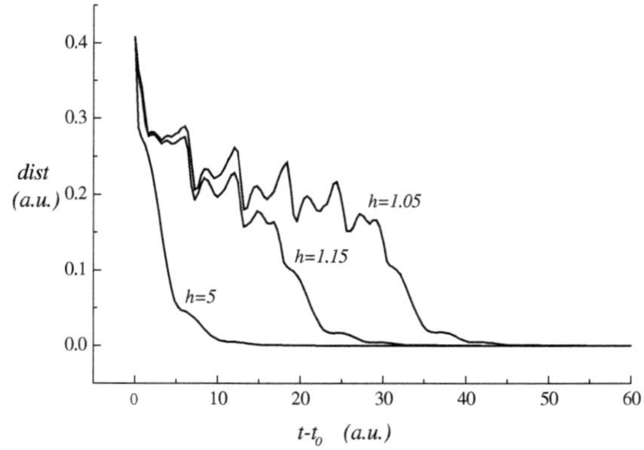

Fig. 6.12. Controlled distance (6.20) as a function of time for three different values of the interfiber interaction. The curves illustrate that synchronization is indeed achieved. Units are arbitrary (a.u.)

for three different values of h. When (6.15) is fulfilled (e.g. $h = 5$) the distance monotonically decreases to a zero value. This is the case of a "rigid" interaction when the system takes either the pulse or the rest state in the second fiber. For lower values of h we have more "elasticity" in the interaction and $\text{dist}(t)$ has rather long-lasting oscillations associated with the appearance of the source of pulses. After some time the source dies out and the distance between the dynamic states in the two fibers tends to zero (Fig. 6.12). In this case mutually synchronized bound states appear. By varying h we have the possibility to control the number of humps or spikes in the pulse train. This number is also sensitive to the values of the other parameters of the system and to changes in the initial conditions.

At variance with the cycle mechanism of pulse reentry in coupled excitable fibers, resulting in the appearance of a sequence of pulses, the initiation of pulse trains or bound states cannot be explained in simple terms of sequential excitation entry from fiber to fiber. Rather, it has to do with the possibility of a stable propagation of trains (bound states) with a variable number of spikes and with the oscillatory properties of a single fiber. In particular, a pulse initiating reentry has a remarkable oscillatory tail whose "oscillations" tend to grow when the interfiber interaction occurs and hence form a number (controlled by h) of spikes in the outcoming pulse trains.

Pulse Driving by Pinned Interfiber Contacts. Let us, finally, consider the case when the interfiber interaction operates with one or a few contacts between the two chains, and hence the vector $\{h_j\}$ has only a few nonzero components (Fig. 6.13). Such contacts can be easily realized with discrete electronic chains.

Single-Pin Contact. Let two identical fibers be connected at a single point, $h_j = h, h_k = 0, \forall k \neq j$, with an h that is strong enough (Fig. 6.13a). It follows from (6.14) that for a strong enough interaction, $h \gg 1$, the difference $\dot{u}_j \sim -2hu_j$ in this pair tends to decrease and, eventually, vanishes. Hence the two coupled units of the different fibers tend to be synchronized (at least their x variables). As in the case of the homogeneous interfiber coupling (Sect. 6.2.3) such a synchronization process takes time, i.e. it is a transient process whose duration depends on the particular value of h. Hence, we have,

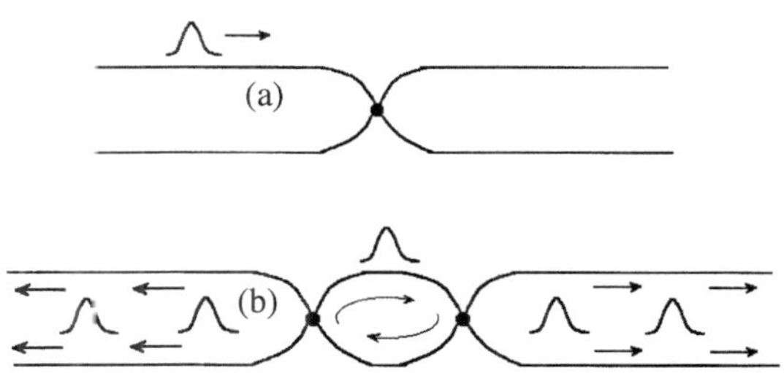

Fig. 6.13. Schematic of two parallel "fibers" coupled by (**a**) a single pin contact and (**b**) a double-pin contact

in fact, a *dynamic contact* between the parallel fibers acting with its own time scale. Two possible results of such a contact are that the pulse reaching the contact in Fig. 6.13a dies out for large enough interfiber coupling, and hence the contact operates as a barrier for excitation transmission, and, alternatively, at the contact point the excitation "enters" the fiber at rest where two new pulses appear. In the latter case, occurring for low enough interfiber coupling, the forward traveling pulse propagates almost synchronously with the original pulse which also propagates in the first fiber. Thus, depending on its strength, the dynamic contact acts in two different ways. It either stops the pulse thus *inhibiting* transmission or allows the pulse to go over, as an *excitatory* action on the next chain element.

The phenomena of pulse driving in the pinned contact can be interpreted in terms of characteristic time scales of the contact action, T_c, and the diffusion action, T_p, accounting for the pulse velocity. Let us look at the interaction process from the spatial site of the chain where the interaction occurs. Roughly, the pulse propagating in the first fiber can be split into several independent "portions" that sequentially bring excitation to those coupled units which are in the rest state. When a portion comes to the site of contact the interaction operates in two steps: (i) the two coupled units become synchronized very rapidly (h has a rather large value); (ii) the synchronous evolution of the coupled pair, defined, approximately, by the dynamics of the single unit, brings the coupled pair to the rest state, and hence tends to kill the original excitation. These two steps account for T_c. Thus, if this time interval is much smaller than the time, T_p, needed for the excitation to reach the units nearest to the site of contact in the two fibers, $T_c \ll T_p$, then the "portions" of the pulse sequentially die at the contact. This inhibits further pulse propagation. But if, when the contact is operating, the excitation has enough time to be transmitted by diffusion to the nearest units, $T_c \gg T_p$, hence to excite them, the pulse overcomes the contact site in the first fiber, and two pulses are excited in the fiber at rest. Note that, for fixed parameter values of the unit, T_p is estimated from the pulse velocity c, $T_p \sim 1/c$. The dependence of c on the diffusion coefficient, D, in discrete reaction–diffusion systems (Chap. 4) is typically $c \sim \sqrt{D}$, hence $T_p \sim D^{-1/2}$. T_c depends only on the value of h and decreases with increasing h to some constant value T_0 at $h \to \infty$, accounting for the second step of the contact action.

It is difficult to quantitatively estimate these time scales because of the complex internal dynamics of the unit. To illustrate the relative action of the two time scales, Fig. 6.14 provides the diagram of pulse driving in the (D, h) parameter plane obtained numerically. In the region below the curve C_1, the contact yields pulse failure. Above the curve C_2 the pulse overcomes the contact and excites (for a certain value of h) the second fiber. The region between C_1 and C_2 corresponds to the complex outcome of the system. Here T_c and T_p become comparable and the result of the contact action crucially

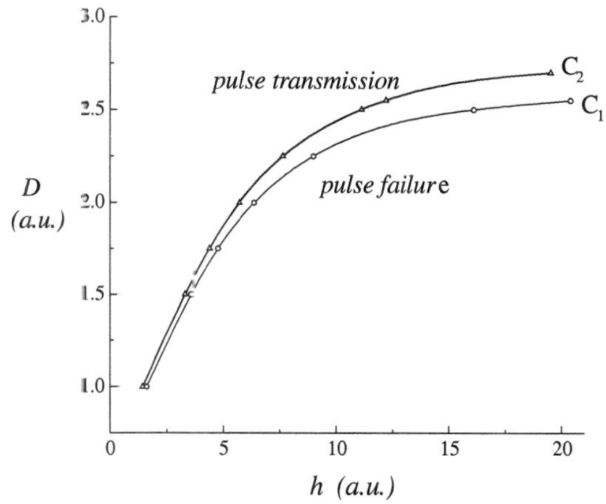

Fig. 6.14. Diagram of pulse driving by single-pin contact in the (D, h) parameter plane. The pulse fails to propagate for values chosen below the curve C_1. Above the curve C_2 the pulse successfully overcomes the contact. Units are arbitrary (a.u.)

depends on the nontrivial dynamics of the unit. In particular, complex wave structures including pulses and wave fronts may appear.

Such behavior of a dynamic contact is not alien to what is known about synaptic linkages connecting an axon with dendrites in another neuron. Similar to synapses, our dynamic contact acts only when an excitation reaches it. Then, the excitation transfer occurs in the dynamic contact with a finite time delay very much like that needed by neurotransmitters to bring an excitation into the post-synaptic cell. The synaptic contacts of nerve cells may be excitatory, hence firing the post-synaptic cell, or inhibitory, hence propagation failure. Our dynamic contact shows both these synaptic-like features.

Double-Pin Contact. Let us now consider the situation when the *excitable* fibers have two dynamic contacts (Fig. 6.13b). In this case the vector $\{h_j\}$ has two nonzero components. Let us assume them to be "elastic" contacts allowing pulse entry (Sect. 6 2.3). Such contacts allow the reentry of pulses from fiber to fiber. Figure 6.15 illustrates the space–time diagram of such a system. The initial pulse put between the contacts is shown by dashed lines. When the pulse passes over the left contact, two new pulses appear in the second fiber. One of them, traveling backward, reaches the right contact, and in turn two other new pulses are excited in the first fiber. Consequently, a sequence of pulse reentries may occur. Two sequences of almost synchronized pulses occur at both ends of the fibers, as shown in Fig. 6.15. Thus, the system represents, in fact, a spatially extended generator of synchronous pulses (Fig. 6.13b).

Note that the outcome of the system (Fig. 6.15) is similar to the cycle reentry observed in FNS fibers, where the sequential reentry of excitation, from fiber to fiber, is defined only by the characteristic refractory period.

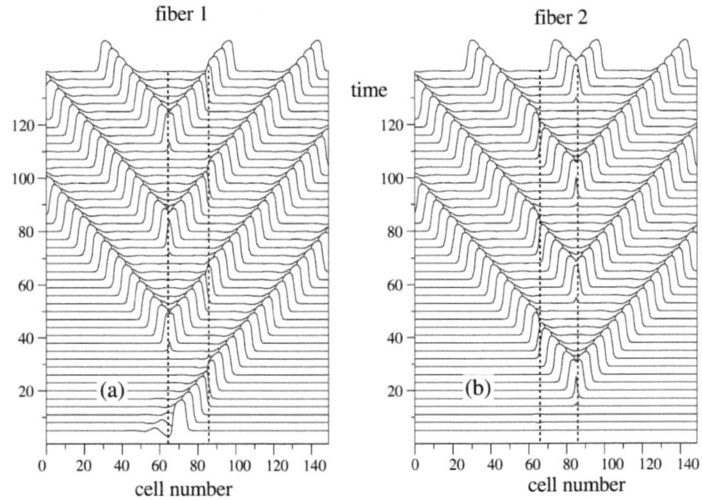

Fig. 6.15. A generator of synchronized pulses when the parallel excitable "fibers" have double-pin contacts. Parameter values: $\alpha = 4.5, D = 2, h_{65} = h_{85} = 5$. The vertical axis accounts for time in arbitrary units

However, in our case the extended generator is stationary and the time lapse or period between the pulses can be controlled at will by suitably placing the dynamic contacts at different sites on the two chains.

6.2.4 Excitable Units

Let us now consider a two-layer, $N \times N$ system composed of two 2D square lattices in which each lattice site is occupied by a Chua's circuit. The units are assumed to be resistively coupled (intralattice coupling coefficient, D). In this case, in (6.1), $l = 1, 2$ and

$$\mathbf{A}_{j,k}^{(l)} = \begin{pmatrix} x_{j,k}^l \\ y_{j,k}^l \\ z_{j,k}^l \end{pmatrix}, \quad \mathbf{F}(\mathbf{A}_{j,k}^{(l)}) = \begin{pmatrix} \alpha\left[y_{j,k}^l - x_{j,k}^l - f(x_{j,k}^l)\right] \\ x_{j,k}^l - y_{j,k}^l + z_{j,k}^l \\ -\beta y_{j,k}^l - \gamma z_{j,k}^l \end{pmatrix},$$

$$\mathbf{D}^{(l)} = \begin{pmatrix} D_l & 0 & 0 \\ 0 & 0 & 0 \\ 0 & 0 & 0 \end{pmatrix}, \quad \mathbf{H}^{(l)} = \begin{pmatrix} h & 0 & 0 \\ 0 & 0 & 0 \\ 0 & 0 & 0 \end{pmatrix},$$

with $f(x)$ satisfying (5.100). Thus the evolution of the two-layer architecture is described by

$$\begin{cases} \dot{x}^1_{j,k} = \alpha\left[y^1_{j,k} - x^1_{j,k} - f(x^1_{j,k})\right] + D_1(\Delta x^1)_{j,k} \\ \qquad +h(x^2_{j,k} - x^1_{j,k}) \,, \\ \dot{y}^1_{j,k} = x^1_{j,k} - y^1_{j,k} + z^1_{j,k} \,, \\ \dot{z}^1_{j,k} = -\beta y^1_{j,k} - \gamma z^1_{j,k} \,, \\ \dot{x}^2_{j,k} = \alpha\left[x^2_{j,k} - y^2_{j,k} - f(x^2_{j,k})\right] + D_2(\Delta x^2)_{j,k} \\ \qquad +h(x^1_{j,k} - x^2_{j,k}) \,, \\ \dot{y}^2_{j,k} = x^2_{j,k} - y^2_{j,k} + z^2_{j,k} \,, \\ \dot{z}^2_{j,k} = -\beta y^2_{j,k} - \gamma z^2_{j,k} \,, \end{cases} \tag{6.21}$$

$$j, k = 1, 2, \ldots, N \,.$$

As already done in Chap. 5, let the values of the parameters of the unit in (6.21) be chosen to provide the bistable and excitable character of the dynamics (5.44).

Identical Layers ($D_1 = D_2 = D$). As for (6.2), using the methodology described in Sect. 6.2.1, it follows that when $D_1 = D_2 = D$ a synchronization manifold exists in the phase space of (6.21) defined by

$$M = \{x^1_{j,k} - x^2_{j,k} = 0,\ y^1_{j,k} - y^2_{j,k} = 0,\ z^1_{j,k} - z^2_{j,k} = 0\} \,.$$

Moreover, for

$$h > \alpha a \tag{6.22}$$

M is *globally* asymptotically stable. Therefore, any initial conditions in the identical coupled layers tend to M, where the synchronized motions are governed by the system describing the dynamics of a single lattice.

Layers with Different Diffusion Coefficients ($D_1 \neq D_2$). Consider (6.21) with a smallness parameter $\mu = 1/h \ll 1$. In this case the motions of (6.21) have both fast and slow features. Accordingly, in the phase space there exists a stable surface of slow motions, and all trajectories of (6.21) after some time become restricted within thin layers (whose thickness is of the order of μ) near this surface. It coincides with M. Then, when $\mu = 0$ the approximate system of slow motions is given by that of a single lattice with the intralattice diffusion coefficient $D^+ = (D_1 + D_2)/2$. Hence, for $h \to \infty$ the motions follow two stages. Any initial conditions in the first (fast) stage rapidly come to the stable surface of slow motions (M). In the second (slow) stage, motions are governed by (5.42), defining the evolution of the single lattice with $D = D^+$, and tend to an attractor of this system (for example, a pattern or a spiral wave).

Pattern Synchronization. When the diffusion coefficients $D_1 = D_2 = D$ belong to the domain D_{ch} (5.47), (6.22) ensures the synchronization of steady patterns in (6.21). The two layers initially containing different steady patterns evolve to a state with a common steady pattern.

Figure 6.16a,b illustrates the result of the synchronization of two different regular patterns. The terminal result is also a regular, steady pattern (Fig. 6.16c,d) identical for both layers. However, the terminal state of each lattice exhibits a new spatial form eventually different from the initial patterns. It is a new "quality" born out of the autonomous evolution of the corresponding two-lattice layered dynamical system. Hence, the pattern synchronization process can be considered as a form of global *self-organization* in the 3D architecture of two coupled lattices.

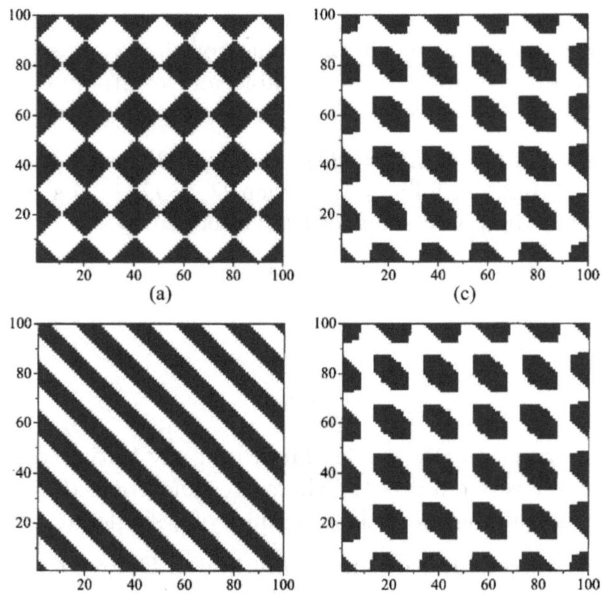

Fig. 6.16. Mutual synchronization of two steady patterns of regular spatial profiles. Parameter values: $D_1 = D_2 = 0.15, h = 3.75$. (**a**–**b**) Initial patterns. (**c**–**d**) Synchronized patterns displaying a new spatial form

6.3 Controlled Patterns and Replication of Form

6.3.1 Bistable Oscillators and Replication

Let us start with a two-layer system (6.10) with units possessing bistability of a limit cycle with a steady state [6.17, 6.25].

Replication. Let the initial conditions be the following: (i) The first lattice produces the in-phase motions with a spatially disordered but steady distribution of the oscillation amplitudes (see, for example, Fig. 6.17). (ii) The

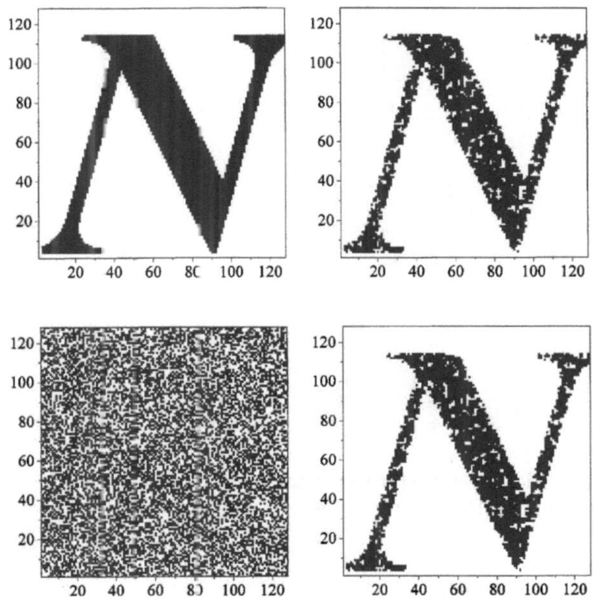

Fig. 6.17. Replication of a patterned stimulus with distortions in the core of the replicated image. Left: initial amplitude distributions; right: terminal patterns. Parameter values: $a = 10$ $d = 0.06, h_{jk} = h = 0.6$

amplitude distribution of in-phase motions in the second lattice is a given pattern (stimulus) which also corresponds to a steady state of the independent lattice. We take as the stimulus the letter "N", with its black core being composed of oscillators which have been excited (having a finite amplitude), while the background outside the letter N contains the oscillators practically at rest (having a vanishing amplitude). Let the interlattice interaction be switched on and the values of coupling coefficients satisfy (6.12). Then, after a transient process the synchronization of oscillations occurs because the manifold M is *globally* stable. But what spatial structure of the terminal amplitude distribution appears in the lattices? Figures 6.17–19 illustrate the result of the interaction by using $d = 0.06$ and $h = 0.6$ for the three cases, while the shape of F (the value of the parameter a) describing the dynamics of a local oscillator is different. In the figures, the left pair of pictures corresponds to initial amplitude distributions, and the right pair to the terminal patterns obtained by numerical integration of the system of amplitude equations originating from (6.10). As stated earlier, the white color in the figures denotes the oscillators close to a state of rest, while the excited

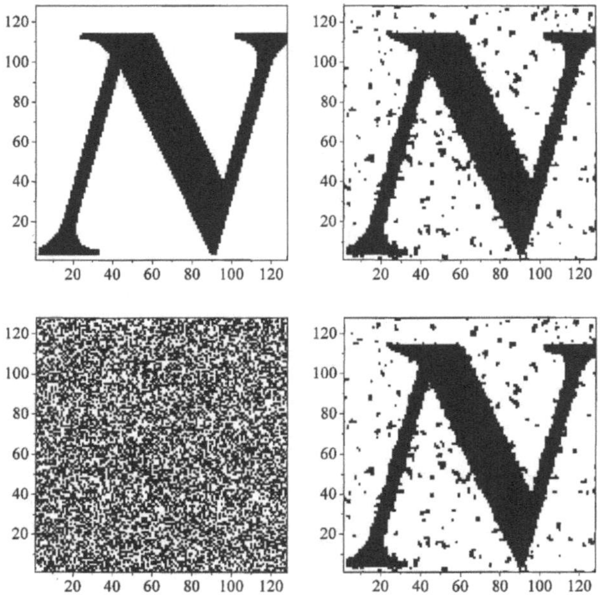

Fig. 6.18. Replication of a patterned stimulus with distortions in the background of the replicated image. Parameter values: $a = 11, d = 0.06, h_{jk} = h = 0.6$

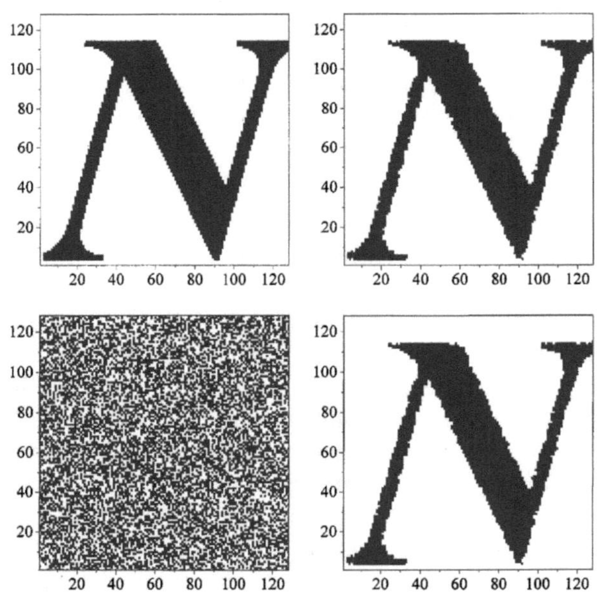

Fig. 6.19. Faithful replication of a patterned stimulus. Parameter values: $a = 10.4$, $d = 0.06, h_{jk} = h = 0.6$

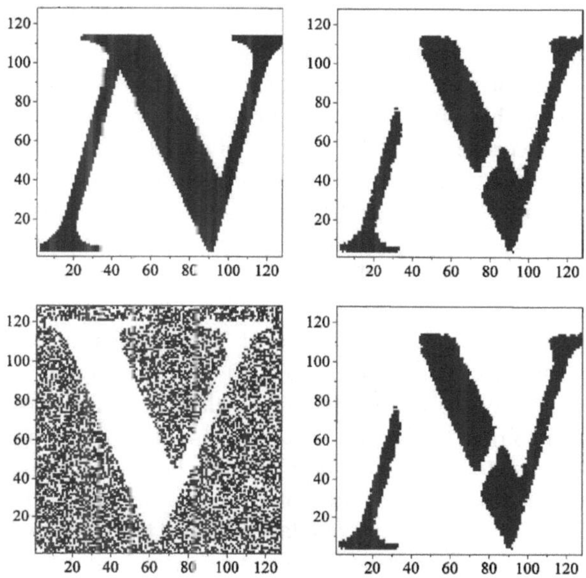

Fig. 6.20. Replication of a patterned stimulus (N) when the initial disorder in the second lattice contains another pattern (V), illustrating competition or cooperation between two patterns. Parameter values: $a = 10.4, d = 0.06, h_{jk} = h = 0.6$

oscillators with maximum amplitude are shown in black. Then in all three figures the initially disordered lattice replicates the shape of the stimulus contained in the other lattice, although the terminal patterns of Fig. 6.17 and 6.18 preserve features of disorder either in the core of the letter (Fig. 6.17) or in the background (Fig. 6.18). The terminal patterns of Fig. 6.19 show the possibility for a rather faithful replication of the stimulus. What scenario and which parameter values control the *quality* or fidelity of the replication? Let us look, first, at Fig. 6.20, where the interaction occurs in the presence of another letter, "V", in the disordered lattice (which, for example, has been earlier replicated from still another lattice), with the parameter values taken to be the same as for the high-fidelity replication shown in Fig. 6.19. We find that the terminal patterns in this case have a significant trace of the letter V in the core of the stimulus. Thus, these four examples show that the fidelity of the replication process is sensitive both to d and to "properties" of the initial distributions. The explanation of why an initially disordered lattice is able to reproduce the pattern of another lattice and to control the quality of this reproduction is given in the following subsection.

Dynamical Origin of Replication. Note, first of all, that the intralattice diffusive coupling, d, between the oscillators in each lattice must be small enough (in the region D_{ch}). The 2^{N^2} possible steady amplitude distributions of in-phase motions yield, in each independent lattice, patterns of widely arbitrary spatial structure. Since a local unit is bistable, the possible amplitude of its oscillations for small d are in the corresponding absorbing domains (Sect. 5.6.1), i.e. in small regions close to the rest state, O^0, and to the excited state, O^+, of each independent oscillator.

Let us consider the stimulus (initial pattern) as a number of clusters. (As earlier a cluster is considered to be a group of a rather large number of neighboring oscillators having the same amplitude, close either to the rest state or to the maximum amplitude state.) For instance, the stimulus in Fig. 6.17–20 (letter "N") has two clusters (core and background). A completely disordered pattern does not have clusters. Then the elements of a cluster are located in the absorbing domains close to the states O^0, if this is a cluster of oscillators at rest, or close to O^+, for oscillators at their maximum amplitude. For instance, the elements of the homogeneous pattern having one cluster are located right on either O^0 or O^+, while the elements of a disordered pattern are located rather far from O^0 or O^+ but within the absorbing domains.

The second point to be noted is that the interlattice couplings, $h_{j,k}$, should be strong enough for the synchronization of the amplitude patterns, i.e. (6.12) is fulfilled. Thus, any unit (j, k) of one lattice has a relatively weak interaction with its neighbors, $d \neq 0$, while it has a relatively strong interaction with the corresponding unit of the other lattice, $h_{j,k} \neq 0$, hence $\inf_{j,k} h_{j,k} \gg d$ is satisfied. Thus the evolution of the amplitude of any pair of oscillators (one taken from the first lattice, r, and the other, ρ, from the second), coupled by $h_{j,k}$, separately can be described by

$$\begin{cases} \dot{r} = -rF(r) - h_{j,k}(r-\rho)\,, \\ \dot{\rho} = -\rho F(\rho) + h_{j,k}(r-\rho)\,. \end{cases} \tag{6.23}$$

Figure 6.21 illustrates the qualitative phase portrait of (6.23). There are two stable steady nodes, $O^{0,0}(0,0)$ and $O^{+,+}(r_0, r_0)$, corresponding, respectively, to the rest and maximum amplitude states of the oscillators, and the saddle point, O^s, whose incoming separatrices are denoted by w_1 and w_2. The initial conditions for this system should be associated with steady patterns of independent lattices. Since we use (6.23) for any pair of oscillators, all possible initial points are located in the absorbing domains, near one of the points $O^{0,0}$, $O^{+,+}$, $O^{0,+}$, $O^{+,0}$ (Fig. 6.21). Taking the initial conditions to be within the absorbing domains, we see the influence of a relatively small d in such a way that the state of all other elements of the lattice only determines the position of the initial point within the domain. Then, if both oscillators from a pair have been near the rest state, or have been excited before they interact (initial conditions should be taken in the domains near $O^{0,0}$ or $O^{+,+}$, respectively), they do not, apparently, change their states (do not leave the corresponding domains) after the interlattice interaction is switched on. If, however, the oscillators have been in different states (the initial point lies within the domains near $O^{0,+}$ or $O^{+,0}$; Fig. 6.21) they evolve, changing their states, and become synchronized as a result of the interlattice coupling. Note that these points are not steady states of (6.23). Possible routes of this process

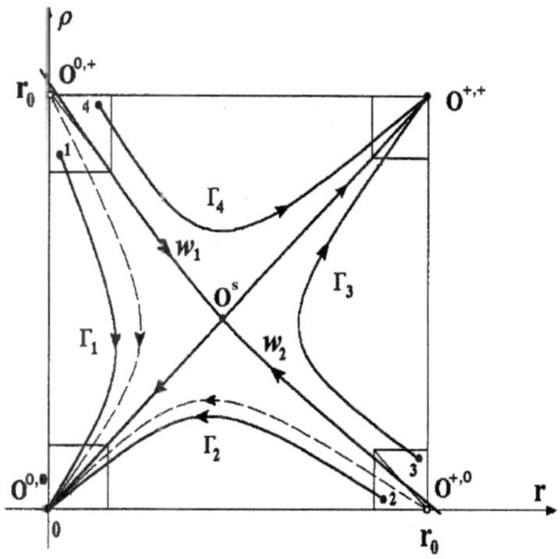

Fig. 6.21. Phase portrait of the auxilary system (6.23). The four rectangles are the domains where the initial conditions for the system should be located. The trajectories $\Gamma_1, \Gamma_2, \Gamma_3, \Gamma_4$ show the possible replication routes as the result of *competition* of a pair of oscillators taken from different lattices.

are shown in Fig. 6.21 by the trajectories $\Gamma_1, \Gamma_2, \Gamma_3$ and Γ_4. They differ by the location of the initial point relative to the separatrices w_1 and w_2. Thus, pattern interaction is reduced to the *competition* of oscillations in the corresponding pairs of oscillators in the coupled lattices.

The replication of a stimulus can also be explained as follows: Let the first lattice, described in (6.23) by r, contain a patterned stimulus, while the second lattice, ρ, is spatially disordered. Consider one oscillator from a cluster at rest of the stimulus and its corresponding excited oscillator in the disordered lattice (point "1" in Fig. 6.21). The excited elements of this pair tend to the motionless state by the route Γ_1. Analogously, the motionless elements of the disordered one in interaction with the excited cluster tend to the excited (maximum amplitude) state (trajectory Γ_3). Thus, as a result of intralattice interaction, the lattice of disordered oscillators *copies* the spatial structure of the stimulus with rather high fidelity.

As earlier mentioned, an oscillator from the disordered pattern stays rather far from the steady points of an independent oscillator, while an oscillator from a cluster is very close to these states. Hence, the initial conditions are located near two *angles* of absorbing domains for a pair of oscillators when one of them is taken within a cluster and another from the disordered state (points "1" and "3" in Fig. 6.21). These angles are associated with the edge of the absorbing domain for one oscillator and the origin of the domain for another. This is the major ingredient in the evolution process, essential for replication. It fails only for oscillators located near the boundaries of the clusters, and hence distortions appear in the replicated images even when the overall replication process is highly faithful.

The second ingredient concerns the relative position of w_1 and w_2 with respect to the absorbing domains. It depends on the shape of the nonlinearity and on the strength of $h_{j,k}$. Faithful replication for all kinds of stimuli is expected when the separatrices lie closest to the bisector lines (Fig. 6.21) of the absorbing domain rectangles (or separate the angles where the initial points are located). If the separatrices intersect the domains near one of the angles, or do not intersect them at all, we have traces of disorder in the cluster at rest (Fig. 6.17) or in the excited cluster (Fig. 6.18) of the replicated image.

Note also that if the initially disordered lattice contains a cluster (letter "V" in Fig. 6.20) the condition for faithful replication is not fulfilled (point "3" stays below the separatrix w_1) for some pairs of oscillators. These pairs do not copy the features of the stimulus but retain the features of the disordered state. Replication in this case is incomplete.

Needless to say, there is no high-fidelity replication for vanishing intralattice diffusion, $d = 0$. In this case the initial conditions are located directly at the points $O^{+,0}$ and $O^{0,+}$ (Fig. 6.20), and no matter what spatial structure of interacting patterns and interlattice coupling exist (no absorbing domains in this case) they are always mapped to the motionless state (as is shown in Fig. 6.21 by the dashed trajectories) or to the excited state (if w_1 and w_2 lie below $O^{+,0}$ and $O^{0,+}$ in Fig. 6.21) for each pair of oscillators. This excludes the possibility of faithful replication.

Rarefied Interlattice Couplings. In the preceding case we considered the mutual interaction of two lattices when all interlattice coupling coefficients, $h_{j,k} = h > 0$, are equal and higher than a threshold value. This is the case of *homogeneous* coupling between the lattice layers. Note that the threshold obtained in Sect. 6.2.2 gives only the estimated upper value of the actual threshold which varies according to the pattern and depends on the lattice parameters and spatial structures of the initial distributions. Let us now see what happens if, as earlier studied in other cases, there is a number of interlattice couplings, $h_{j,k}$, below threshold or, simply, there are broken bonds.

Let us first turn to Fig. 6.22, which illustrates the results of the numerical integration of (6.10) for the case of rarefied couplings while all other parameters are taken the same as for the faithful replication shown in Fig. 6.19. The coefficients of the matrix $h_{j,k}$ are randomly distributed within the interval $[0, 0.6]$. Thus, we have some couplings above threshold, others below threshold and a number of broken bonds. Actually the replication still occurs here but the replicated image is rather imperfect relative to the case of homogeneous coupling (Fig. 6.19).

Following the arguments given in Sect. 6.3.1 we can say the following: The elements at site (j, k), for which $h_{j,k}$ exceeds the threshold, interact as described for the case of homogeneous coupling, i.e. the disordered oscillators in each lattice tend to the state defined by the elements of the stimulus. Pairs of oscillators with broken or too weak bonds cannot be described by (6.10), because the condition $h_{j,k} \gg d$ is not fulfilled, and due to the significant role

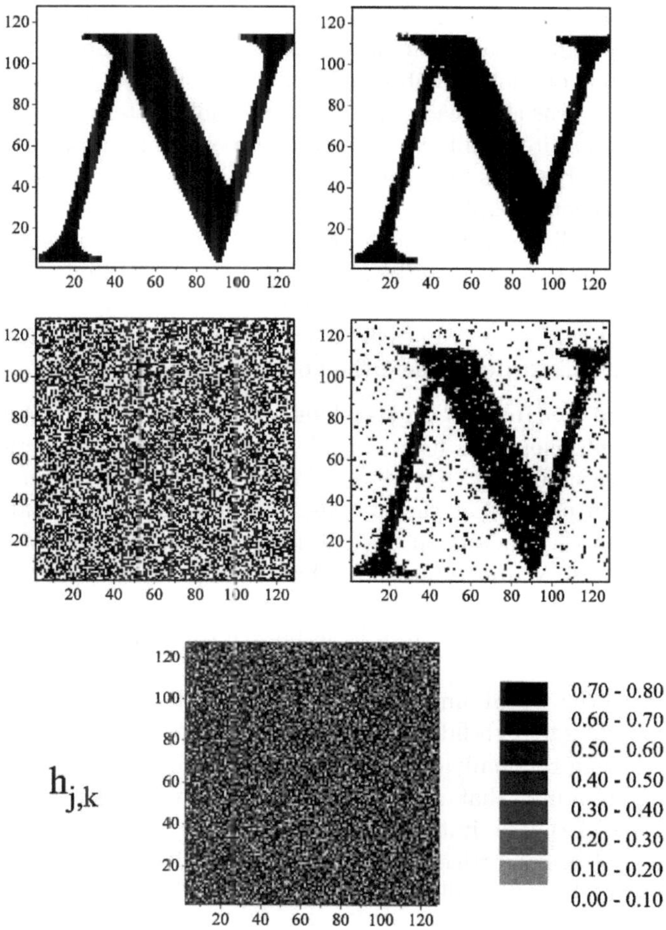

Fig. 6.22. Replication of a patterned stimulus with *rarefied* coupling between the lattices. The values of $h_{j,k}$ are randomly distributed (bottom panel) and corresponding grey scale indicates $0 < h_{j,k} < 0.8$. Parameter values: $a = 10.4, d = 0.06$

played by the intralattice coupling they cannot be considered independent of the other lattice elements in their own layer. If before they interact the oscillators of such a pair have been in different states (near the state of rest or near the excited state) they cannot change their states because d is too weak (the magnitude of d has been chosen, namely, to provide the possibility of existence of all $2^{(N-1)^2}$ steady states of the lattice for fixed state of a given oscillator). Thus, at (j, k), with drastically weak or broken coupling $(h_{j,k} \to 0)$, the replicated image has distortions which appear in the terminal patterns of Fig. 6.22. The oscillators of the disordered lattice, for which $h_{j,k}$ is not vanishing but does not exceed the threshold, can both synchronize with the stimulus by means of the cooperative action of a weak

d and a not too strong $h_{j,k}$. Either they replicate features of the stimulus or they tend to a state located somewhere in between the state of rest and the excited state but outside the absorbing domains of these states. Note, that the terminal patterns are steady states for the whole system (the two coupled lattices as a composite architecture) because the system is gradient. However, they are not steady states for each lattice taken separately, in contrast to the case of homogeneous coupling.

Thus, the rarefied coupling also provides the ability to replicate a given patterned stimulus in the disordered lattice through the mutual interaction of the two lattices. Since the system cannot "overcome" the broken coupling by the action of neighbors (d in each lattice is too weak), the replicated image will always have distortions where the bonds are broken.

Replication of a Stimulus on Lattices of Oscillators with Arbitrarily Distributed Phases. We have considered the mutual interaction of two coupled lattices assuming that all oscillators in either lattice are in-phase. It has also been noted that the in-phase motions are locally stable, i.e. all small disturbances imposed on the phases in the process of evolution will asymptotically decay to the phase-synchronization mode. Here we illustrate how replication occurs through the synchronization of oscillator *amplitudes*, when the phases of the oscillators have been initially randomly distributed along the lattices.

Let us take the initial amplitude patterns and the parameters of the system corresponding to high-fidelity replication for in-phase motions (Fig. 6.19), their phases being randomly distributed within the interval $(-1, 1)$. Integrating (6.10) it is found that amplitude patterns synchronize rather quickly ($t \approx 12$) (Fig. 6.19). The initially disordered lattice reproduces the stimulus given by the patterned lattice. The phases of the oscillators evolve on a longer time scale, and Fig. 6.23 illustrates the stages of this evolution and how the system reaches a terminal state with all oscillators close to the phase-synchronization mode (the last pair of pictures in Fig. 6.23). First, there is *phase clustering* (as earlier, a phase cluster is a group of neighboring oscillators with equal phase) which grows following the intralattice interaction (smaller clusters are absorbed by bigger ones, like the Ostwald ripening of drops and crystal growth) and, finally, we have all oscillators in-phase. Note that at intermediate stages the phase patterns copy the main features (the contours of the letter "N"; Fig. 6.23) of the amplitude patterns already synchronized (Fig. 6.19). Thus, the coupled lattices of oscillators whose phases have been initially arbitrarily distributed are able to replicate a patterned stimulus carried by the amplitudes, while the phases of all oscillators tend to the phase-synchronization mode.

Image Transfer by Replication. To ilustrate the process of pattern replication we take as the stimulus the black-and-white portrait of a young lady (Fig. 6.24). The initial condition in the first layer of the 3D architecture when ($h = 0$) is associated with this picture in the following way: Let the

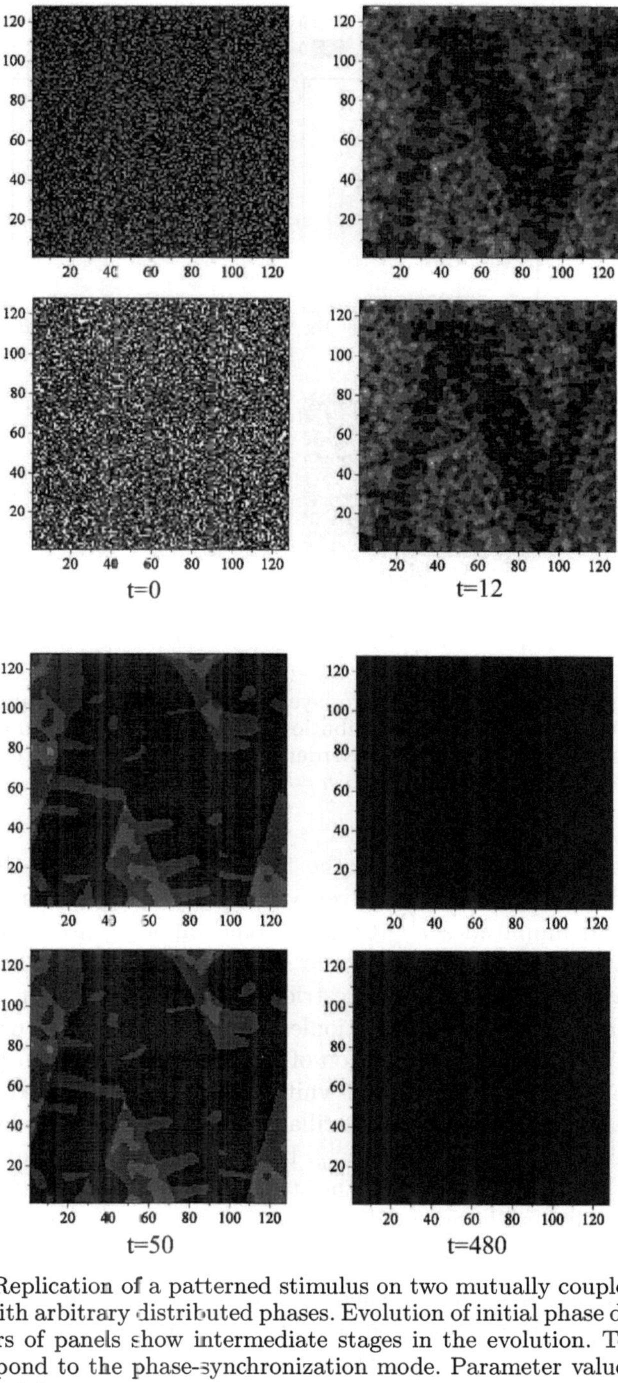

Fig. 6.23. Replication of a patterned stimulus on two mutually coupled lattices of oscillators with arbitrary distributed phases. Evolution of initial phase distributions. Vertical pairs of panels show intermediate stages in the evolution. Terminal patterns correspond to the phase-synchronization mode. Parameter values: $a = 10.4$, $d = 0.06, h = 0.6$

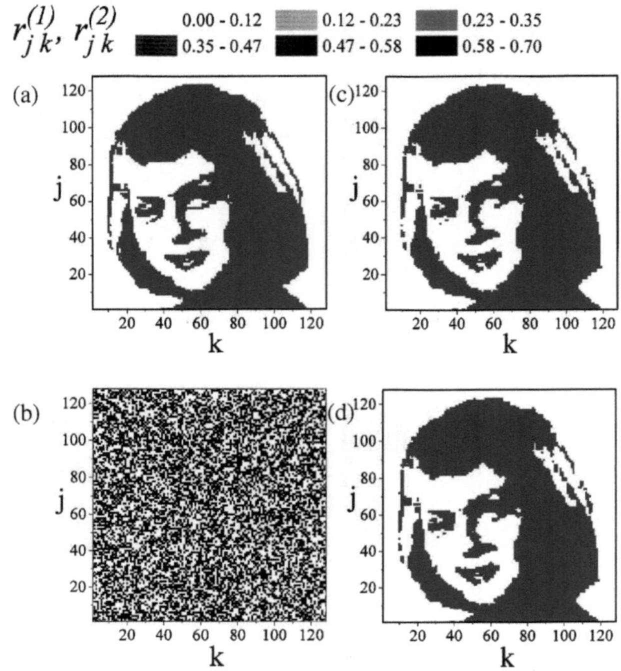

Fig. 6.24. Image replication in a two-layer system. (**a–b**) Initial amplitude patterns; (**c–d**) terminal amplitude distribution in the layers. (**a**) A young lady's face encoded as amplitude pattern; (**b**) disordered pattern – "pure" state of the layer. Parameter values: $a = 10.65, D = 0.06, h = 0.7$

initial phases of the oscillators in the first layer, $\varphi^{(1)}_{j,k}(0)$, be chosen to be arbitrarily distributed around a given value [for instance, near $\varphi^{(1)}_{j,k}(0) = 0$] and the initial amplitudes $r^{(1)}_{j,k}(0)$ have some inhomogeneous distribution. As earlier mentioned, in this case the system evolves to the coherent (in-phase) mode with the amplitudes restricted within small regions (absorbing domains) near the unperturbed motionless state, $r = 0$, and the excited state, $r = r_0$. Hence, the steady distribution of the amplitudes, $r^{(1)}_{j,k}$, in $\{\mathbf{Z}^2, \mathbf{R}\}$ has a *bistable* character. Any black-and-white image can be encoded as a steady amplitude pattern of the coherent oscillation of the lattice layer by a suitable choice of the initial conditions for $r^{(l)}_{j,k}$. In our simulations we take the white color when the oscillators are near the state of rest, and the black color when they are near the oscillatory mode.

Replication in a Two-Layer System. Let us start considering the assembly of two layers ($M = 2$ in system 1) such that when $h = 0$ the first layer contains the amplitude pattern associated with the stimulus (Fig. 6.24a) while the second layer is in a spatially disordered, seemingly chaotic state. This state is given by the amplitude distribution which is also steady in time but random in space, corresponding to a pseudo-random sequence generated by computer

(Fig. 6.24b). Thus, the second layer can be considered as structureless, "raw" material.

Two Copies of the Original Image. When the interlayer interaction is switched on and becomes strong enough, $h > h_1^*$ (Sect. 6.2.2), after a rather short transient process the system tends to the synchronized state (Fig. 6.24c,d). It is also a steady amplitude pattern which is a faithful copy of the initial image. Thus, the oscillators of the disordered layer become *self-organized* according to the "template" proposed by the stimulus. This "template" does not strictly force the disordered layer because there is mutual (bidirectional) interaction between the two layers. The replicated copies, or off-springs, always have some imperfection or distortions relative to the original image. Hence, our system does not work like a photocopying machine or a printing press but actually operates as a dynamically *self-regulating* machine.

Note, however, that replicated "misprints" do not occur at random. Figure 6.25 illustrates the misprint pattern built as the difference signal between the original image and its copy. As this pattern clearly selects the *contours* of the lady's face, we can say that it contains the key features of the stimulus. This is what may be of potential interest for this redundancy reduction functions as a data compression tool like in perceptual neural networks. Let us take the difference signal in each unit as the output of the system. A general rule for efficient data compression is to minimize the "mutual information" between the output and the input given with the original stimulus. Loosely, the "mutual information" has a high value if the output unit acts synchronously with the input and vanishes if the output state is different. In our case this measure for the output (Fig. 6.25) and the input (Fig. 6.24a) tends to have a rather low value. In other words, only a few spatial sites, about 1% (points labeled by black dots in Fig. 6.25), are needed to preserve the key features of the stimulus. In fact, we have a 100 : 1 ratio in the data compression as the result of the self-organization process of the two interconnected layers.

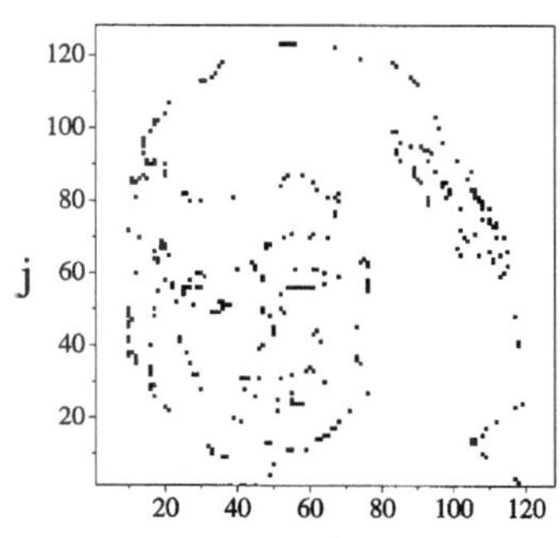

Fig. 6.25. Differential image obtained as the misprint signal between replicated copies (Fig. 6.24c,d) and the original image (Fig. 6.24a)

The Quality of Replication. To estimate the quality of the copies, and hence the fidelity to the original, we introduce a quantity, Δ, called the *replication quality factor*,

$$\Delta(\%) = \frac{N_1}{N^2} \times 100\%\,,$$

where N_1 is the number of oscillators (pixels in the pictures of Fig. 6.24) correctly replicating the stimulus in a given point of the lattice. N^2 is the number of oscillators in each layer. The ratio $N^2 : (N^2 - N_1)$ characterizes the degree of compression accurately described with Δ. Hence, in what follows we deal only with Δ to quantify the fidelity of the replication process.

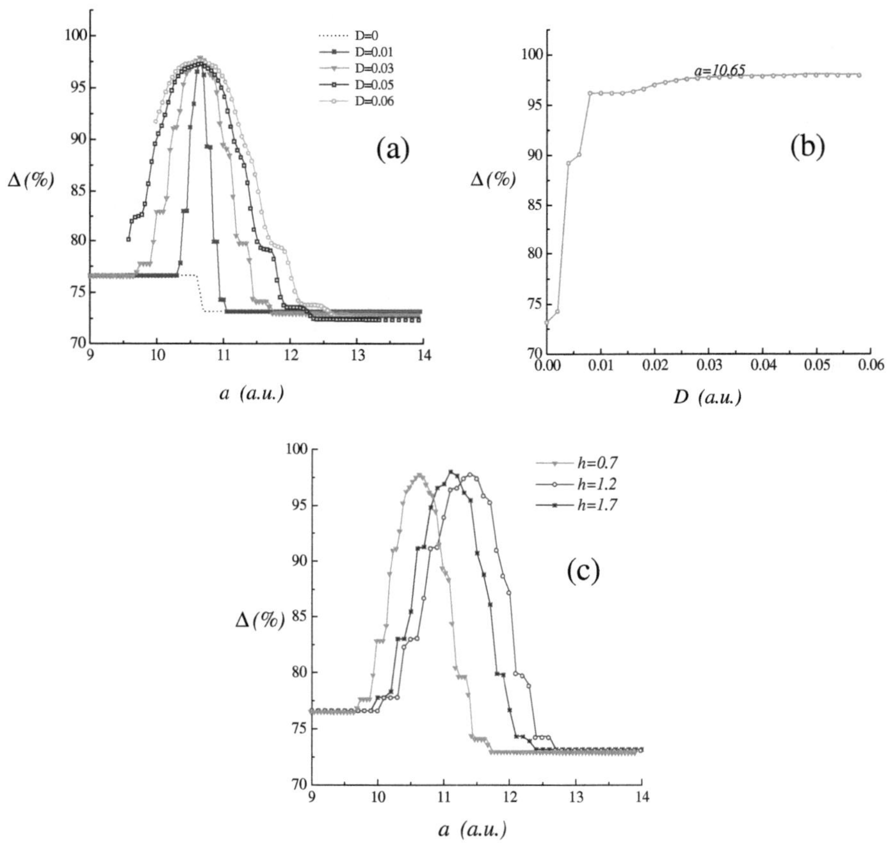

Fig. 6.26. Replication quality factor Δ for pattern replication in a two-layer system. (**a**) Dependence of Δ on the excitability parameter a for $h = 0.7$. Different curves correspond to different fixed values of the intralattice diffusion D. (**b**) Evolution of the peak value of Δ at $a = 10.65$ with increasing D. (**c**) Shift of the "resonance" quality with the parameter h. Units are arbitrary (a.u.)

Figure 6.26a illustrates the replication quality factor as a function of the control parameter a for fixed h and different values of the intralayer diffusion D. The values of the parameter a define the *excitability* of the unit. In particular, high values of a make the oscillator more easily "excitable" in the sense that a small perturbation of its motionless state brings it to the oscillatory mode. Thus, the parameter, a, of the bistable unit regulates the basins of attraction of the stable steady point and the stable limit cycle in the phase plane. When $D = 0$ the replication quality factor, depicted by the dashed curve in Fig. 6.26a, takes only two values $\Delta \approx 76{,}6\%$ and $\Delta \approx 73{,}2\%$. The value of a at the discontinuity of the curve corresponds to the equal basins of attraction of the motionless and oscillatory states. The replicated image in this case has rather significant distortions either in its core (black part of the pattern) or in its background (white part). The behavior of Δ for different values of $D \neq 0$ is shown in Fig. 6.26a. Each curve has an apparent maximum at $a \approx 10.65$. This maximum does not appear instantaneously, although quite rapidly with increasing the D. For that value of a, the dependence of Δ on D is shown in Fig. 6.26b. Note that the quality factor at the maximum point is very high ($\Delta \approx 98\%$). The replicated image is shown in Fig. 6.26c,d. It looks very much like the original. Values of a on either side of the maximum provoke distortions of the copies either in the core of the image or in its background.

It follows that the role of a rather small but nonzero D is significant. The system, in fact, acquires a kind of *selectivity* property. Actually, the curve of replication quality for $D \neq 0$ looks very much like a resonance curve in frequency selection systems. But in our case we have to do with a specific kind of spatial selection. It can be called a *synergetic image selection* when the system replicates any regular spatial image independently of its concrete structural context. For instance, with nearly the same quality the system will process an image of a man's face or a form of capital letters (see Sect. 6.3.1). Note that our synergetic resonance has nothing to do with the time behavior (frequencies) of the local units. It deals only with the steady amplitude patterns and has no oscillatory dynamics.

Figure 6.26c shows how changing the strength of h results in the *shift* of the resonance peak corresponding to the highest value in the replication quality. The three curves are calculated for a fixed value of D and for three different values of h.

Image Transfer in Multilayer Assemblies.

One Original Leading to Many Copies. Let us consider now a multilayer system composed of three lattice layers ($M = 3$). As for $M = 2$, when $h = 0$ the first layer contains the original black-and-white image of the lady's face encoded as a steady amplitude pattern. The other two layers are in spatially chaotic states given by two different disordered patterns. Figure 6.27a illustrates the initial states of the three layers in $\{\mathbf{Z}^2, \mathbf{R}\}$ space. As for the two-layer system a strong enough interlayer interaction, $h > h^*$, leads to the mutually synchronized steady patterns. Figure 6.27b shows the snapshot

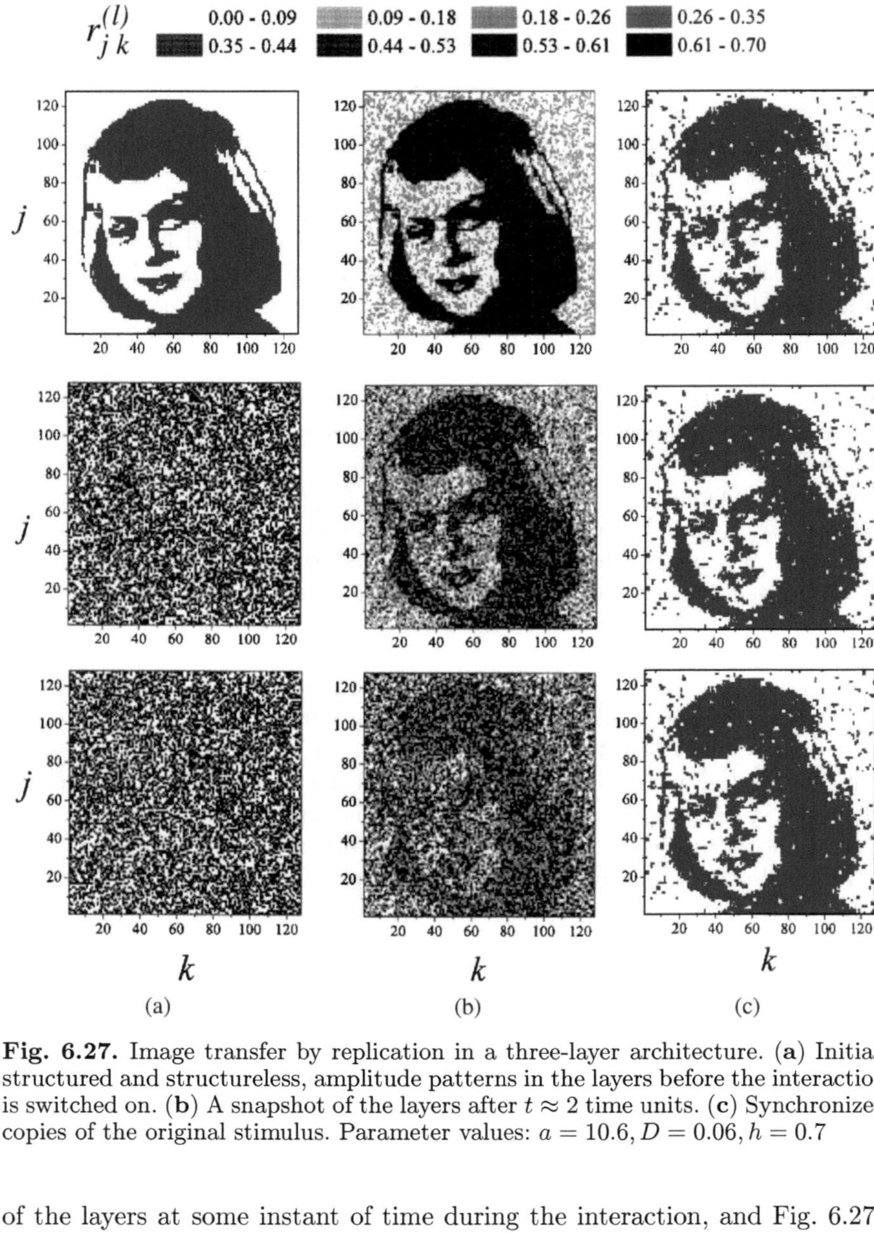

Fig. 6.27. Image transfer by replication in a three-layer architecture. (**a**) Initial, structured and structureless, amplitude patterns in the layers before the interaction is switched on. (**b**) A snapshot of the layers after $t \approx 2$ time units. (**c**) Synchronized copies of the original stimulus. Parameter values: $a = 10.6, D = 0.06, h = 0.7$

of the layers at some instant of time during the interaction, and Fig. 6.27c the terminal, synchronized patterns. We obtain three self-replicated patterns which are rather faithful copies of the original image.

Replication as a Process of Image Transfer. Let us analyze the replication process in the three-layer architecture as a competition between the states of three interacting oscillators, each one taken at the same relative site (j, k). The approximate system describing the amplitudes of three strongly coupled oscillators ($h > h^*$) is of the third order. The analysis of this system which would allow us to predict the "winning" and "slaved" units in this competition is very difficult and represents a separate problem. We rather consider the replication in multilayer systems from another point of view.

Note that the replicated images in the second and third layers do not appear simultaneously, but rather sequentially (see Fig. 6.27b). As we are using nearest-neighbor interaction, first the lady's face appears in the layer nearest to the stimulus and subsequently in the next one. Therefore, the replication process can be considered, approximately, as a sequence of independent acts of *image transfer* from layer to layer. As the connection between the layers is of a diffusive type (6.1), the process of appearance of the image in the disordered layers looks like *image diffusion* in the direction of the "space coordinate" l in a real 3D space. Each "act" of transfer represents the image replication in two interacting lattices and has been described in Sect. 6.3.1. It is clear that distortions or misprints of "self-transferred" images will accumulate during the multiple acts of the replication process, with foreseen quality losses in the terminal pattern relative to the original figure (Fig. 6.27).

Quality of a Tranferred Image in the Multilayered Architecture. To find the conditions for the highest quality of pattern transfer in the three-layer system we analyze the replication quality function Δ (Sect. 6.3.1). Figure 6.28 shows the dependence of Δ on the parameter of excitability a. The curves are calculated for a fixed value of h and for different values of the intralattice diffusion, D. Analogous to Fig. 6.26a the curves exhibit "resonance" characteristics of the three-layer architecture as a synergetic image transfer system.

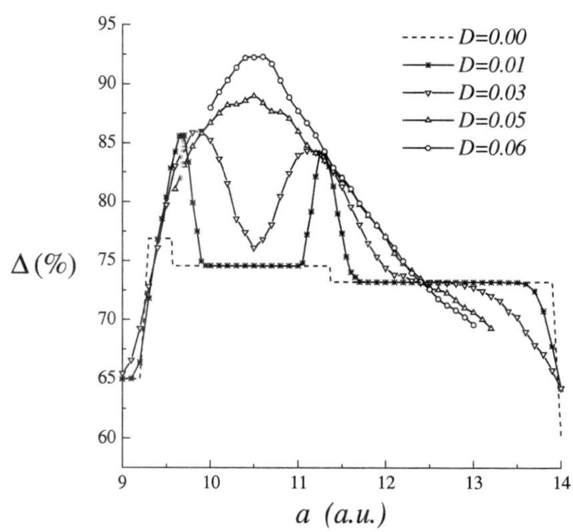

Fig. 6.28. Replication quality factor, Δ, as a function of parameter a for fixed $h = 0.7$ and different values of D in a three-layer image transfer system. Units are arbitrary (a.u.)

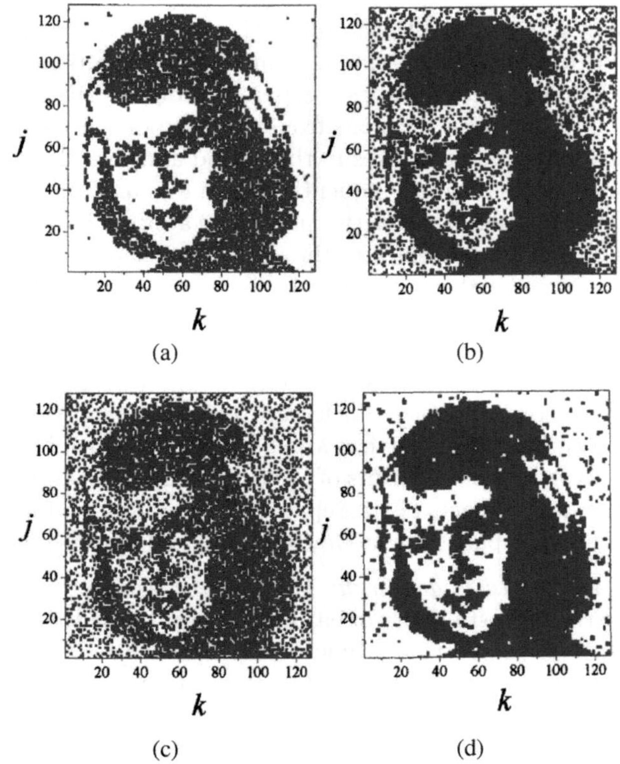

Fig. 6.29. Distorted copies of a stimulus replicated in a three-layer architecture or anatomy for different points of the "resonance" curves of Fig. 6.26a obtained for fixed $h = 0.7$. (**a**) $a = 9.85, D = 0.03$; (**b**) $a = 11.15, D = 0.03$; (**c**) $a = 10.6$, $D = 0.03$; (**d**) $a = 10.6, D = 0.06$

The behavior of the curves looks rather different to that obtained for the two-layer system. First, for smaller values of D the function has two maxima ("peaks") corresponding to nearly the same quality (about 85%) of the copies. However, the distortions or misprints in the replication are rather different for the two "peaks". The transferred images for the two values of a are shown in Fig. 6.29a,b and the distortions appear either in the core or in the background of the image. Figure 6.29c,d illustrate the distortions of a transferred image for fixed a and different values of D. With increasing values of D the two maxima "merge" into one of higher value (see Fig. 6.28). Comparing the curve for $D = 0.06$ with the corresponding one from Fig. 6.26a, we find that the maximum value of Δ becomes lower for the three-layer system, but it stays rather high (about 93%) hence providing a faithful transfer of the stimulus (Fig. 6.27).

Spatial Resolution of the Image Replicating System and Fidelity. A warning should now be given. In the replication process discussed above, the characteristic spatial scale of the input image relative to the lattice constant is crucial for achieving an acceptable copying quality. For instance, let us, consider a system with just two layers and take as the input image given to one of them the chessboard of Fig. 6.30a, whose spatial scale is δ. Then, decreasing δ, the replication quality factor, $\Delta(\delta)$, shows that below a certain value, δ_c, the replication process fatally fails and the original chessboard fades way. In the particular case considered, we have $\delta_c = 6 \div 10$ units (Fig. 6.30b). It is the spatial resolution of the replicating system. Thus, those features of the input image whose characteristic spatial scale is below δ_c cannot be replicated with an acceptable degree of quality $[\Delta \geq \Delta(\delta_c)]$.

Stability of the Image Replication Process with Regard to Noisy Disturbances. Let us now consider the influence of noise on the dynamic replication of images in the multilayer system. Let us assume that each unit is independently perturbed by a weak external noise. Since the replication process only deals with amplitude distributions of oscillators synchronized in phase, we only introduce the noise in the amplitude equations:

$$\dot{r}_{j,k}^{(l)} = f\left(r_{j,k}^{(l)}\right) + d\left((\Delta r)_{j,k}^{(l)}\right) + h\left(r_{j,k}^{(l+1)} + r_{j,k}^{(l-1)} - 2r_{j,k}^{(l)}\right) + \xi_{j,k}^{(l)}(t) \quad (6.24)$$

where $\xi_{j,k}^{(l)}(t)$ is a "white" Gaussian noise of zero mean and variance σ^2,

$$\langle\xi_{j,k}^{(l)}(t)\rangle = 0\,, \quad \langle\xi_{j,k}^{(l)}(t)^2\rangle = \sigma^2\,,$$
$$\langle\xi_{j,k}^{(l)}(0)\xi_{j',k'}^{(l')}(t)\rangle = \delta_{jj'}\delta_{kk'}\delta_{ll'}\delta(0)\,.$$

For simplicity, we only consider a two-layer architecture. The parameter values and initial amplitude distributions are the same as in Fig. 6.24, corre-

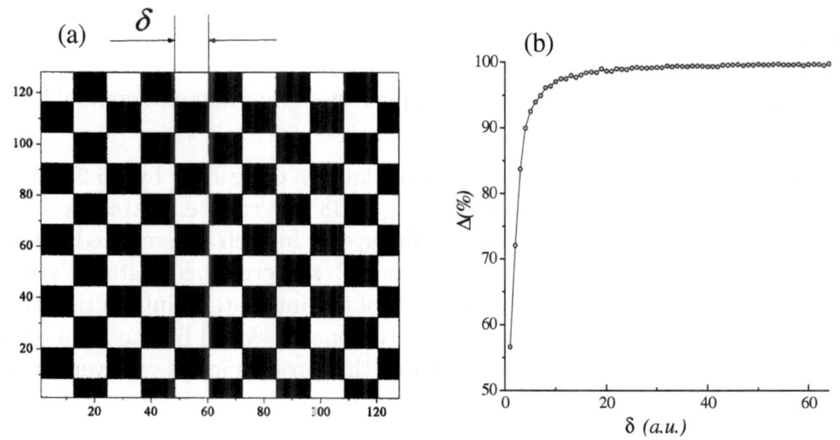

Fig. 6.30. (**a**) Pattern used to test the spatial resolution of the image replication system. (**b**) Dependence of the replication quality factor on the relative characteristic spatial scale of the test image δ; $a = 10.6, D = 0.06, h = 0.7$. Units are arbitrary (a.u.)

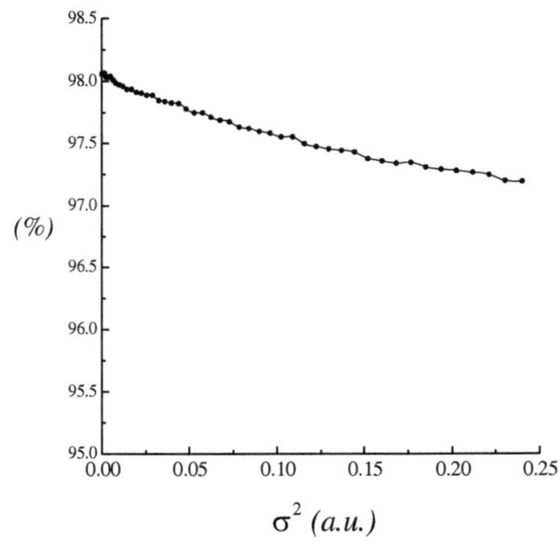

Fig. 6.31. Dependence of the replication quality factor, Δ, on the variance, σ^2, of a Gaussian noise for a two-layer architecture forced with noise. The parameter values and the initial conditions coincide with the values used in the noise-free case shown in Fig. 6.24. Units are arbitrary (a.u.)

sponding to a noise-free case. It appears that in the presence of weak noise the initial stimulus (the lady's face) is still replicated rather faithfully in the spatially disordered second layer (Fig. 6.24). Figure 6.31 illustrates the dependence of Δ on σ^2. Increasing the noise intensity to $\sigma^2 = 0.25$, the value of Δ goes down only by 1% relative to the noise-free case.

6.3.2 Excitable Units

Let us, finally, consider the problems of pattern replication and control in the two-layer system (6.21) [6.21].

Replication of Form. When a single lattice described by (6.21) is in a steady, spatially disordered pattern (Fig. 5.38b) it can be treated as a kind of "raw material". Let us assume that one layer is in such a disordered pattern while the second layer carries an "encoded" structure as a stable steady pattern, given as a stimulus. As the result of the interlattice interaction there is *replication of form.* The disordered state appears slaved by the form of the patterned stimulus. Figure 6.32 illustrates this process in a two-layer lattice. Figure 6.32a,b shows the initial states of the layers. The off-springs, replicated and hence synchronized patterns, are shown in Fig. 6.32c,d. They are quite faithful copies of the initial regular form.

At variance with the reaction diffusion systems of the FNS type the lattice of Chua's circuits has more complex, three-"species" reaction kinetics, where the bistability is provided by the oscillatory stable states (foci). However,

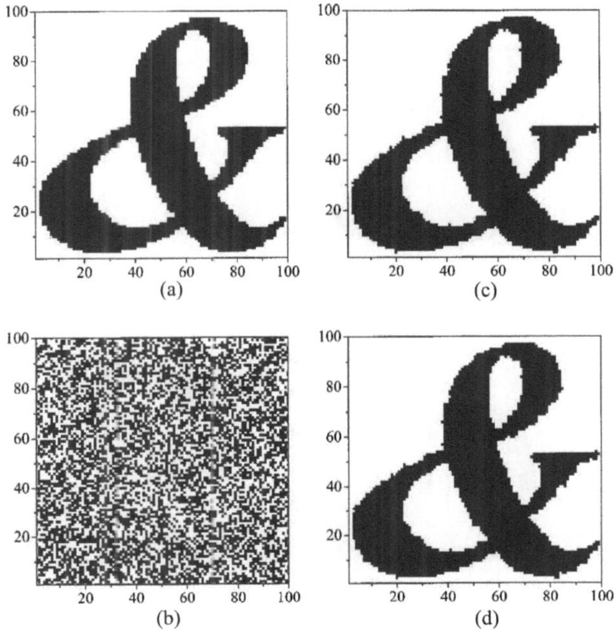

Fig. 6.32. Replication of form in two coupled lattices. Parameter values: $D_1 = D_2 = 0.14, h = 3.75$. (**a**) Patterned stimulus. (**b**) Disordered pattern. (**c,d**) Synchronized state. As time proceeds both layers attain a rather faithful common copy of the stimulus

when weak diffusion and strong enough interlayer interaction are the main processes, the two-layer architecture of Chua's circuits represents an appropriate network realization for replication of form.

Controlled Spiral Waves. Let us consider (6.21) when each layer, taken separately, exhibits different spiral wave patterns (Chap. 5).

Let the first layer of (6.21) exhibit a "bright" excitable spiral wave (Fig. 6.33) while all elements of the second layer are in the motionless state, P^-. At a given instant of time we switch on the interlayer coupling. Its strength is chosen high enough to provide synchronization between the layers. Note that (6.22) gives an upper value for the coupling coefficient. The actual (numerically obtained) value appears to be smaller and depends on the kind of synchronizing motions. The interaction provides entry of the excitation into the initially motionless layer and back to the layer with a rotating spiral. The synchronization process occurs rather quickly. The sequence of snapshots given in Fig. 6.33 shows the development of the spiral wave in the synchronized layers. As with interacting steady patterns, the interlayer interaction yields a wave pattern of a new form. It represents a large-scale spiral wave of the "oscillatory" type (Chap. 5). Hence, the discrete medium,

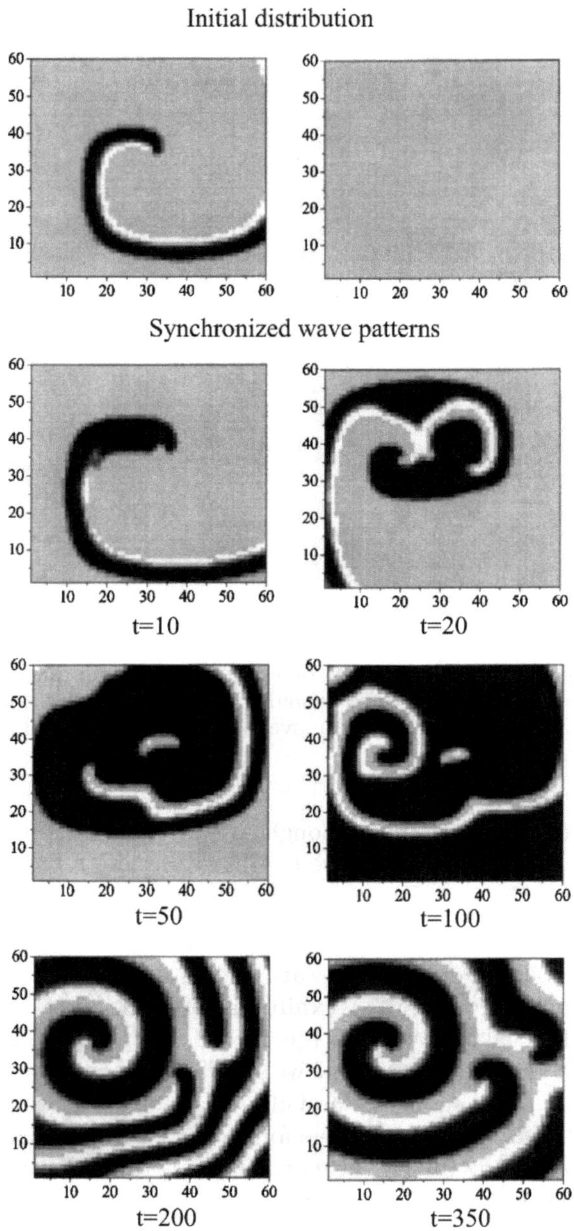

Fig. 6.33. Time sequence illustrating the appearance of an "oscillatory" spiral wave as a result of the mutual interaction of an "excitable" spiral wave with the "unexcited" homogeneous state. Parameter values: $D_1 = D_2 = 0.8, h = 1$

in fact, changes its properties from excitable to oscillatory "reaction kinetics" as a result of the interlattice interaction.

The second example to illustrate wave pattern reentry is the interaction of the large-scale "oscillatory" spiral in the first layer with the homogeneous motionless state in the second. Figure 6.34 shows the initial distribution and snapshots of the synchronized motions in the layers. The resulting wave pattern represents a bright excitable spiral. Thus, again there is "inversion" of the properties of the medium as the elements initially oscillatory (Fig. 5.41b,c) return to excitable behavior (see Fig. 5.40b,c).

The interaction of bright and oscillatory spirals (see Fig. 6.35) leads to the appearance of a synchronized, "dark" spiral wave pattern.

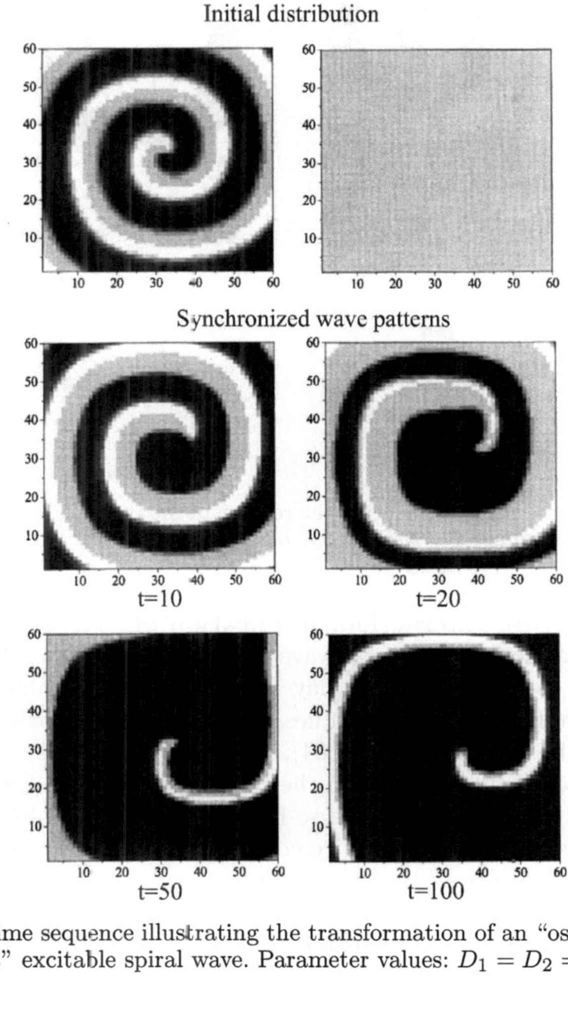

Fig. 6.34. Time sequence illustrating the transformation of an "oscillatory" spiral into a "bright" excitable spiral wave. Parameter values: $D_1 = D_2 = 0.8, h = 1$

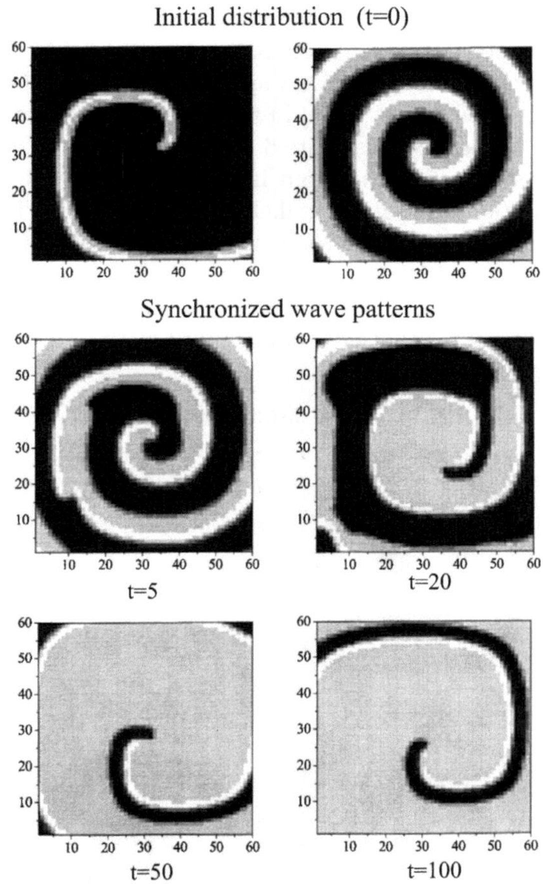

Fig. 6.35. Time sequence illustrating the reentry of the two different types of spiral waves. Parameter values: $D_1 = D_2 = 0.8, h = 1$

Pattern and Wave Interaction. Replication of Spiral Waves. Let us consider the interaction of a spiral wave in one layer with an inhomogeneous steady pattern in the other. This may occur when the intralattice diffusion coefficients have largely different values. To have a steady pattern in a layer, diffusion must be weak enough (5.104), while traveling waves occur for high enough values of D. It follows from the results presented in Sect. 6.2.3 that to have mutual synchronization of two layers, with largely different diffusion coefficients, a very strong interlattice coupling, $h \to \infty$, is required.

Figure 6.36 illustrates the interaction of a disordered steady pattern (Fig. 6.36a) and a "dark" spiral wave (Fig. 6.36b) for $h = 10$. The layers become almost synchronized and exhibit the same spiral waves. This situation is similar to the replication of steady patterns in identical layers. But now the lattice initially in the disordered steady state replicates the spiral wave which

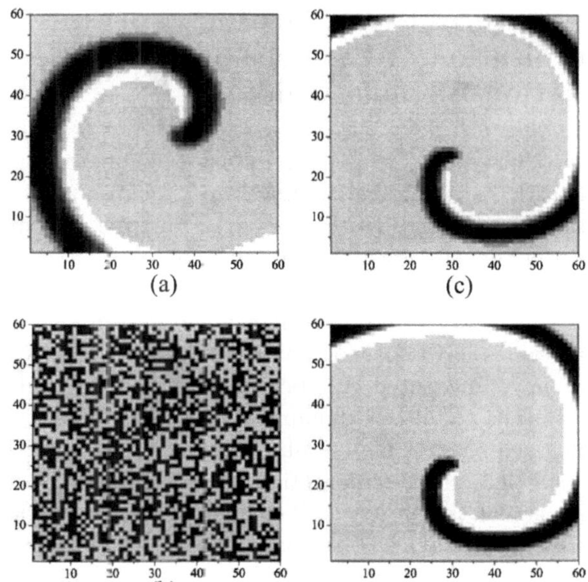

Fig. 6.36. Replication of a spiral wave. (**a**) Initial "dark" excitable spiral wave for $D_1 = 2.5$. (**b**) Steady, spatially disordered state for $D_2 = 0.14$. (**c,d**) Synchronized state ($h = 10$) representing the spiral wave of the same type as the initial stimulus which behaves like the spiral in the lattice with $D \approx 1.3$

is a traveling, rotating pattern. The resulting spirals differ from the original wave. They become "narrower" (Fig. 6.36b,c) as the approximate system describing the synchronized state (Sect. 6.2.3) has the diffusion coefficient $D^+ = (D_1 + D_2)/2$, smaller than that for the initial wave.

What about the dynamic properties of the layers? Before they interact we have two different layers. The first one with rather strong diffusion describes a medium "suitable" only for traveling wave solutions, while the second one with weak diffusion could exhibit only steady patterns. The synchronized layers represent, in fact, a single lattice with D^+, which loses the property to support inhomogeneous steady patterns but keeps, however, the capability for spiral waves. Thus, appropriately choosing D_1 and D_2 in each layer, we can control the dynamics of the resulting, synchronized discrete medium. It is possible, for example, by decreasing D_1 in the first layer, to cause propagation failure of the spiral in the synchronized state when D^+ is not strong enough to support travelling waves.

6.4 Salient Features of Replication Processes via Synchronization of Patterns and Waves with Interacting Bistable Units

We have shown that when intralattice diffusion competes with interlattice coupling, two lattice systems can manage to synchronize through mutual interaction to finally display identical behavior in time and space. Different spatial patterns which are steady structures of each independent lattice, hence taken separately, through mutual interaction evolve to identical steady patterns whose structure can be very diverse and depends on the initial conditions and the values of the parameters of the system. The synchronization process that we have illustrated here proves the ability of a lattice system to reproduce itself as the result of a dynamical process. It is for this reason that the replication process is not merely like that of a printing press.

We have investigated the interaction between patterns, and waves or pulses, in a system of two coupled fiber-like chains. In particular we have considered fibers or (linear) chains composed of electronic cells resistively coupled. We have analytically shown and numerically verified that for strong enough interfiber coupling the possibility of mutual synchronization of all motions is actually realized. We have also numerically illustrated how the synchronization can lead to nontrivial, synergetic effects in the spatio-temporal dynamics of the coupled fibers. Synchronization of spatial disorder, spatial intermittency, the reverse in the propagation of a wave front, pulse reentry and pulse failure, overcoming of propagation failure and the appearance of sources of pulses are a few spectacular examples of the results found. Thus, interfiber synchronization allows us to excite and control different types of spatio-temporal patterns in coupled fibers. We have also shown the results of interaction between parallel fibers with few (one and two) interfiber connections. In particular, pinned coupling between fibers with *excitable* kinetics brings controlled transfer of an excitation from fiber to fiber. Such a contact behaves qualitatively very much like the synaptic connections between neurons. Fibers with two dynamic contacts can operate as a spatially extended generator of synchronous sequences of pulses with controlled time delay between them.

We have also shown that, for strong enough coupling strength between lattices, the spatial pattern on one lattice controls completely the image pattern replicated on another. Such control occurs even when the number of connections between layers in the 3D architecture is less and even much less than the number of units in each lattice. The observed phenomena also occur when the spatial distribution of interlattice connection is disordered in space, hence exhibiting rather randomly distributed values. The robustness of the observed control and synchronization phenomena permits us to safely say that similar results are expected for other models.

The main features of the replication process described above are as follows:

(i) The replication of a patterned stimulus by a disordered lattice occurs through mutual interaction of strongly coupled lattices with weak but non-vanishing diffusion in each lattice ($0 < d \ll h_m$).
(ii) The interlattice interaction can be described as a kind of *competition* of the amplitude states of each pair of strongly coupled oscillators taken independently at a site (j, k) of both lattices. The initial condition for this competition is defined uniquely by the spatial structure of the image and the disorder, for nonvanishing intralattice coupling.
(iii) The second lattice, or "raw material" ready for replication, should be quite disordered or at least it should not contain clusters of oscillators. This provides an advantage to a stimulus relative to a disordered distribution in the competition and hence gives the possibility of faithful replication.

We have shown how dynamic replication of images can be achieved with a controllable degree of fidelity in a multilayer architecture of diffusively coupled *bistable* oscillators. Each layer forms and preserves black and white images or two-level encoded forms or functions as 2D steady patterns of amplitudes of coherent, time-bound oscillations. Spatially chaotic, disordered patterns are used as "raw" frames capable of being "filled" according to the form of any stimulus.

We have shown that, with appropriate inter-layer interaction, the system in the process of evolution can replicate in multiple copies an original image imposed as a stimulus in one of its layers. Thus, the multilayer architecture may mimic nature as, for example, it shows *cloning*-like dynamics, and it is capable of repeating many times at least the key features of an original. The misprints of replicated images arising as a result of pattern competition can be treated as some additional "degrees of freedom" in the copying system. The distorted copies may acquire new (useful) features, new information appearing, for example, from the accumulation of misprints in multiple acts of image replication or as the result of specific action of some fluctuations.

By studying the replication phenomenon, we have found that it can be considered as a process of image transfer (image diffusion) from layer to layer inside the multilayered architecture. The quality of the transferred image given by the replication quality function, Δ, has "resonance" features. The system possesses somewhat image (synergetic) selectivity. The high-quality transfer occurs only for specific values of the parameter a responsible for the "excitability" of the units. By taking the difference signal between the input stimulus and the self-replicated image, we have illustrated that processes in layered lattices provide efficient information compression.

Mutual synchronization of two lattices of excitable and bistable units yields the possibility of replicating steady, regular and irregular patterns as well as wave-like behavior, including oscillatory and excitable spiral waves, a phenomenon typical of reentry processes in 3D reaction–diffusion systems.

To conclude, in view of the contents of our chapter, let us emphasize the importance of reading the interesting publications of Heagy, Caroll and Pecora [6.6–9], Hopfield [6.10, 6.11], and Hoppensteadt and Izhikevich [6.12, 6.13], who have discussed the possible applications of active lattice systems for information processing and storage.

7. Spatio-Temporal Chaos in Bistable Coupled Map Lattices

7.1 Introduction and Motivation

In this chapter we consider lattice models with discrete time, i.e. dynamical systems with a discrete time, discrete space and continuous state and hence coupled map lattices (CML). CML describe the collective dynamics of a finite or infinite number of interacting finite-dimensional local dynamical systems (elements, units or cells) placed on the sites of a spatial lattice (the "physical" space). The dynamics of a single element is described by a point map. Thus, CML are characterized not only by their internal dynamics, i.e. that of the local elements, but also by the dynamics in the "physical" space. CML have been used to model a variety of phenomena including pattern formation, nonlinear waves, topological spatial chaos, distribution of defects in the spatial structures of nonlinear fields, reentry initiation in coupled parallel fibers, etc. [7.1–3, 7.6, 7.7, 7.9–16, 7.18, 7.20–22, 7.24, 7.25–31].

The contents of this chapter has been divided into two parts. The first part is devoted to the investigation of unbounded multidimensional CML systems and the second part to bounded systems having a 3D architecture. The systems of the first type are used for investigating the phenomena of spatial chaos and chaos of traveling waves. Such phenomena cannot be considered in the frame of bounded systems.

We start the investigation of unbounded CMLs from the discussion of a constructive method for studying the linear stability of bounded motions. Then, using this method we study the phenomenon of spatial chaos (1D and 2D cases) and chaotic traveling waves in a bistable CML with diffusive coupling between the units.

In the second part of the chapter we consider CML composed of two interacting lattices. Each lattice may be treated as a layer (Fig. 1.5d). A layer represents a 2D lattice of diffusively coupled cubic maps. First, we study the dynamics of a single lattice (layer). By constructing invariant domains in the phase space, we obtain the conditions for pattern formation in such a lattice. The distribution of the "species" in the lattice space has a binary character, hence the layer is capable of storing binary images as stationary patterns or fixed points in the phase space. We obtain the conditions of global interlayer synchronization and study the phenomena of pattern synchronization and pattern replication. Let the first CML layer contain an image encoded as a stationary pattern and the second layer be in a disordered or spatially chaotic state. As a result of interlayer synchronization of the CML, the stimulus is replicated in both layers with some degree of fidelity. Moreover, the difference between the original and replicated pattern may select the contour of the processed image. Earlier, we have studied the replication phenomenon in layered systems with continuous time (see Chap. 6). Similar phenomena have been found and studied in systems of coupled chaotic maps. We consider the possible instabilities of the synchronization manifold. Instability typically occurs through riddling and blowout bifurcation leading to off-diagonal attractors with specific properties (for example, bubbling and on–off intermittency). In the case of the layered CML the synchronizing subsystems (layers) may have very complex time dynamics (large number of fixed points, limit cycles, chaotic attractors), while the space dynamics concerns pattern formation and pattern replication. We discuss some characteristic features of loss of pattern synchronization by using a transverse instability factor and a density function. [7.4, 7.5, 7.32, 7.33]

7.2 Spectrum of the Linearized Operator

Let us return to Chap. 1, where we presented the CML in the following form:

$$\mathbf{U}(n+1) = \mathbf{F}\big(\mathbf{U}(n)\big) \quad \text{or} \quad u_j(n+1) = F(\{u_j(n)\}^s)\,, \tag{7.1}$$

where $\mathbf{U}_j \in E^p, j \in \mathbf{Z}^d$ and F is subjected to imposed conditions (1.8–11,14). The system (7.1) is a classical dynamical system (generated by a continuous map on a complete metric space). However, the operator F is not differentiable in the sense of Fréchèt but only in the sense of Gateaux. Thus the stability of a solution of (7.1) in linear approximation does not generally imply the stability of this solution relative to the original system. Nevertheless, information on the linear stability is very useful. Indeed, on the one hand linear stability is essential for actual stability. On the other hand, for lattice models that consist of a finite rather than infinite number of subsystems, linear stability implies stability relative to a nonlinear system, so that at the heuristic level we need not justify the correspondence between linear stability and stability relative to the original system. Finally, certain correspondence does exist thanks to the "Lipschitz version" of stability theory and owing to the fact that the evolution operator is differentiable in the sense of Gateaux.

7.2.1 Linear Operator

Let $\mathbf{U}^*(n)$ be a known bounded solution of (7.1),

$$|u_j^*(n)| \le M_1\,, \quad j \in \mathbf{Z}^d\,, \tag{7.2}$$

and let $\boldsymbol{\xi}(n)$ be a perturbation,

$$\boldsymbol{\xi}(n) = \mathbf{U}(n) - \mathbf{U}^*(n)\,.$$

Then we have

$$\boldsymbol{\xi}(n+1) = A(n)\boldsymbol{\xi}(n) + G(\boldsymbol{\xi}(n)\mathbf{U}^*(n))\,,$$

where $A(n)\boldsymbol{\xi}(n)$ is the linear part,

$$\left(A(n)\boldsymbol{\xi}(n)\right)_j = L_j(\{\xi_j(n)\}^s) \equiv \sum_{|i-j|\le s} \left.\frac{\partial F}{\partial u_i}\right|_{\{u_j^*(n)\}^s} \xi_i(n)\,,$$

and G is the nonlinear part. Hence due to (7.2) the following inequality holds:

$$|L_j(\{\xi_j(n)\}^s)| \le c \max_{|i-j|\le s} \{|\xi_i(n)|\}\,,$$

where c depends on j. Therefore, the operator $A(n)$ is bounded:

$$||A(n)||_q < \infty\,.$$

Then if the spectrum of $A^*(n)A(n)$ belongs to a set lying strictly inside the unit circle, the following inequality holds:

$$||A^*(n)A(n)||_q < 1 - \Delta(n)\,,$$

where $0 < \Delta(n) < 1$. Consequently, the solutions $\mathbf{U}^*(n)$ will be linearly stable if $\Delta(n) \ge \Delta_0 > 0$.

7.2.2 A Finite-Dimensional Approximation of the Linear Operator

Let $A : B_q \to B_q$ be a linear bounded operator given by

$$(A\xi)_j = L_j(\{\xi_j\}^s)\,, \tag{7.3}$$

where $\boldsymbol{\xi} \in B_q$ and $j \in \mathbf{Z}^d$. Then $L_j : (E^p)^{(2s+1)^d} \to E^p$ is a linear map such that

$$|L_j(\{\xi_j(n)\}^s)| \le c \max_{|i-j|\le s} \{|\xi_i(n)|\}\,, \tag{7.4}$$

and c does not depend on j.

Note that with A we mean $A^*(n)A(n)$. Let us denote by $A_N : B_q \to B_q$ an operator having the following form:

$$\begin{aligned}
(A_N\xi)_j &= L_j(\{\xi_j\}^s)\,, & |j| &\le N - s\,,\\
(A_N\xi)_j &= 0\,, & |j| &> N\,,\\
(A_N\xi)_j &= L_j^b(\{\xi_j\}_N^s)\,, & N - s &< |j| \le N\,,
\end{aligned}$$

where $\{\xi\}_N^s = \{\xi_i | |i| \le N, |i-j| \le s\}$ and L_j^b are linear maps such that

$$|L_j^b(\{\xi_j\}_N^s)| \le c \max_{|i-j|\le s, |i|\le N} \{|\xi_i|\}\,, \tag{7.5}$$

where c is the constant from (7.4). Then A_N is a bounded operator which is, in fact, a finite dimensional approximation of A (see Fig. 7.1).

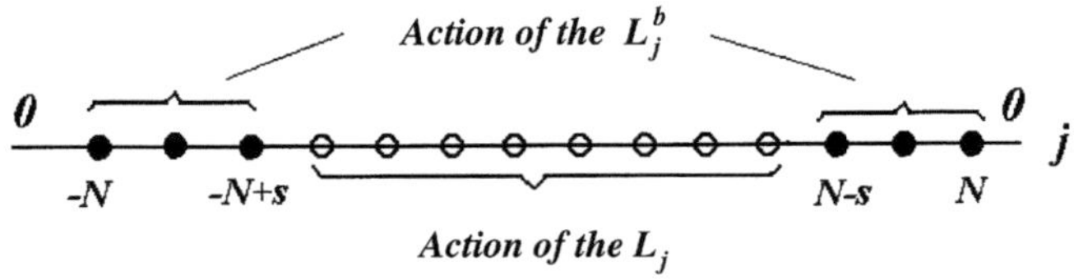

Fig. 7.1. Schematic of the construction a finite-dimensional approximation of the linearization operator

7.2.3 Methodology to Obtain the Linear Spectrum

We take

$$\Lambda_N = \operatorname{spec} A_N\,, \quad \Lambda = \operatorname{clos}\left\{\bigcup_{N=N_0}^{\infty} \Lambda_N\right\},$$

where $N_0 \gg 1$ is a fixed integer, and assume that a certain orthogonal basis is given in E^p. Then the orthogonal basis in $(E^p)^{(2N+1)^d}$ is readily found corresponding to the scalar product

$$\langle \eta, \zeta \rangle_{N,q} = \sum_{|j|\le N} \frac{(\eta_j, \zeta_j)}{q^{|j|}}\,,$$

where $\eta, \zeta \in (E^p)^{(2N+1)^d}$ and $\tilde{A}_N : (E^p)^{(2N+1)^d} \to (E^p)^{(2N+1)^d}$ corresponds to A_N. Let

$$\tilde{\Lambda}_N = \operatorname{spec} \tilde{A}_N\,.$$

The following properties of A_N hold:

- $\lim\limits_{N\longmapsto\infty} ||A_N \boldsymbol{\xi} - A\xi||_q = 0$ for any $\boldsymbol{\xi} \in B_q$;
- $\Lambda_N = \tilde{\Lambda}_N \bigcup\{0\}$;
- $\tilde{\Lambda}\bigcup\{0\} = \Lambda$, where $\tilde{\Lambda} = \operatorname{clos}\left(\bigcup\limits_{N\ge N_0} \tilde{\Lambda}_N\right)$.

Using these properties we can find the criteria to obtain the spectrum of the linearized operator.

If A is a normal operator and $A_n(\tilde{A}_N)$ is a normal operator for an arbitrary $N \ge N_0$ then

$$\operatorname{spec} A \subset \Lambda\,.$$

Note that so far A has been assumed to be equal to $A^*(n)A(n)$ and, hence self-conjugate (and, all the more, normal). Thus it can be assumed that

$A_N(\tilde{A}_N)$ is equal to $A^*_N(n)A_N(n)\left(\tilde{A}^*_N(n)\tilde{A}_N(n)\right)$, where $A_N(n)$ is a "finite-dimensional" operator approximating $A(n)$, $A^*_N(n)$ is the operator conjugate to $A_N(n)$, and $\tilde{A}^*_N(n)$ is the operator conjugate to $\tilde{A}^*_N(n)$ in the space $(E^p)^{(2N+1)^d}$ with $\langle\cdot,\cdot\rangle_{N,q}$ representing a scalar product.

We take into account the fact that $\tilde{A}^*_N(n)$ does not coincide with the transposed operator. Indeed

$$\begin{aligned}\langle \tilde{A}_N u, v\rangle_{N,q} &= \sum_{|i|\le N} \frac{1}{q^{|i|}} \sum_{j\le N} (a_{ij}u_j, v_i) \\ &= \langle u, \tilde{A}^*_N v\rangle = \sum_{|j|\le N} \frac{1}{q^{|j|}} \sum_{|i|\le N} (u_j, b_{ji}v_i)\,.\end{aligned}$$

Since

$$b_{ji} = \frac{1}{q^{|i|-|j|}} a^T_{ij}\,,$$

where $a^T_{ij} : E^p \to E^p$ is the matrix transpose of a matrix $a_{ij} : R^p \to R^p$ for an arbitrary $i, j \in \mathbf{Z}^d, |i| \le N, |j| \le N$.

Thus we have an effective method for the estimation of the linearized operator.

7.2.4 Gershgorin Disks

We need below in the estimation of the spectrum of $\tilde{A}_N$ the help of Gershgorin disks (see Appendix G). They are defined in the following way: To be concrete we assume that $d = 2$. Let a linear operator $\tilde{A}_N$ be written in the coordinate form as follows

$$\xi_{jk}(n+1) = \sum_{|r-j|\le s, |l-k|\le s} a_{jkrl}\xi_{rl}(n)\,, \tag{7.6}$$

where $\xi_{jk} \in R$, $j, k \in \{-N \dots N\}$.

We consider on the complex plane the set of disks

$$D_{jk} : |a_{jkjk} - \lambda| \le \sum_{(r,l)\ne(j,k)} |a_{jkrl}|\,.$$

The spectrum of $\tilde{A}_N$ belongs to the union $\bigcup_{j,k=-N}^{N} D_{jk}$. The proof of this statement is the same as for the case $d = 1$ (see Appendix F). Indeed, let $\{\xi^0_{jk}\}$ be an eigenvector of $\tilde{A}_N$ corresponding to the eigenvalue λ_0. Then

$$(a_{jkjk} - \lambda_0)\xi^0_{jk} = -\sum_{(r,l)\ne(j,k)} a_{jkrl}\xi^0_{rl}\,, \quad \forall j, k\,.$$

Assume that $|\xi^0_{rl}| = \max_{j',k'} |\xi^0_{j',k'}|$. Then,

$$|a_{jkjk} - \lambda_0| \leq \sum_{(r,l)\neq(j,k)} |a_{jkrl}| \,.$$

7.2.5 An Alternative Way to Obtain the Stability Criterion

In what follows we consider examples in which the linearization operator is symmetric but not self-conjugate in B_q. In order to avoid cumbersome calculations connected with finding a conjugate matrix and its product and the original matrix, we show another way to estimate the norm of an operator L. Let us consider the case $d = 2$. Let $\tilde{L}$ be an infinite "matrix" which determines the map

$$\bar{\xi}_{jk} = \sum_{|r-j|\leq s, |l-k|\leq s} a_{jkrl}\xi_{r\rho} \,, \tag{7.7}$$

where $\xi_{jk} \in R$, $|a_{jkrl}| \leq M < \infty$ and $a_{jkrl} = a_{rljk}$ for arbitrary $j, k, l, r \in \mathbf{Z}$. Let $\Delta^\Gamma(\tilde{L}) = \bigcup_{j,k} D_{jk}(\tilde{L})$ be the union of Gershgorin disks for the matrix $\tilde{L}$. Note that the Gershgorin disks are well defined for arbitrary j, k due to the finiteness of the number of "diagonals" of $\tilde{L}$. We set

$$r\left(\Lambda^\Gamma(\tilde{L})\right) \equiv \sup\{|\lambda| | \lambda \in \Delta^\Gamma(\tilde{L})\} \,.$$

If $r < 1$, then there exists $q^* > 1$ such that for any $q, 1 < q < q^*$, the inequality $\|L\|_q < 1$ holds. Here $L : B_q \to B_q$ is the operator determined by the "matrix" $\tilde{L}$.

Thus, if q is close enough to unity then for stability it is sufficient to estimate the set $\Delta^\Gamma(\tilde{L})$ and to determine parameters for which $r\left(\Delta^\Gamma(\tilde{L})\right) < 1$.

7.3 Spatial Chaos in a Discrete Version of the One-Dimensional FitzHugh–Nagumo–Schlögl Equation

7.3.1 Spatial Chaos

The complexity of motions in lattice systems in many cases (in particular, for gradient-like models) is related to the existence of a large (or infinite) number of stable steady solutions randomly located along spatial coordinates, hence spatial chaos. Before defining spatial chaos we remind the reader what a translational dynamical system (TDS) is. In the space B (Sect. 1.4) the translation group $S_j, j \in \mathbf{Z}^d$, acts as follows: $(S_{j_0}\mathbf{u})_j = u_{j+j_0}$ if $(\mathbf{u})_j = u_j$. Each element $S_{j_0}, j_0 \in \mathbf{Z}^d$, is a linear bounded operator, and for any $\{u_j\} \in B$

$$\sum_{j\in\mathbf{Z}^d}\frac{|u_{j+j_0}|^2}{q^{|j+j_0|}}=\sum_{j\in\mathbf{Z}^d}\frac{|u_{j+j_0}|^2}{q^{|j+j_0|}}q^{|j+j_0|-|j|}\le q^{|j_0|}||\mathbf{u}||_q\,.$$

Thus, $||S_{j_0}||\le q^{|j_0|/2}$. Clearly, each map S_{j_0} commutes with the evolution map F.

The dynamical system $(\{S_j\},B)$, $j\in\mathbf{Z}^d$, *will be referred to as a translational dynamical system.*

If the lattice dynamical system is an adequate model of a nonequilibrium medium, then the spatio-temporal behavior of the dynamics of the medium must be described rigorously by the dynamical properties of these two systems – translational as well as evolutionary.

Let us give a rigorous definition of this phenomenon. It is, in principle, the same as defined by Bunimovich and Sinai but in different words.

A lattice dynamical system is said to have spatial chaos if there exists a stable set of stationary states on which a TDS behaves stochastically (for example, has positive topological entropy).

7.3.2 A Discrete Version of the One-Dimensional FitzHugh–Nagumo–Schlögl Equation

Consider the following 1D lattice dynamical system with diffusion coupling:

$$u_j(n+1)=u_j(n)+d[u_{j+1}(n)-2u_j(n)+u_{j-1}(n)]+\alpha\left[f\left(u_j(n)\right)\right]\,,\quad(7.8)$$

where $j\in\mathbf{Z},n\in\mathbf{Z}_+$ and $u_j(n)\in R$ are variables describing the medium, d is the diffusion coefficient, α is a parameter of the medium taken positive, and the function $f(u)$ is

$$f(u)=u(u-a)(1-u)\,,\ 0<a<1\quad\text{if}\quad u\in[-R,R],\ R\gg 1$$

with $|f'(u)|\le M,|f'(u)|<M,u\in R$. The system (7.8) can be treated as a discrete version of the FNS model.

7.3.3 Steady States

For the steady solutions of (7.8), $u_j(n)\equiv\psi_j$, the following equation holds:

$$d(\psi_{j+1}-2\psi_j+\psi_{j-1})+\alpha[f(\psi_j)]=0\,.\quad(7.9)$$

Under the substitution $\psi_{j-1}=x_j,\psi_j=y_j$ this equation can be represented as a system,

$$x_{j+1}=y_j\,,\quad y_{j+1}=2y_j-\frac{\alpha}{d}f(y_j)-x_j\,.\quad(7.10)$$

This system is, in fact, a TDS on the set of stationary states. Indeed, the action of the TDS is determined by the relationship $(S_{j_0}\mathbf{u})_j=u_{j+j_0}$. In our case

$S_{j_0} = S_1$, so TDS $\{S_j\}_{j\in\mathbf{Z}}$ is generated by the shift map $S_1 : (S_1\mathbf{u})_j = u_{j+1}$ if $(\mathbf{u})_j = u_j$. Any bounded trajectory $(\dots(x_j, y_j), (x_{j+1}, y_{j+1}), \dots)$ corresponds to a bounded solution $\Psi = (..., \psi_j = y_j, \psi_{j+1} = y_{j+1}, ...)$ of (7.9) which belongs to $B \equiv B_q$. Thereby we have the map $h : F \to B$, where F is the set of points of all bounded trajectories of (7.10) such that $S_1 \circ h = h \circ S$. Here S is a Hénon-like map $(x, y) \to \left(y, 2y - \frac{\alpha}{d} f(y) - x\right)$ which generates (7.10). It can be easily shown that h is a one-to-one and continuous map.

First, we note that S has three fixed points: $O_1(0,0), O_2(a,a)$ and $O_3(1,1)$. The points O_1 and O_3 are saddles with positive multipliers, and O_2 is either a saddle with negative multipliers or it has complex multipliers. Thus, (7.8) has three spatially homogeneous steady states corresponding to these fixed points. Moreover, (7.10) is conservative and has numerous bounded trajectories, each corresponding to a steady state of (7.8). We want to separate the stable states using symbolic dynamics. Let us construct a rectangle in which S acts like a Smale's horseshoe map. Then we impose hyperbolicity conditions on the invariant set in this rectangle and show the conjugation of S on this set with the shift map in the topological Bernoulli scheme with two symbols.

Smale's Horseshoe Map. Consider the rectangle

$$\Pi = \{(x, y : |x| \leq \beta, |y| \leq \gamma)\}$$

with boundaries

$$\begin{aligned} \Gamma_\pm &= \{(x, y) : y = \pm\gamma, |x| \leq \beta\}, \\ B_\pm &= \{(x, y) : x = \pm\beta, |y| \leq \gamma\}, \end{aligned}$$

while the images of the boundaries are

$$\begin{aligned} S\Gamma_\pm &= \{(x, y) : x = \pm\gamma, |y - \phi(\pm\gamma)| \leq \beta\}, \\ SB_\pm &= \{(x, y) : y = \phi(x) \mp \beta, |x| \leq \gamma\}, \end{aligned}$$

where $\phi(t) \equiv 2t - \frac{\alpha}{d} f(t)$.

We require $S\Pi$ to intersect with Π as Smale's horseshoe map, i.e. we assume the following:

(i) the y-coordinate of any point S_+ is greater than γ, and the $\phi(x)$ the y-coordinate of any point S_- is smaller than $(-\gamma)$;

(ii) the absolute value of the x-coordinate of any point of $S\Pi$ is smaller than β;

(iii) the maximum of the function plotted at $SB_\pm$ is greater than γ, while the minimum of the function plotted at $SB_\mp$ is smaller than γ.

These conditions lead to the inequalities

$$\begin{aligned} &\gamma < \beta, \quad \gamma > 1/3(1 + a + \sqrt{1 - a + a^2 - 6d/\alpha}), \\ &d/\alpha < 6/(1 - a + a^2), \quad \phi(\gamma) - \beta - \gamma > 0, \\ &\beta < \min\{-\gamma + \phi(x_1); -\gamma - \phi(x_2)\}, \end{aligned} \tag{7.11}$$

where $x_{1,2} = \frac{1}{3}\left(1 + a \mp \sqrt{1 - a + a^2 - 6d/\alpha}\right)$. If these conditions are satisfied, then $S\Pi$ has the form shown in Fig. 7.2, and the intersection $S\Pi \bigcap \Pi$ has three components: Ω_1, Ω_2 and Ω_3. Let

$$\omega_i = S^{-1}\Omega_i\,, \quad i = 1, 2, 3\,.$$

Let us show that the hyperbolicity conditions hold in the set $\omega_1 \bigcup \omega_3$ (Fig. 7.3). Note that for maps like

$$(x, y) \mapsto \big(P(x, y), Q(x, y)\big)$$

these conditions can be written in the following form:

$$\|P'_x\| < 1\,, \quad \|(Q'_y)^{-1}\| < 1\,,$$

$$1 - \|P'_x\|\,\|(Q'_y)^{-1}\| > 2\sqrt{\|(Q'_y)^{-1}\|\,\|Q'_x\|\,\|(Q')_y^{-1}P'_y\|}\,,$$

$$1 - \big(\|P'_x\| + \|(Q'_y)^{-1}\|\big) + \|P'_x\|\,\|(Q'_y)^{-1}\| > \|Q'_x\|\,\|(Q'_y)^{-1}\|P'_y\|\,,$$

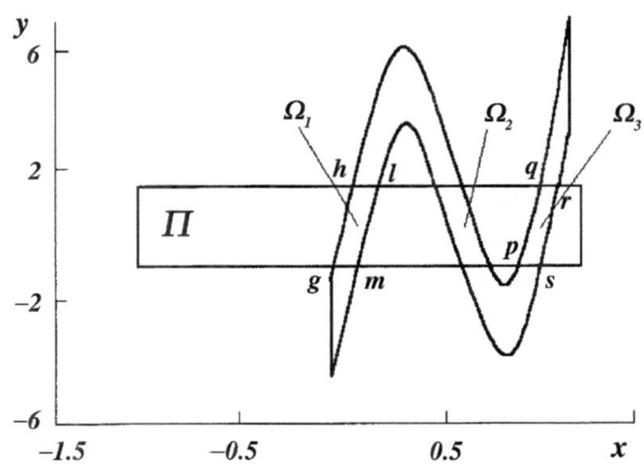

Fig. 7.2. Mapping of the rectangle Π is due to the iteration of map S

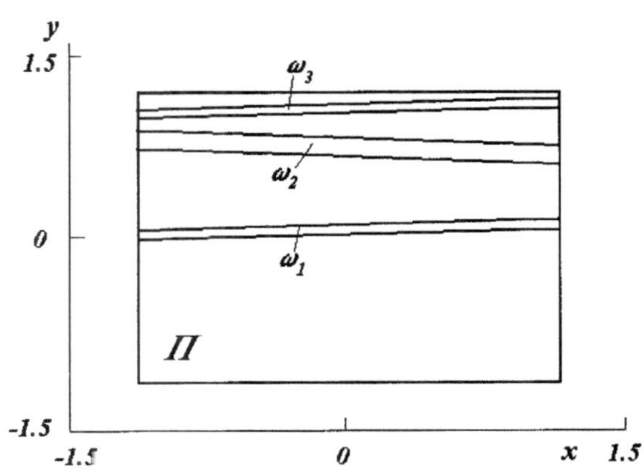

Fig. 7.3. The action S^{-1} on Π

where $\|\cdot\| = \sup_{(x,y)\in\omega_1\bigcup\omega_3} |\cdot|$. In our case we have $|P'_x| = 0, |P'_y| = |Q'_x| = 1$. Then the hyperbolicity conditions follow from the inequality

$$\sup_{(x,y)\in\omega_1\cup\omega_3} \frac{1}{|2-\alpha/df'(y)|} < \frac{1}{2},$$

which is satisfied if $x_l < x_3, x_p > x_4$, where x_l, x_p are the x-coordinates of the point (l,p) (see Fig. 7.2), and $x_{3,4} = \frac{1}{3}\left(1+a\mp\sqrt{1-a+a^2}\right)$. Since x_l and x_p are the roots of the equations $\phi(x)-\beta=\gamma$ and $\phi(x)+\beta=-\gamma$, these conditions are equivalent to

$$\beta < -\gamma+\phi(x_3), \quad \beta < -\gamma-\phi(x_4). \tag{7.12}$$

Let us demand that (7.11–12) be simultaneously satisfied:

$$\frac{d}{\alpha} < \begin{cases} \min\left\{\dfrac{2a-(1+a)x_3}{6(1-x_3)}; \dfrac{2a^2-5a+2}{18}\right\} & \text{if } a<1/2 \\ \dfrac{[(1+a)x_4-2a]x_4}{6(1+x_4)} & \text{if } a\geq 1/2. \end{cases} \tag{7.13}$$

Hence, when (7.11) is satisfied, in the (β,γ) plane there exists a nonempty region, b_0, with length and height equal to those of the rectangle Π which is mapped in detail in Smale's horseshoe (see Fig. 7.4, where b_0 is illustrated for the case $a \geq 1/2$).

Let Λ be a set of points of all the trajectories that lie completely in $(\omega_1\bigcap\Omega_1)\bigcup(\omega_3\bigcap\Omega_3)$. Employing symbolic dynamics we find that $S|_\Lambda$ is topologically conjugate to the shift-map of the topological Bernoulli scheme with two symbols (in particular, $h_{top}(S|_\Lambda) = \log 2$, therefore $h_{top}(S_1|_{h_\Lambda} = \log 2$.

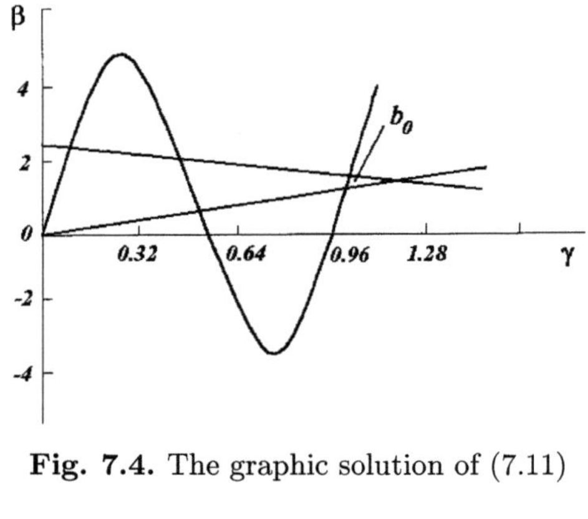

Fig. 7.4. The graphic solution of (7.11)

7.3.4 Stability of Spatially Steady Solutions

Spatially Homogeneous Steady Solutions. We show first how to find stability conditions for spatially homogeneous steady solutions of (7.8) $\{u_j = 0\}$ and $\{u_j = 1\}$. The equation for perturbations of each of these solutions has the following form:

$$\xi_j(n+1) = [1 - \alpha f'(u^0) - 2d]\xi_j(n) + d[\xi_{j+1}(n) + \xi_{j-1}(n)] \,.$$

with $u^0 = 0$ or $u^0 = 1$

In accordance with our method let us consider the following $(2N+1) \times (2N+1)$ matrices:

$$\mathbf{a}_N = \begin{pmatrix} b & d & 0 & \cdots & & 0 \\ d & b & d & \cdots & & 0 \\ 0 & d & b & d & \cdots & 0 \\ \cdots & \cdots & \cdots & \cdots & \cdots & \\ & & & & d & b \end{pmatrix},$$

$$\mathbf{a}_N^* = \begin{pmatrix} b & d/q & & & \cdots & 0 \\ dq & b & d/q & & \cdots & 0 \\ 0 & dq & b & d/q & \cdots & 0 \\ \cdots & \cdots & \cdots & \cdots & \cdots & \\ & & & & dq & b \end{pmatrix},$$

where $b = 1 - \alpha f'(u^0) - 2d$. The matrix $\mathbf{a}_N^*$ is conjugate to the matrix $\mathbf{a}_N$ in the $\langle\cdot,\cdot\rangle_{N,q}$ scalar product. Then we can define the operator A_N through the use of the product $\tilde{A}_N = \mathbf{a}_n^* \mathbf{a}_N$:

$$\begin{pmatrix} b^2 + \frac{d^2}{q} & db + \frac{db}{q} & \frac{d^2}{q} & 0 & 0 & 0 & \cdots \\ m & d^2q + b^2 + \frac{d^2}{q} & db + \frac{db}{q} & \frac{d^2}{q} & 0 & 0 & \cdots \\ d^2q & m & d^2q + b^2 + \frac{d^2}{q} & db + \frac{db}{q} & \frac{d^2}{q} & 0 & \cdots \\ \cdots & \cdots & \cdots & \cdots & \cdots & \cdots & \\ \cdots & \cdots & \cdots & \frac{d^2}{q} & m & d^2q + b^2 + \frac{d^2}{q} & db + \frac{db}{q} \;\cdots \\ 0 & & & 0 & d^2q & m & k^2q + b^2 \end{pmatrix},$$

with $m = kbq + kb$. Thus, we have the following five Gershgorin disks:

$$|\lambda - b^2 - \frac{d^2}{q}| \le d|b| + \frac{d|b|}{q} + \frac{d^2}{q},$$

$$|\lambda - d^2q - b^2 - \frac{d^2}{q}| \le d|b|q + d|b| + \frac{d|b|}{q} + \frac{d^2}{q},$$

$$|\lambda - d^2 q - b^2 - \frac{d^2}{q}| \le d^2 q + d|b|q + d|b| + d|b| + \frac{d|b|}{q} + \frac{d^2}{q}, \quad (7.14)$$

$$|\lambda - d^2 q - b^2 - \frac{d^2}{q}| \le d^2 q + d|b|q + d|b| + d|b| + \frac{d|b|}{q},$$

$$|\lambda - d_q^2 - b^2| \le d^2 q + d|b|q + d|b| .$$

The fulfillment of the following condition is sufficient to obtain the linear stability of a spatially homogeneous steady solution,

$$2d^2 q + b^2 + \frac{2d^2}{q} + 2d|b| + d|b|q + \frac{d|b|}{q} < 1 . \quad (7.15)$$

Assume that

$$d < \frac{1 + \alpha f'(u^0)}{2} . \quad (7.16)$$

Then (7.15) can be rewritten as

$$d\frac{(q-1)^2}{q} < \frac{-\alpha f'(u^0)\left[2 + \alpha f'(u^0)\right]}{\left[1 + \alpha f'(u^0)\right]} . \quad (7.17)$$

Thus for arbitrary admissible values of the parameters d, α and a there exists an interval $1 < q < q_*$ such that spatially homogeneous steady solutions are linearly stable in Bq for every q belonging to this interval. Then $q_* \ge \frac{1}{2}\left[2 + \phi/d + \sqrt{(4+\phi/d)\phi/d}\right]$, where ϕ equals the right-hand side of (7.17).

Spatially Inhomogeneous Steady Solutions. Let $\mathbf{u}^* = \{u_j^*\}$ be one of the bounded steady solutions of (7.8). In addition let

$$\xi_j(n+1) = \sigma_j \xi_j(n) + d\left[\xi_{j-1}(n)\xi_{j+1}(n)\right] \quad (7.18)$$

be the linear equation for perturbations of $\mathbf{u}^*$, where $\sigma_j = 1 - 2d + \alpha f'(u_j^*)$. Its corresponding linear operator, L, is symmetric (and so the "matrix" $\tilde{L}$). Therefore we can use the results of Sect. 7.2.5 to find the conditions under which $\|L\|_q < 1$. We consider the set $\Delta^\Gamma(\tilde{L})$ of Gershgorin disks D_j,

$$|\lambda - \sigma_j| \le 2d ,$$

on the complex plane. Then if $\sigma_j > 0$, we have $r\left(\Delta^\Gamma(\tilde{L})\right) \le \sup_{u_j^*}\left(1 + \alpha f'(u_j^*)\right)$.

Since $(u_{j-1}^*, u_j^*) \in (\Omega_1 \bigcap \omega_1) \bigcup (\Omega_3 \bigcap \omega_3)$ then $f'(u_j^*) \le f'(x_0) < 0$, where x_0 equals either x_l or x_p and $x_l(x_p)$ is the x-coordinate of point $l(p)$ (see Figs. 7.2 and 7.5). It follows that if

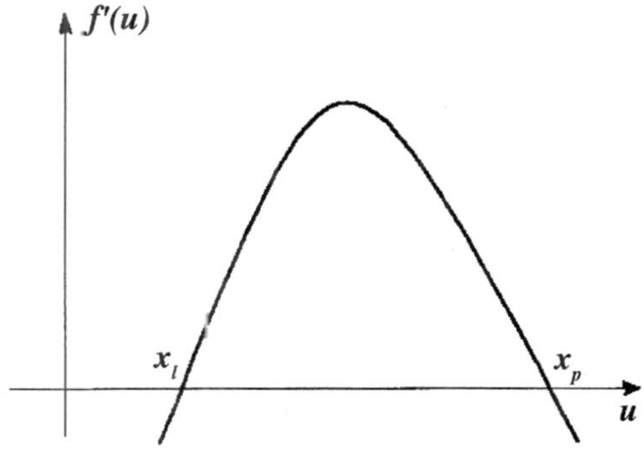

Fig. 7.5. Graphic illustration of how the value $f'(u_j^*)$ is found

$$\gamma < \gamma_0 \equiv \left[-(1+a) + \sqrt{1 - a + a^2 + \frac{3(1-2d)}{\alpha}} \right] \Big/ 3 \tag{7.19}$$

then $\sigma_j > 0$ (because $|u_j^*| \leq \gamma$). Inequalities (7.13) and (7.19) are compatible if

$$\frac{d}{\alpha} < \frac{1}{2\alpha} + \frac{5+3a}{2} . \tag{7.20}$$

Consequently, if (7.20) holds, then $r\left(\Delta(\tilde{L})\right) < 1$ and $\|L\|_q < 1$ for values of q close to unity. Hence, for the points of the parameter region restricted by (7.13) and (7.20) (see Fig. 7.6, where this region is shown for $a = -1/2$), every steady solution of (7.8), corresponding to the elements of the set Λ, is linearly stable. Moreover, $\mathbf{u}^*$ is stable also with respect to the nonlinear system (7.8) in the following sense. Let $\boldsymbol{\xi} = \{\xi_j\}$ be a perturbation of $\mathbf{u}^*$ such that $(u_{j-1}^* + \xi_{j-1}, u_j^* + \xi_j)$ belongs to $(\Omega_1 \bigcap \omega_1) \bigcup (\Omega_3 \bigcap \omega_3)$ for each $j \in \mathbf{Z}^d$. Then $\|F^n \boldsymbol{\xi}\|_q \to 0$ as $n \to \infty$, where F is an evolution operator determined by (7.8) and is close enough to unity.

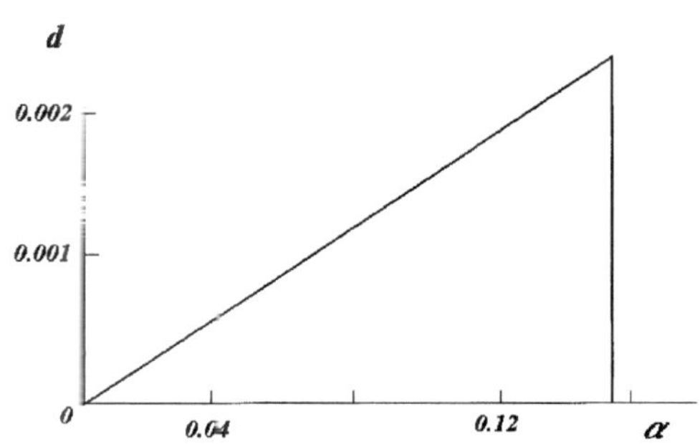

Fig. 7.6. Parameter region corresponding to the spatial chaos in the discrete FNS equation (7.8)

In conclusion, (7.8) has a continuum set of steady states. These states are stable and hence a highly multistable situation. The profiles of the steady states "along" the spatial coordinate j are extremely diverse. They include a simple homogeneous distribution (represented by the spatially homogeneous states $\{u_j(n) = 0\}$ and $\{u_j(n) = 1\}$; the state $\{u_j(n) = a\}$ is unstable for this parameter set), a periodic distribution (corresponding to the periodic trajectories of S), and "soliton"-type profiles (corresponding to the homoclinic trajectories of S), etc. Moreover, it is possible to realize the profiles [of course, also corresponding to the trajectories of S from $(\Omega_1 \bigcup \Omega_3) \bigcap (\omega_1 \bigcup \omega_3)$] of any degree of complexity, because there always exists a trajectory of S which is associated with an arbitrary double infinite sequence of two symbols in the topological Bernoulli scheme. Thus, there is spatial chaos in (7.8).

7.4 Chaotic Traveling Waves in a One-Dimensional Discrete FitzHugh–Nagumo–Schlögl Equation

In this section we search for traveling waves for the discrete FNS equation (7.8).

7.4.1 Traveling Wave Equation

For simplicity we study traveling waves with unit velocity, i.e. waves of the form $u_j(n) = \psi(j+n)$. Then, the function $\psi(j+n)$ must satisfy

$$\psi(k+2) = \psi(k+1) + \alpha f\left(\psi(k+1)\right) + d[\psi(k) - 2\psi(k+1) + \psi(k+2)]\,, \tag{7.21}$$

where $k \equiv j - 1 + n$ is the "traveling coordinate". Equation (7.21) can be rewritten as

$$\psi(k+2) = \frac{1}{1-d}\{(1-2d)\psi(k+1) + \alpha f\left(\psi(k+1)\right) + k[\psi(k)]\}\,. \tag{7.22}$$

Under the substitutions

$$\psi(k) = x_k \quad \text{and} \quad \psi(k+1) = y_k\,,$$

(7.22) becomes

$$x_{k+1} = y_k\,, \quad y_{k+1} = \frac{d}{1-d}x_k + \frac{1}{1-d}[(1-2d)y_k + \alpha f(y)]\,. \tag{7.23}$$

The system (7.23) is generated by the Hénon-like map

$$H : (x, y) \to \left(y, \frac{d}{1-d}x + \frac{1}{1-d}[(1-2d)y + \alpha f(y_k)]\right)\,,$$

which is a diffeomorphism for $0 < d < 1$.

7.4.2 Existence of Traveling Waves

As in Sect. 7.2.3 we can show that there exists a rectangle $\Pi = \{(x,y) \| |x| \leq \beta, |y| \leq \gamma\}$ on which the map H acts like a Smale's horseshoe map if the following inequalities hold:

$$\beta > \gamma, \quad \beta < -\frac{2-3d}{d} - \frac{\alpha}{d} f(\gamma), \quad 3\gamma > 1 + a + \sqrt{1 - a + a^2 + \frac{3(1-2d)}{\alpha}},$$
$$\beta < -\frac{1-d}{d} - \frac{1}{d}\left[(1-2d)x_{min} + \alpha f(x_{min})\right], \tag{7.24}$$
$$\beta < -\frac{1-d}{d} + \frac{1}{d}\left[(1-2d)x_{max} + \alpha f(x_{max})\right],$$

where

$$x_{min} = \frac{1}{3}\left(1 + a - \sqrt{1 - a + a^2 + \frac{3(1-2d)}{\alpha}}\right),$$
$$x_{max} = \frac{1}{3}\left(1 + a + \sqrt{1 - a + a^2 + \frac{3(1-2d)}{\alpha}}\right).$$

These inequalities are satisfied if

$$6d < \alpha(1 - a + a^2) + 3, \quad 2d > 1 - \alpha a. \tag{7.25}$$

Let Ω_1 be the left-hand, Ω_2 the middle and Ω_1 the right-hand components of $H(\Pi) \bigcap \Pi$ (see Fig. 7.7), and let $\omega_i = H^{-1}(\Omega_i), i = 1,2,3$, and denote with A_i, B_i, C_i and D_i the "angular" points of the domain Ω_i (Fig. 7.7). Let a_i, b_i, c_i and d_i be their x-coordinates. We now look for the hyperbolicity conditions on the set $\omega_1 \bigcup \omega_3$. If these conditions are met, each trajectory of (7.23) that belongs to $\omega_1 \bigcup \omega_3$ will be hyperbolic. We denote by Δ the set of such trajectories. For (7.23) the hyperbolicity conditions (see Sect. 7.2.3) take the form

$$\sup_{(x,y)\in\omega_1\cup\omega_3} \frac{1}{|1 - 2d + \alpha f'(y)|} < 1. \tag{7.26}$$

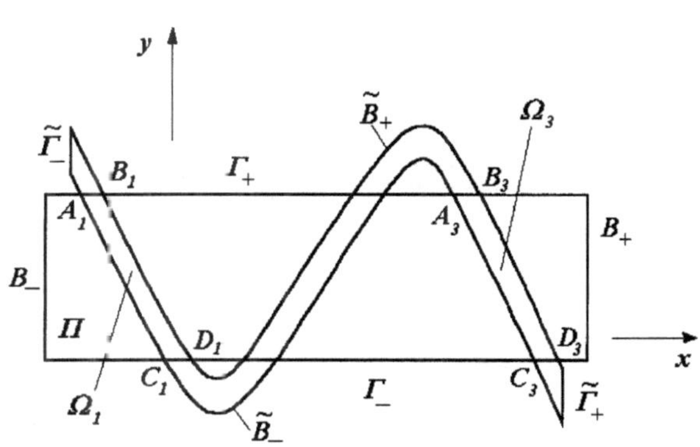

Fig. 7.7. Smale's horseshoe for the Hénon-type map: images and pre-images

Thus the supremum is attained when $y = d_1$ for ω_1 and when $y = d_3$ for ω_3 (Fig. 7.8). Consequently, (7.26) is satisfied if

$$\begin{aligned} d_1 < d_1^0 &\equiv \frac{\alpha(1+a) - \sqrt{\alpha^2(1-a+a^2) + 6\alpha(1-d)}}{3\alpha}, \\ a_3 > a_3^0 &\equiv \frac{\alpha(1+a) + \sqrt{\alpha^2(1-a+a^2) + 6\alpha(1-d)}}{3\alpha}. \end{aligned} \tag{7.27}$$

Let us introduce the function $G(x) \equiv d\beta/(1-d) + [(1-2d)x + \alpha f(x)]/(1-d) + \gamma$. The zeros of this function are the x-coordinates of the points of intersection of $\tilde{B}_+$ and Γ_-. Then the inequality in (7.27) holds if $G(d_1^0) < 0$ or

$$\beta < -\frac{1-d}{d}\gamma + \frac{1}{d}[\alpha f(d_1^0) + (1-2d)d_1^0]\,. \tag{7.28}$$

Analogously, the second inequality in (7.27) is satisfied if $G(a_3^0) > 0$ or

$$\beta < -\frac{1-d}{d}\gamma + \frac{1}{d}[\alpha f(a_3^0) + (1-2d)a_3^0]\,. \tag{7.29}$$

Hence, (7.28–29) guarantee the hyperbolicity of the trajectories in Δ. The inequalities (7.24,28–29) do not have solutions β and γ for all values of the parameters d, α and a. We will show that such values exist and also satisfy (7.25). Let $d = 1/2$. Then (7.25) is satisfied. The other inequalities, after simple but tedious calculations, can be written in the following forms:

$$\begin{aligned} -f(x_{min}) &> \frac{\gamma^0(\alpha)}{\alpha}, & f(x_{max}) &> \frac{\gamma^0(\alpha)}{\alpha}, \\ -f(d_1^0) &> \frac{\gamma^0(\alpha)}{\alpha}, & f(a_3^0) &> \frac{\gamma^0(\alpha)}{\alpha}, \end{aligned} \tag{7.30}$$

where

$$\gamma^0 \equiv \frac{1+a}{2} + \sqrt{\frac{(1-a)^2)}{4} + \frac{1}{\alpha}}\,.$$

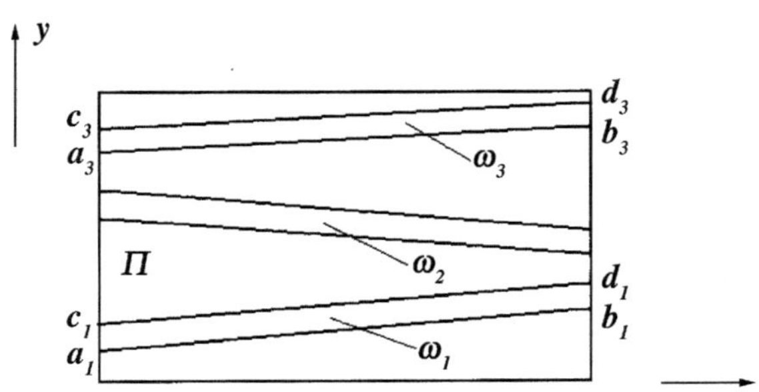

Fig. 7.8. Geometry for deriving the supremum conditions

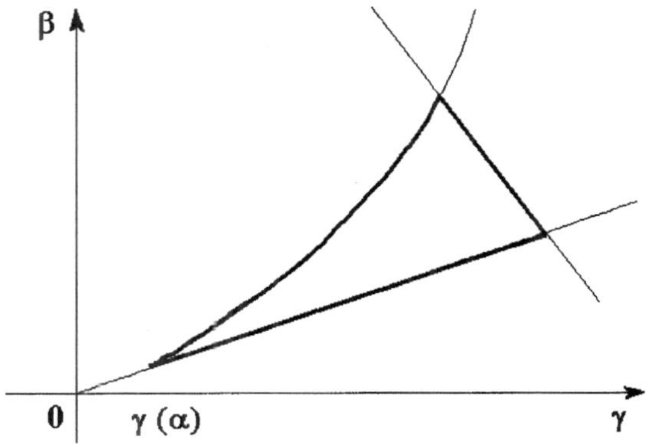

Fig. 7.9. Region of existence of a hyperbolic set

If (7.30) holds, then there exists on the (γ, β) plane a nonempty region for which the conditions of the existence of Smale's horseshoe and the hyperbolicity conditions are satisfied (in Fig. 7.9 this region is depicted for $a \leq 1/2$). Accordingly (7.30) holds for $d = 1/2$ and $\alpha > \alpha^*$, where α^* is the root of $-f(d_1^0) = \gamma^0(\alpha)/\alpha$ for $a \leq 1/2$ and of $f(a_3^0) = \gamma^0(\alpha)/\alpha$ for $a \geq 1/2$.

Thus, in the space of the parameters d, α and a there exists a nonempty region for which there is a Smale's horseshoe on the rectangle Π and a set Δ of hyperbolic trajectories that lie completely in $(\Omega_1 \cap \omega_1) \cup (\Omega_3 \cap \omega_3)$. Using symbolic dynamics we find that $H|_\Delta$ is topologically conjugated to the shift-map on the set of infinite sequence of two symbols (the topological Bernoulli scheme of two symbols). Since each trajectory from Δ corresponds to a stationary wave, this ensures the existence of traveling waves of a chaotic profile.

7.4.3 Stability of Traveling Waves

We pass now to the study of the stability of traveling waves. Let $u_j(n) = \psi(j+n)$ be one of the solutions of (7.8) that corresponds to a trajectory Δ moving with velocity $c = 1$ (see Sect. 1.4). The linear equation for perturbations has the following form:

$$\xi_k(n+1) = d\xi_{k-2}(n) + \sigma_{k-1}\xi_{k-1}(n) + d\xi_k(n)\,, \tag{7.31}$$

where $k = j + n$ and $\sigma_j \equiv 1 - 2d + \alpha f'(\psi_j), \forall j$. In fact, (7.31) determines the operator $A : B_q \to B_q$ for any $q > 1$. We estimate its q norm from the estimation of the spectrum of the operator $A^*(n)A(n)$. Following the method earlier described we supplement (7.31) with the boundary conditions

$$\xi_{-N-2}(n) = \xi_{-N-1}(n) \equiv 0\,. \tag{7.32}$$

Then (7.31–32) determine the set of approximating matrices $\mathbf{a}_N$ of the form

$$\mathbf{a}_N = \begin{pmatrix} d & 0 & \cdots & \cdots & 0 \\ \sigma_{-N} & d & 0 & \cdots & 0 \\ d & \sigma_{-N+1} & d & \cdots & 0 \\ \cdots & \cdots & \cdots & \cdots & \cdots \\ 0 & 0 & \cdots & d\sigma_{N-1} & d \end{pmatrix},$$

$$\mathbf{a}_N^* = \begin{pmatrix} d & \frac{\sigma_{-N}}{q} & \frac{d}{q^2} & 0 & \cdots & & 0 \\ 0 & d & \frac{\sigma_{-N+1}}{q} & \frac{d}{q^2} & & \cdots & 0 \\ \cdots & \cdots & \cdots & \cdots & \cdots & \cdots & \cdots \\ & \cdots & \cdots & \cdots & \cdots & d\frac{\sigma_{N-1}}{q} & \frac{d}{q^2} \\ & \cdots & \cdots & \cdots & \cdots & 0 & d \end{pmatrix}.$$

The matrix approximating $A^*(n)A(n)$ can be represented as follows:

$$\mathbf{a}_N^*\mathbf{a}_N =$$

$$\begin{pmatrix} d^2+\frac{\sigma_{-N}^2}{q}+r & \frac{d\sigma_{-N}}{q}+\frac{ds}{q^2} & r & 0 & \cdots & \cdots & 0 \\ \sigma_{-N}d+\frac{sd}{q} & d^2+\frac{\sigma_{-N+1}^2}{q}+r & \frac{sd}{q}+r & r & \cdots & & \\ d^2 & ds+\frac{\sigma_{-N+2}d}{q} & \cdots & \cdots & \cdots & 0 & \\ \cdots & & \cdots & \cdots & d^2+\frac{\sigma_{N-2}^2}{q}+r & \frac{\sigma_{N-1}}{q}+r & r \\ \cdots & & \cdots & d^2 & d\sigma_{N-2}+\frac{\sigma_{N-1}d}{q} & d^2+\frac{\sigma_{N-2}^2}{q} & \frac{d\sigma_N}{q} \\ \cdots & & \cdots & 0 & d^2 & d\sigma_{N-1} & d^2 \end{pmatrix},$$

where $r = \frac{d^2}{q^2}$ and $s = \sigma_{-N+1}$. To find the eigenvalues of the matrix we limit $|\sigma_j|$ from above by the cut-off σ, which does not depend on j:

$$\sigma = \sup_{\left(\psi(j-1),\psi(j)\right)\in\left((\Omega_1\cap\omega_1)\cup(\Omega_3\cap\omega_3)\right)} |1-2d-\alpha f'(\psi_j)|\,.$$

If the point $(\psi(j-1),\psi(j)) \in \Omega_1 \cap \omega_1$, then $a_1 < \psi(j-1) < d_1$, $-\gamma < \psi(j) < \gamma$, and if $(\psi(j-1),\psi(j)) \in \Omega_3 \cap \omega_3$, then $a_3 < \psi(j-1) < d_3$ and $-\gamma < \psi(j) < \gamma$, where $a_{1,3}$ and $d_{1,3}$ are x-coordinates of the points A_1 and D_3, respectively (Fig. 7.7). It follows from this that

$$\sigma < \omega = 3\alpha\gamma^2 + 2\alpha(1+a)\gamma + \alpha a + 2d - 1\,.$$

We estimate now the position of the spectrum of the matrix $\mathbf{a}_N^*\mathbf{a}_N$ by means of Geshgorin disks (Appendix G). This spectrum lies on the complex plane inside a union of disks bounded by the disk

$$(x-a_0)^2 + y^2 \le b_0^2\,.$$

Here

$$a_0 = d^2 + \frac{\omega^2}{q} + \frac{d^2}{q^2} \quad \text{and} \quad b_0^2 = d^2 + \frac{d^2}{q^2} + \sigma\left(d + \frac{d}{q}\right) + \sigma\left(\frac{d}{d} + \frac{d}{q^2}\right) + \frac{\omega^2}{q},$$

and hence the disk shrinks into

$$(x-d^2)^2+y^2 \leq d^2+\sigma d \tag{7.33}$$

as $q \to \infty$. Accordingly, the spectrum of the matrix $\mathbf{a}_N^* \mathbf{a}_N$ lies in a set which does not depend on N strictly inside the unit circle if q is large enough and d is small enough. It is possible to estimate more precisely the critical values of d and q and to show than (7.23) and (7.25) are compatible.

In conclusion, we have shown that (7.8) has an infinite set of stationary waves moving with unit velocity. The profiles of these waves are linearly stable to perturbations moving with the same velocity. Hence chaotic traveling waves occur in the discrete version of a 1D FNS equation.

7.5 Two-Dimensional Spatial Chaos

We now consider the system

$$\begin{aligned} u_{jk}(n+1) = u_{jk}(n) - \alpha f\big(u_{jk}(n)\big) + d\big[u_{j+1,k}(n) - 4u_{jk}(n) \\ + u_{j-1,k}(n) + u_{j,k-1}(n) + u_{j,k+1}(n)\big] \,, \end{aligned} \tag{7.34}$$

where $j,k \in \mathbf{Z}, n \in \mathbf{Z}_+$, α and d are positive parameters, $f(u) = u(u-a)(1-u)$ if $|u| \leq R \gg 1$ and $|df/du| \leq M, |f''(u)| \leq M$ for any u. This system is a discrete version of the two–dimensional nonlinear diffusion FNS model. Let us search for steady solutions of (7.34). Let us also show that, under suitable conditions, (7.34) exhibits spatial chaos.

7.5.1 Invariant Domains

Let us denote by Φ the evolution operator determined by (7.34) and consider it in the space B_q introduced in Chap. 1. Let $Q_r = \{|u_{jk}| \leq r\}$. We show now that Q_r is a Φ-invariant set. Let us prove a few statements about the properties of the solutions of (7.34):

(i) *For any fixed $a \in (0,1)$ there exist $\alpha_0 > 0$ and $d_0(\alpha)$ such that if $0 < \alpha < \alpha_0$ and $0 < d < d_0(\alpha)$ then there exists a number r, $0 < r < R$, for which $\Phi Q_r \subset Q_r$.*

Indeed, if $|u_{jk}(0)| \leq r < R$ then it follows from (7.34) that

$$|u_{jk}(1)| \leq \big|\varphi\big(u_{jk}(0)\big)\big| + 8dr \,, \tag{7.35}$$

with $\varphi(x) \equiv x + \alpha f(x) = -\alpha x^3 + \alpha(1+a)x^2 + (1-\alpha a)x$. The function ϕ has the following extremum points:

$$x_{min} = \frac{1}{3}\left(1 + a - \sqrt{1 - a + a^2 + 3/\alpha}\right),$$

$$x_{max} = \frac{1}{3}\left(1 + a + \sqrt{1 - a + a^2 + 3/\alpha}\right).$$

Assume that

$$\begin{aligned} x_{max} < r < \tfrac{1}{2}\left(1+a+\sqrt{(1-a)^2+4/\alpha}\right), \\ \tfrac{1}{2}\left(1+a-\sqrt{(1-a)^2+4/\alpha}\right) < -r < x_{min}. \end{aligned} \tag{7.36}$$

If (7.36) is valid then

$$\varphi_0 \equiv \sup_{x\in[-r,r]} |\varphi(x)|$$

is reached either for $x = x_{min}$ or for $x = x_{max}$ whatever the value of r chosen from the intervals shown in (7.36). Then (7.36) is satisfied if

$$x_{max} < r < \frac{1}{2}\left[\sqrt{(1-a)^2+4/\alpha} - (1+a)\right], \tag{7.37}$$

and (7.37) is satisfied for some r if the parameters of the system satisfy the following conditions:

$$\begin{aligned} &\alpha(5+16a+5a^2) < 1, \quad \alpha a < 1, \\ &4 - \alpha(45+38a+15a^2) + \alpha^2(15a^3+34a^2+15a) > 0. \end{aligned} \tag{7.38}$$

These conditions hold if α is small enough (for instance, $\alpha < 1/26$).

We return now to (7.35) and claim that

$$\left|\varphi\big(u_{jk}(0)\big)\right| + 8kr < r. \tag{7.39}$$

The inequality (7.39) follows from the condition $\varphi_0 + 8kr < r$ or

$$8k < 1, \quad r > \frac{\varphi_0}{1-8k}. \tag{7.40}$$

The inequalities (7.37) and (7.40) are compatible if

$$\begin{aligned} &\varphi_0(1-8d)^{-1} < \tfrac{1}{2}\left[\sqrt{(1-a)^2+4/\alpha} - (1+a)\right] \quad \text{or} \\ &d < d_0(\alpha) \equiv \tfrac{1}{4}\left\{\tfrac{1}{2} - \phi_0/\left[\sqrt{(1-a)^2+4/\alpha} - (1+a)\right]\right\}. \end{aligned} \tag{7.41}$$

Then $\lim_{\alpha\to 0} d(\alpha) > 0$; therefore, statement (i) is true for any r satisfying (7.37) and (7.40).

The set Q_r can be treated as an invariant "domain" for (7.34) [note that (7.39) is strictly so]. We now show that inside this set there are many Φ-invariant "domains" of the following form:

$$I^0_{jk}(\nu) = \{u_{jk} \in Q_r,\ |u_{jk}| \le \nu\},$$

$$I^1_{jk}(\epsilon) = \{u_{jk} \in Q_r,\ |u_{jk} - 1| \le \epsilon\}.$$

(ii) *For sufficiently small values of $\alpha, d,$ and α/d, there exist values of ϵ and ν such that the subsets $I^0_{jk}(\nu)$ and $I^1_{jk}(\epsilon)$ are Φ-invariant for any $(j,k) \in \mathbf{Z}^2$.*

The function $\varphi(x)$ monotonically increases in small neighborhoods of the points $x = 0$ and $x = 1$. Therefore

$$-\nu + \alpha f(-\nu) \le x + \alpha f(x) \le \nu + \alpha f(\nu) \tag{7.42}$$

if $|x| \le \nu$ and

$$1 - \epsilon + \alpha f(1-\epsilon) \le x + \alpha f(x) \le 1 + \epsilon + \alpha f(1+\epsilon) \tag{7.43}$$

if $|x - 1| \le \epsilon$ for sufficiently small values ν and ϵ. As

$$\varphi\big(u_{jk}(0)\big) - 8dr \le |u_{jk}(1)| \le \varphi\big(u_{jk}(0)\big) + 8dr\,,$$

then for $|u_{jk}(0)| \le \nu$ we have from (7.42)

$$\varphi(-\nu) - 8dr \le |u_{j,k}(1) \le \varphi(\nu) + 8dr$$

and for $|u_{j,k}(0) - 1| \le \epsilon$ we have from (7.43)

$$\varphi(1-\epsilon) - 8dr \le |u_{jk}(1)| \le \varphi(1+\epsilon) + 8dr\,.$$

Thus

$$\begin{array}{ll} \varphi(\nu) + 8dr < \nu\,, & \varphi(-\nu) - 8dr > -\nu\,, \\ \varphi(1+\epsilon) + 8dr < 1+\epsilon\,, & \varphi(1-\epsilon) - 8dr > 1-\epsilon\,, \end{array} \tag{7.44}$$

and we obtain the following inequalities:

$$\nu^3 - (1+a)\nu^2 + a\nu - \frac{8dr}{\alpha} > 0\,, \tag{7.45}$$

$$\epsilon^3 - (2-a)\epsilon^2 + (1-a)\epsilon - \frac{8dr}{\alpha} > 0\,. \tag{7.46}$$

Then for sufficiently small values of the ratio d/α these inequalities have positive solutions. They can be estimated as follows:

$$\frac{8r}{a}\frac{d}{\alpha} - \ldots < \nu < a - \frac{8r}{a(1-a)}\frac{d}{\alpha} + \ldots\,,$$

$$\frac{8r}{1-a}\frac{d}{\alpha} + \ldots < \epsilon < (1-a) - \frac{8r}{a(1-a)}\frac{d}{\alpha} + \ldots$$

Hence, statement (ii) is valid.

7.5.2 Existence of Steady Solutions

(iii) *Under conditions (i) and (ii), (7.34) has an infinite set of steady solutions such that any infinite matrix* $(a_{jk}), (jk) \in \mathbf{Z}^2, a_{jk} \in \{0,1\}$, *corresponds to one of these solutions,* $\mathbf{u}^* = \{u^*_{jk}\}$, $\mathbf{u}^* \bigcap_{j,k} I^{a_{jk}}_{jk}$.

Indeed, let us fix an arbitrary infinite matrix $(a)_{jk}$ and consider the intersection $J = \bigcap_{j,k} I^{a_{jk}}_{jk}$. The set J is convex because it is a product of intervals on the coordinate axes. Also J is a compact set in q-norm; because for an arbitrary $\delta > 0$ there exists $N_0 \in \mathbf{Z}_+$ such that $\sum_{|(j,k)| \geq N_0} |u_{jk}|^2 / q^{|(j,k)|} < \delta$ for any $\{u_{j,k}\} \in J$. It follows from condition (ii) that $\Phi(J) \subset J$. Indeed, since $J \subset I^{a_{jk}}_{jk}$ then $\Phi(J) \subset \Phi\left(I^{a_{jk}}_{jk}\right) \subset I^{a_{jk}}_{jk}$ for any $j,k \in \mathbf{Z}$, i.e. $\Phi(J) \subset \Phi\left(\bigcap_{j,k} I^{a_{jk}}_{jk}\right) = J$. Applying the Schauder–Tikhonov theorem we find that the map Φ has a fixed point $\mathbf{u}^*$ in J, i.e. statement (iii) is satisfied.

Note that if $a_{j,k} = 0$ then u^*_{jk} is *close* to 0 and if $a_{jk} = 1$ then u^*_{jk} is close.

7.5.3 Stability of Steady Solutions

Let us now study the stability of the steady solutions of (7.34). Let $\mathbf{u}^* = \{u^*_{jk}\}$ be one of these solutions belonging to $\bigcap_{j,k} I^{a_{j,k}}_{j,k}$. In addition, let

$$\xi_{jk}(n+1) = \sigma_{jk}\xi_{jk}(n) + d\left[\xi_{j+1k}(n) + \xi_{jk+1}(n) + \xi_{j-1k}(n) + \xi_{jk-1}(n)\right] \tag{7.47}$$

be the linear equation for perturbations of $\mathbf{u}^*$, where $\sigma_{jk} = 1 - 4d + \alpha f'(u^*_{jk})$. Its corresponding linear operator, L, is symmetric (and so the "matrix" $\tilde{L}$). Therefore, we can use the earlier introduced alternative way for obtaining the stability criterion (see Sect. 7.2.4) in order to find the conditions under which $||L||_q < 1$. We consider the set $\Delta^\Gamma(\tilde{L})$ of Gershgorin disks D_{jk}

$$|\lambda - \sigma_{jk}| \leq 4d \tag{7.48}$$

on the complex plane.

(iv) *For any fixed* $a \in (0,1)$ *there exists* $\alpha^* > 0$ *and* $d^*(\alpha) > 0$ $[\alpha^* \leq \alpha_0$, $d^*(\alpha) \leq d_0(\alpha)]$ *such that if* $0 < \alpha < \alpha^*$ *and* $0 < d < d^*(\alpha)$, *then each steady solution* $\mathbf{u}^*$ *is linearly stable for any value of* q *which is close enough to unity.*

Indeed, it follows from (7.48) that all Gershgorin disks lie strictly inside the unit circle if

$$f'(u^*_{jk}) < 0\,, \quad \alpha f'(u^*_{jk}) > -2 + 8d\,. \tag{7.49}$$

The first inequality is fulfilled if

$$\begin{aligned} \nu < \nu_0 &= \frac{1}{3}\left(1 + a - \sqrt{1 - a + a^2}\right), \\ \epsilon < \epsilon_0 &= \frac{1}{3}\left(2 - a - \sqrt{1 - a + a^2}\right), \end{aligned} \tag{7.50}$$

because $\mathbf{u}^* \in J = \bigcap_{jk} I_{jk}^{\imath_{jk}}$. From the second inequality in (7.49) we have

$$\begin{aligned} &\frac{1}{3}\left[1 + a - \sqrt{1 - a + a^2 + 6(1 - 4d)/\alpha}\right] < u_{jk}^* \\ &\quad < \frac{1}{3}\left[1 + a + \sqrt{1 - a + a^2 + 6(1 - 4d)/\alpha}\right]. \end{aligned} \tag{7.51}$$

Since $\mathbf{u}^* \in J$, then (7.51) is satisfied if

$$\begin{aligned} \nu &< \frac{1}{3}\left[\sqrt{1 - a + a^2 + 6(1 - 4d)/\alpha} - (1 + a)\right], \\ \epsilon &< \frac{1}{3}\left[a - 2 + \sqrt{1 - a + a^2 + 6(1 - 4d)/\alpha}\right]. \end{aligned} \tag{7.52}$$

As ν and ϵ have to be positive, we must have

$$0 < d < \frac{1}{8}(2 - \alpha a), \quad 0 < d < \frac{1}{8}[2 - \alpha(1 - a)]. \tag{7.53}$$

Consequently, under (7.53) there exist values of ν and ϵ such that all Gershgorin disks lie strictly inside a unit circle. Inequalities (7.45–46,53) are consistent if

$$\begin{aligned} \frac{d}{\alpha} &< \frac{1}{8r}[\nu_0^3 - (1 + a)\nu_0^2 + a\nu_0], \\ \frac{d}{\alpha} &< \frac{1}{8r}[\epsilon_0^3 - (2 - a)\epsilon_0^2 + (1 - a)\epsilon_0]. \end{aligned} \tag{7.54}$$

Then, there exists $\alpha^* > 0$ and $d^*(\alpha) > 0$ such that for $d < d^*(\alpha)$ and $\alpha < \alpha^*$ values ν and ϵ can be estimated in the following way:

$$\begin{aligned} &\frac{8r}{a}\frac{d}{\alpha} + \ldots < \nu < \min\left\{\nu_0, \frac{1}{3}\left[\sqrt{1 - a + a^2 + 6(1 - 4d)/\alpha} - (1 + a)\right]\right\}, \\ &\frac{8r}{1 - a}\frac{d}{\alpha} + \ldots < \epsilon < \min\left\{\epsilon_0, \frac{1}{3}\left[\sqrt{1 - a + a^2 + 6(1 - 4d)/\alpha} + (a - 2)\right]\right\}, \end{aligned}$$

if d/α is small enough. Hence, if (7.54) holds then $r\left(\Delta^\Gamma(\tilde{L})\right) < 1$ and $||L||_q < 1$ for values of q close to one. Thus, all solutions $\mathbf{u}^*$ are linearly stable.

7.5.4 Two-Dimensional Spatial Chaos

From the spectrum of the linearized operator (see Sect. 7.1) it is possible to estimate the q-norm of the operator $d\Phi(\mathbf{u})$ for any $\mathbf{u} \in J$ and to show that if d/α is small enough and q is close enough to unity then all solutions $\mathbf{u}^*$ are linearly stable. In addition, they are stable with respect to (7.34) in the following sense: For each steady solution $\mathbf{u}^* = \{u^*_{jk}\}$ there exists a set $\tilde{J} = \bigoplus \tilde{I}^{a_{jk}}_{jk} \subset I$ with $\tilde{I}^{a_{jk}}_{jk} \subset \{|u_{jk}| \le \nu\} \bigcup \{|u_{jk} - 1| \le \epsilon\}$ an open interval such that $\|\Phi^n \mathbf{v} - \Phi^n \mathbf{u}\|_q \to 0$ as $n \to \infty$ for any $\mathbf{v} \in \tilde{J}$.

Let us denote by Λ_0 the set of steady solutions $\{\mathbf{u}^*\}$. Then, $S_{j_0 k_0}\Lambda_0 = \Lambda_0$ for any $(j_0, k_0) \in \mathbf{Z}^d$, i.e. the set Λ_0 is invariant with respect to the TDS. Let Λ^2 be the set of all infinite matrices, $\Lambda^2 = \{\mathbf{a}\}$, $\mathbf{a} = (a_{ij}), a_{ij} \in \{0, 1\}$, $(i, j) \in \mathbf{Z}^2$ with the metric

$$\mathrm{dist}(\mathbf{a}, \mathbf{a}') = \sum_{i,j} \frac{|a_{i,j} - a'_{i,j}|}{2^{|i|+|j|}}$$

and $\{\sigma_{i_0,j_0}\}_{i_0,j_0} \in \mathbf{Z}^2$ be the full Bernoulli shift on the set. As a corollary of the considerations given earlier we have, in fact, the existence of a homeomorphism $h : \Lambda^2 \to \Lambda_0$ such that $S_{j_0 k_0} \circ h = h \circ \sigma_{j_0 k_0}$ for any $(j_0, k_0) \in \mathbf{Z}^2$. Accordingly, (7.34) admits spatial chaos.

7.6 Synchronization in Two-Layer Bistable Coupled Map Lattices

7.6.1 Layered Coupled Map Lattices

We consider the system [7.33]

$$\begin{cases} u_{j,k}(n+1) = u_{j,k}(n) + \alpha f\big(u_{j,k}(n)\big) + d\big(\Delta u(n)\big)_{j,k} - h\left[u_{j,k}(n) - v_{j,k}(n)\right] , \\ v_{j,k}(n+1) = v_{j,k}(n) + \alpha f\big(v_{j,k}(n)\big) + d\big(\Delta v(n)\big)_{j,k} - h\left[v_{j,k}(n) - u_{j,k}(n)\right] , \end{cases}$$

$$j, k = 1, 2, \dots, N , \tag{7.55}$$

where $f(z) = z(1-z)(z-a), /, 0 < a < 1$, $\alpha > 0$ and h and d are coupling coefficients characterizing the coupling strength between the two layers (interlayer coupling) and between elements in each layer (intralayer coupling). The pair (j, k) defines a lattice site; $\big(\Delta z(n)\big)_{j,k}$ is the discrete Laplace operator

$$\begin{aligned} \big(\Delta z(n)\big)_{j,k} &= \big(\Delta z(n)\big)_j + \big(\Delta z(n)\big)_k - 4z_{j,k}(n) , \\ \big(\Delta z(n)\big)_j &= z_{j+1,k}(n) + z_{j-1,k}(n) , \\ \big(\Delta z(n)\big)_k &= z_{j,k+1}(n) + z_{j,k-1}(n) . \end{aligned}$$

We impose on (7.55) von Neumann boundary conditions:

$$z_{0,k} = z_{1,k} , \quad z_{j,0} = z_{j,1} , \quad z_{j,N+1} = z_{j,N} , \quad z_{N+1,k} = z_{N,k} . \tag{7.56}$$

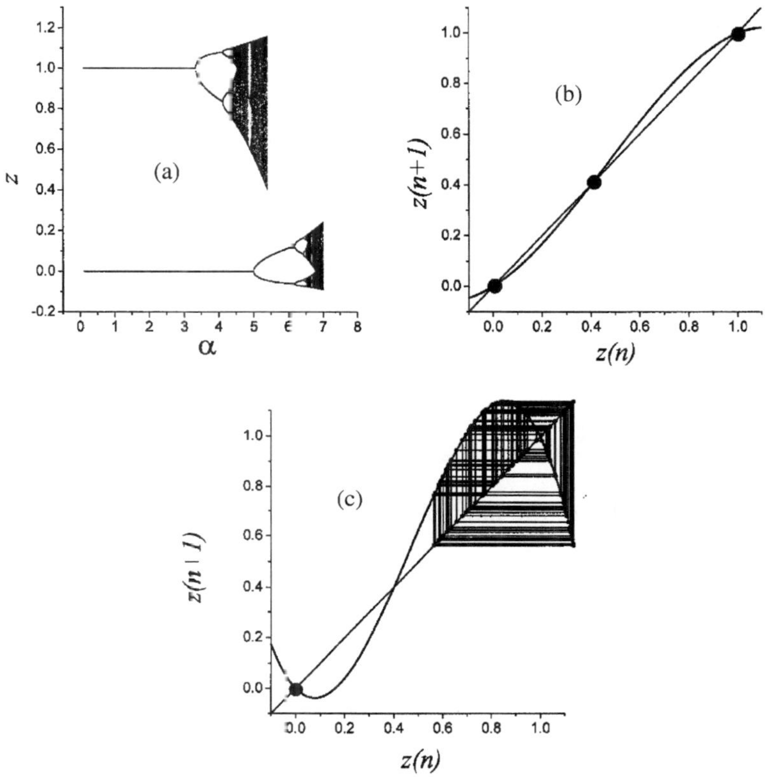

Fig. 7.10. (**a**) Bifurcation diagram of the first-order map (7.57). (**b**) Regular bistable map (7.57). (**c**) Chaotic map (7.57)

For $h = 0$, (7.55) describes the dynamics of a single layer. The composite CML for $h \neq 0$ represents a discrete version of a two-component "reaction–diffusion" medium with two diffusive components where $f(z)$ defines the "reaction kinetics".

The dynamics of a CML element is given by the first-order map

$$z(n+1) = z(n) + \alpha f\big(z(n)\big)\,. \tag{7.57}$$

Taking α as a control parameter we obtain the bifurcation diagram shown in Fig. 7.10a. It follows that the dynamics of the local map (7.57) can be either regular or chaotic. Figure 7.10a illustrates one of the possible regular modes of (7.57) when it has two stable fixed points whose basins of attraction are separated by an unstable fixed point. The chaotic dynamics appears according to the Feigenbaum period-doubling scenario [7.8, 7.19, 7.23], as shown in Fig. 7.10c.

Spatially Homogeneous States. The spatially homogeneous states of the lattice correspond to the fixed points of (7.55) whose coordinates do not depend on indices j and k. Such fixed points are defined by the solutions of the nonlinear system of equations

$$\begin{cases} \alpha f(u) - h(u-v) = 0\,, \\ \alpha f(v) - h(v-u) = 0\,. \end{cases} \tag{7.58}$$

The solution of (7.58) is given by the intersection of the two curves shown in Fig. 7.11a. For all α and $h > 0$, (7.58) has three symmetric solutions corresponding to three symmetric fixed points of (7.55)

$$O_0\left(u_{j,k}(n) = v_{j,k}(n) = 0\right), \quad O_1\left(u_{j,k}(n) = v_{j,k}(n) = a\right), \\ O_2\left(u_{j,k}(n) = v_{j,k}(n) = 1\right),$$

$$j, k = 1, 2, \ldots, N\,.$$

Using (F.16) we obtain the multipliers of these fixed points:

$$\mu_{s,l}^{(1),(2)}(O_i) = 1 + \beta_i - 2d\left(2 + \cos\frac{s\pi}{N} + \cos\frac{l\pi}{N}\right) - h \pm h\,, \\ s, l = 1, 2, \ldots, N\,, \quad i = 0, 1, 2\,, \tag{7.59}$$

with

$$\beta_0 \equiv -\alpha a\,, \quad \beta_1 \equiv \alpha(1-a)\,, \quad \beta_2 \equiv -\alpha(1-a)\,.$$

It follows from (7.59) that the fixed point O_1 is always unstable, while O_0 and O_2 are stable for the parameter values defined by

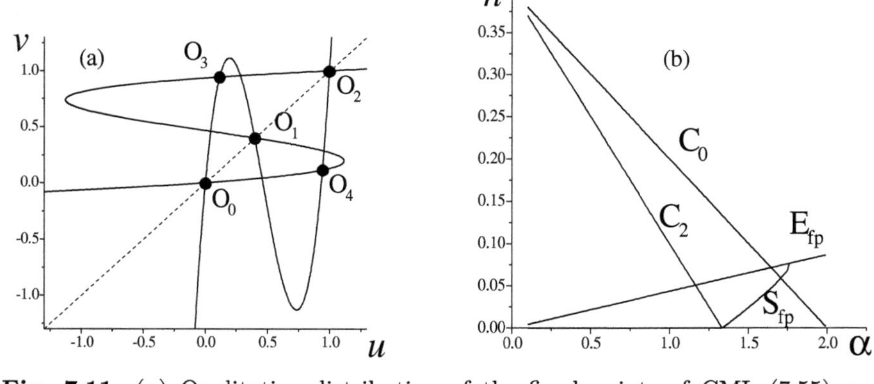

Fig. 7.11. (**a**) Qualitative distribution of the fixed points of CML (7.55) corresponding to spatially homogeneous states defined by the solutions of (7.58). (**b**) Parameter regions of existence and stability of spatially homogeneous states of the CML (7.55)

$$
\begin{aligned}
h &< \frac{1}{2}\left[2 - \alpha a - 4d\left(1 + \cos\frac{\pi}{N}\right)\right] \quad \text{for} \quad O_0\,, \\
h &< \frac{1}{2}\left[2 - \alpha(1-a) - 4d\left(1 + \cos\frac{\pi}{N}\right)\right] \quad \text{for} \quad O_2\,.
\end{aligned}
\tag{7.60}
$$

The stability boundaries defined by (7.60) are shown in Fig. 7.11b. Thus, the points O_0 and O_2 are stable in the regions below C_0 and C_2, respectively. Along with the symmetric fixed points system, (7.55) can have asymmetric fixed points whose coordinates are given by the solutions of (7.58) with $u \neq v$. The maximum number of such solutions is six (Fig. 7.11a) and all of them exist in the region below the curve E_{fp} shown in Fig. 7.11b. Using (F.16) we find that the fixed points O_3 and O_4, shown in Fig. 7.11a, are stable for the parameter region D^s_{fp} located between the curves E_{fp} and S_{fp} (Fig. 7.11b) while all other asymmetric fixed points are unstable. Thus, depending on the parameter values, the lattice (7.55) exhibits either two or four stable spatially homogeneous states.

Invariant Domain of the Layered Coupled Map Lattices. Consider (7.55) when $1 - \alpha a > 0$ is satisfied. In the phase space of (7.55) we introduce the domain

$$
Q = \{\mathbf{u}, \mathbf{v} \mid -r_2 \le u_{j,k} \le r_1, -r_2 \le v_{j,k} \le r_1\}\,,
$$

with $r_{1,2} > 0$. Let us show that there exist values of r_1 and r_2 for which Q is an invariant domain in the phase space. It means that any trajectory started in Q does not leave it for any $n > 0$. Let $(u_{j,k}(0), v_{j,k}(0)) \in Q$, $j,k = 1,2,\ldots,N$. Let us find the conditions on the parameters of (F.1) and on the "size" of the domain Q, i.e. on r_1 and r_2, for $(u_{j,k}(1), v_{j,k}(1)) \in Q$.

(i) Let us take the parameters α and a in the region

$$
D_\alpha = \{\alpha, a \mid 0 < a < 1, \alpha < \min 1/a, 1/(1-a)\}
$$

shown in Fig. 7.12a below the curve $\alpha = \alpha_1(a)$. If $(\alpha, a) \in D_\alpha$, then the function $\varphi(z) \equiv z + \alpha f(z)$ has two extrema (Fig. 7.12b) and the following properties:

$$
\begin{aligned}
& z_{\substack{\max\\ \min}} = \frac{\sqrt{1 - a + a^2 + 3/\alpha} \pm (1+a)}{3}\,, \qquad z_{max} > z_{min}\,, \\
& \varphi(-z_2) = \varphi(z_{max})\,, \quad \varphi(z_1) = \varphi(z_{min})\,, \\
& z_2 = z_{max} - (1+a)\,, \quad z_1 = 2z_{min} + (1+a)\,, \quad z_2 < z_1\,, \\
& z_2 > z_{max} \quad \text{if} \quad \alpha < \alpha_2(a)\,, \\
& z_2 \le z_{max} \quad \text{if} \quad \alpha \ge \alpha_2(a)\,,
\end{aligned}
\tag{7.61}
$$

with $\alpha_2(a) \equiv (a^2 + 3a + 1)^{-1}$.

(ii) Let the values of r_1 and r_2 satisfy

$$
z_{max} < r_1 < z_1\,, \qquad -z_2 < -r_2 < -z_{min}\,,
\tag{7.62}
$$

which define the rectangle Π on the plane (r_1, r_2) shown in Fig. 7.13.

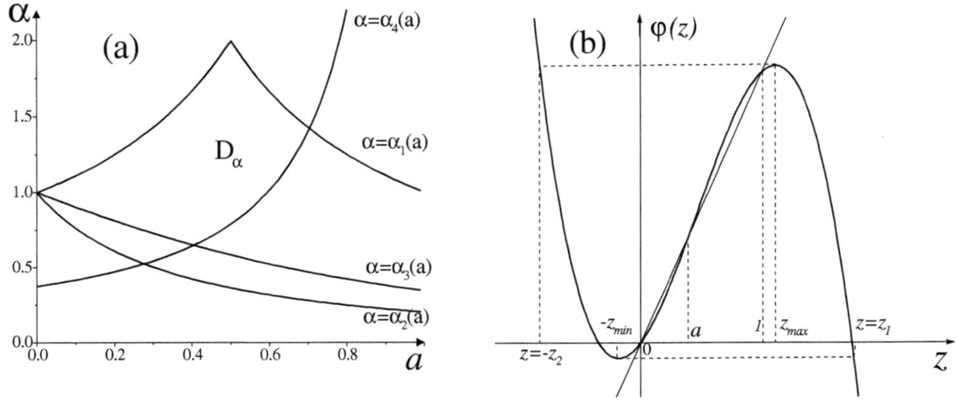

Fig. 7.12. (**a**) Parameter regions in the (α, a) plane defining the existence of the invariant domain Q in the phase space. (**b**) Qualitative shape of the function $\varphi(z)$ and relative distribution of auxiliary points

(iii) Let us take $\alpha < \alpha_2(a)$ (Fig. 7.12a). In this case the line $r_1 = r_2$ divides the rectangle Π into two parts, Π_1 and Π_2 (Fig. 7.13a). Let $(r_1, r_2) \in \Pi_1$ and, therefore, $r_2 < r_1$. Then, we obtain from (7.55)

$$\begin{aligned} |u_{j,k}(1)| = \Big| \varphi\big(u_{j,k}(0)\big) + d \Big[\big(\Delta u(0)\big)_j + \big(\Delta u(0)\big)_k - 4u_{j,k}(0) \Big] \\ -h\big[u_{j,k}(0) - v_{j,k}(0)\big] \Big| \leq \big|\varphi\big(u_{j,k}(0)\big)\big| + (8d + 2h)r_1 \,. \end{aligned} \tag{7.63}$$

It follows from (7.61) that the function $\varphi\big(u_{j,k}(0)\big)$ takes its maximum value for $u_{j,k}(0) = z_{max}$. This value does not depend on r_1 and r_2. Therefore,

$$|u_{j,k}(1)| \leq \varphi(z_{max}) + (8d + 2h)r_1 \,. \tag{7.64}$$

We require that the point $\{u_{j,k}(1), v_{j,k}(1)\} \in Q$, $j, k = 1, 2, \ldots, N$. This condition is satisfied if

$$r_2 > r_1(8d + 2h) + \varphi(z_{max}) \,. \tag{7.65}$$

Let us impose conditions on the parameter values such that within region Π_1 there are some points satisfying (7.65). Such values are restricted by (Fig. 7.13a)

$$8d + 2h < \frac{z_2 - \varphi(z_{max})}{z_2} \,. \tag{7.66}$$

When (7.66) is fulfilled there exists a subregion (triangle ABC in Fig. 7.13a) in Π for which (7.65) is satisfied and, therefore, Q is an invariant domain.

(iv) Let $\alpha \geq \alpha_2(a)$. In this case the whole rectangle Π lies inside the region $r_2 < r_1$ (Fig. 7.13b) and the points satisfying (7.65) always exist if

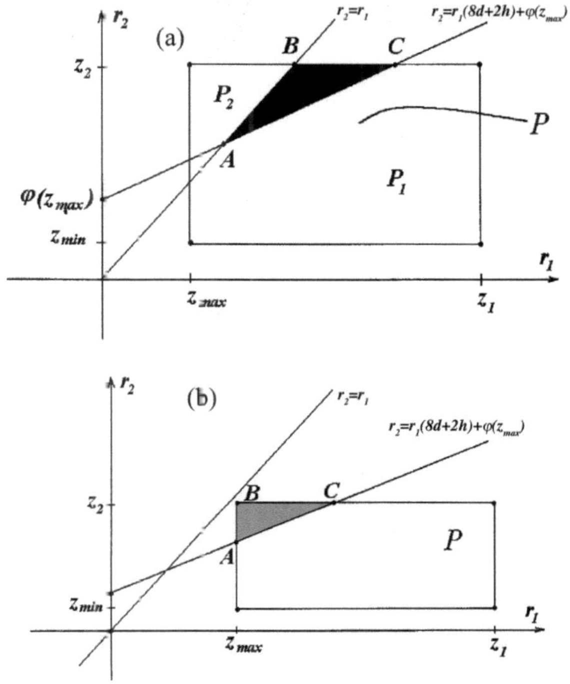

Fig. 7.13. Auxiliary plane (r_1, r_2). Values of r_1 and r_2 taken within the triangle ABC define the "size" of the invariant domain Q. (**a**) $\alpha < \alpha_2(a)$. (**b**) $\alpha_2(a) < \alpha < \alpha_3(a)$

$$8d + 2h < \frac{z_2 - \varphi(z_{max})}{z_{max}} \tag{7.67}$$

is satisfied. Since the parameters d and h are positive definite, (7.67) is satisfied if $z_2 - \varphi(z_{max}) > 0$. This inequality defines the region in the (α, a) plane restricted between $\alpha = \alpha_2(a)$ and $\alpha = \alpha_3(a)$ as shown in Fig. 7.12a.

Thus, (7.55) has in its phase space an invariant domain Q if the parameters satisfy (7.66) for $\alpha < \alpha_2(a)$ and (7.67) for $\alpha_2(a) \le \alpha \le \alpha_3(a)$.

7.6.2 Dynamics of a Single Lattice (Layer)

Let $h = 0$, and hence the layers are independent. Their evolution taken separately is described by the system

$$\begin{aligned} &u_{j,k}(n+1) = u_{j,k}(n) + \alpha f\big(u_{j,k}(n)\big) + d\big(\Delta u(n)\big)_{j,k}, \\ &j, k = 1, 2, \ldots, N. \end{aligned} \tag{7.68}$$

Note that (7.68) represents (7.34) (see Sect. 7.4) with boundary conditions (7.56). By taking $h = 0$ in (7.66–67) we obtain the parameter values for which 7.68) has an invariant domain

$$Q_u = \{\mathbf{u} \mid -r_2 \le u_{j,k} \le r_1\}.$$

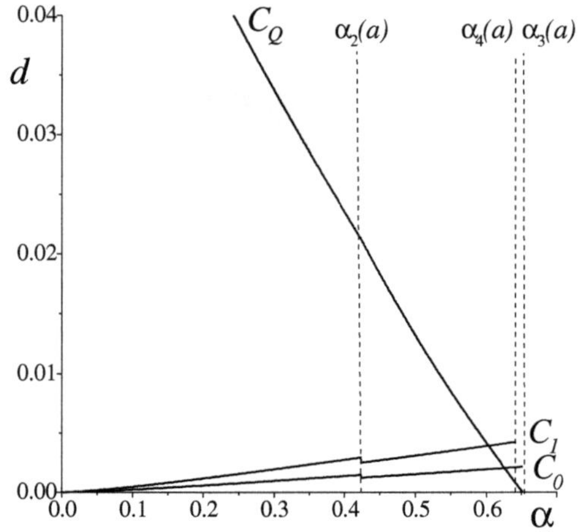

Fig. 7.14. Parameter region corresponding to high multistability and spatial disorder in a single layer of the CML. The curve C_Q delineates the region of existence of the invariant domain Q_u (region below). C_0 and C_1 provide the region of existence of the narrower invariant domains $U^0_{j,k}$ and $U^1_{j,k}$, respectively, for $a = 0.4$

Figure 7.14 illustrates the corresponding parameter region located below the curve C_Q in the (α, d) plane.

Let us construct "narrower" invariant domains, which are crucial for understanding the dynamics of the lattice.

Narrower Invariant Domains.

First Family of Invariant Domains. Consider in the phase space of (7.68) the following regions:

$$U^1_{j,k}(\varepsilon) = \{u_{j,k} \in Q_u,\ |u_{j,k} - 1| < \varepsilon\},$$

where ε satisfies the inequality

$$0 < \varepsilon < z_{max} - 1. \tag{7.69}$$

Let $u_{j,k}(0) \in U^1_{j,k}(\varepsilon)$. Since $U^1_{j,k}(\varepsilon) \subset Q_u$, then from (F.14) we obtain

$$u_{j,k}(1) \le \varphi\big(u_{j,k}(0)\big) + 8dr_1. \tag{7.70}$$

The function $\varphi(u)$ increases monotonically for the points of $U^1_{j,k}(\varepsilon)$, and hence

$$\varphi(1-\varepsilon) \le \varphi\big(u_{j,k}(0)\big) \le \varphi(1+\varepsilon). \tag{7.71}$$

From (7.70–71) we obtain

$$\varphi(1-\varepsilon) - 8dr_2 \le u_{j,k}(1) \le \varphi(1+\varepsilon) + 8dr_1. \tag{7.72}$$

Then, each of the regions $U^1_{j,k}(\varepsilon)$ is an invariant domain if $u_{j,k}(1) \in U^1_{j,k}(\varepsilon)$. This condition is satisfied if

$$\begin{aligned} \varphi(1+\varepsilon) + 8dr_1 &< 1+\varepsilon\,, \\ \varphi(1-\varepsilon) - 8dr_2 &> 1-\varepsilon\,. \end{aligned} \tag{7.73}$$

These inequalities can be rewritten as conditions on the "size" of the region Q_u

$$r_1 < \frac{1+\varepsilon-\varphi(1+\varepsilon)}{8d}\,, \quad r_2 < \frac{\varphi(1-\varepsilon)-1+\varepsilon}{8d}\,. \tag{7.74}$$

Let us show that there exists a nonempty subregion in the triangle ABC in the (r_1, r_2) plane, shown in Fig. 7.13 with $h = 0$, with points where (7.74) is satisfied.

Consider first $\alpha < \alpha_2(a)$. In this case such a subregion exists if

$$\begin{aligned} \frac{1+\varepsilon-\varphi(1+\varepsilon)}{8d} &> r_A \equiv \frac{\varphi(z_{max})}{1-8d}\,, \\ \frac{\varphi(1-\varepsilon)-1+\varepsilon}{8d} &> \frac{\varphi(z_{max})}{1-8d}\,, \end{aligned} \tag{7.75}$$

where r_A is the abscissa of A shown in Fig. 7.13a. The inequalities (7.75) impose the conditions on ε, i.e. on the "sizes" of regions $U^1_{j,k}(\varepsilon)$. Analyzing (7.75) we find that it has a solution in the region

$$d < \frac{\alpha\Phi(\varepsilon_{max})}{8[\varphi(z_{max})+\alpha\Phi(\varepsilon_{max})]}\,, \tag{7.76}$$

with

$$\Phi(\varepsilon) \equiv \varepsilon^3 - (2-a)\varepsilon^2 + (1-a)\varepsilon\,,$$

$$\varepsilon_{max} \equiv \frac{2-a-\sqrt{1-a+a^2}}{3}\,.$$

Moreover, ε must satisfy (7.69). The inequalities (7.75) and (F.15) are simultaneously fulfilled if, for example, $z_{max} - 1 > \varepsilon_{max}$. Thus, we obtain one more condition on the parameters of the system

$$\alpha < \min_{a\in(0,1)} \{\alpha_2(a), \alpha_4(a)\}\,, \tag{7.77}$$

with

$$\alpha_4(a) \equiv \frac{3}{4(2-a)\left(2-a-\sqrt{1-a+a^2}\right)}\,.$$

Finally, (7.76–77) provide the existence of the narrower invariant domains $U^1_{j,k}(\varepsilon)$.

For $\alpha_2(a) < \alpha < \alpha_3(a)$, the construction of domains $U^1_{j,k}(\varepsilon)$ is similar. In this case these domains exist for the parameter region

$$d < \frac{\alpha\Phi(\varepsilon_{max})}{8z_{max}}, \tag{7.78}$$

$$\alpha_2(a) < \alpha < \min_{a\in(0,1)} \{\alpha_4(a), \alpha_3(a)\} . \tag{7.79}$$

In the (α, d) parameter plane, the region of existence of the narrower invariant domains $U^1_{j,k}(\varepsilon)$ is located below curve C_1 (Fig. 7.14).

Second Family of Invariant Domains. Consider the regions

$$U^0_{j,k}(\nu) = \{u_{j,k} \in Q_u, \ |u_{j,k}| \le \nu\},$$

where ν satisfies

$$0 < \nu < z_{min}. \tag{7.80}$$

Following (7.70–73) we obtain the following restrictions on the "size" of Q_u:

$$r_1 < \frac{\nu - \varphi(\nu)}{8d}, \quad r_2 < \frac{\nu + \varphi(-\nu)}{8d}. \tag{7.81}$$

For $\alpha < \alpha_2(a)$, (7.80) is satisfied. Accordingly, (7.81) defines a nonempty subregion in the triangle ABC (Fig. 7.13a) if

$$\frac{\nu - \varphi(\nu)}{8d} > \frac{\varphi(z_{max})}{1-8d}, \quad \frac{\nu + \varphi(-\nu)}{8d} > \frac{\varphi(z_{max})}{1-8d}. \tag{7.82}$$

The inequalities (7.82) can be simultaneously solved if the parameters of (7.68) satisfy

$$d < \frac{-\alpha f(u_{min})}{8[\varphi(z_{max}) - \alpha f(u_{min})}, \tag{7.83}$$

where u_{min} is the coordinate of the minimum of $f(u)$,

$$u_{min} \equiv \frac{1 + a - \sqrt{1 - a + a^2}}{3}.$$

For $\alpha_2(a) < \alpha < \alpha_3(a)$ the conditions on r_1 and r_2 are (Fig. 7.13b)

$$\frac{\nu - \varphi(\nu)}{8d} > z_{max}, \quad \frac{\nu + \varphi(-\nu)}{8d} > 8dz_{max} + \varphi(z_{max}). \tag{7.84}$$

From (7.80) and (7.84), we find that a solution exists if (7.79) and

$$d < \frac{-\alpha f(u_{min})}{8z_{max}} \tag{7.85}$$

are satisfied.

Thus, the second family of invariant domains, $U^0_{j,k}(\nu)$, when $\alpha_2(a) < \alpha < \alpha_3(a)$ exists if (7.78) is satisfied or (7.79) and (7.85) are simultaneously satisfied. In the (α, d) parameter plane the region of existence of the narrower invariant domains, $U^0_{j,k}(\varepsilon)$, is located below the curve C_0 (Fig. 7.14).

Multistability of a Single Layer. Let us choose the parameters of (7.68) such that all invariant domains, $U^0_{j,k}(\nu)$ and $U^1_{j,k}(\varepsilon)$, simultaneously exist. It is provided by (7.76–77), (7.79), (7.83) and (7.85). The corresponding parameter region in the (α, d) plane for $a = 0.4$ is shown in Fig. 7.14. By construction it lies below the curve C_0. Let us fix an arbitrary matrix (a_{jk}) consisting of the symbols "0" and "1" and consider the intersection $J = \cap U^{a_{j,k}}_{j,k}$ corresponding to this matrix. The set J is convex as it is represented by the direct product of the segments of the coordinate axes. Besides, J is a compact set. Since $J \subset U^{a_{jk}}_{jk}$, $S(J) \subset J$, where S is the point map given by (7.68) (see Sect. 7.5.4). Thus, any given matrix c_{jk} corresponds to an invariant domain J. Therefore, there exist at least 2^{N^2} attractors located in each domain, J, and (7.68) displays a high multistability. Using the Schauder–Tikhonov theorem we find that S has a fixed point $\{u^*_{j,k}\}$ in J. The numerical solution of (7.68) shows that, for suitable parameter values, the fixed point in J is the only attractor, and hence it is stable. In this case, any fixed point defines a stationary pattern in the space $\{\mathbf{Z}^2, R\}$ realized for appropriate initial conditions. Note that for these values of the parameters there exist stationary patterns of highly diverse spatial configurations ranging from spatially homogeneous ($a_{j,k}$ consists of only 0 or 1) to irregular or spatially chaotic (0 and 1 are arbitrarily distributed in $a_{j,k}$).

Figure 7.15a1 illustrates a possible stationary pattern obtained in the region of high multistability. It evolves from the initial distribution shown in Fig. 7.15a2. The space–time diagram shown in Fig. 7.15a3 illustrates the transient process. It follows that the distribution of "species" in the lattice space acquires bistable character. Hence, the single-layer CML for the chosen parameters can be a possible tool for information storage encoded as stationary patterns. As has been noted in Sect. 7.5.1 the unit of the CML can exhibit more complex dynamics (Fig. 7.10). For illustration, we choose the parameters to yield in each site a stable fixed point and a time-dependent attractor (limit cycle or chaotic set). In this case, a small enough intralayer diffusion also provides the existence of bistable patterns. Figure 7.15b shows one of them for the same initial distribution as in Fig. 7.15a but for complex dynamics of the unit. The lattice comes to almost the same stable spatial configuration. However part of the units (dark ones) oscillate in the neighborhood of the attractor of a single unit.

7.6.3 Global Interlayer Synchronization

Sufficient Conditions for Global Synchronization. Consider the possibility of interlayer synchronization in the CML (7.55). Let us change variables:

$$x_{j,k}(n) = u_{j,k}(n) - v_{j,k}(n)\,, \quad y_{j,k}(n) = u_{j,k}(n) + v_{j,k}(n)\,. \qquad (7.86)$$

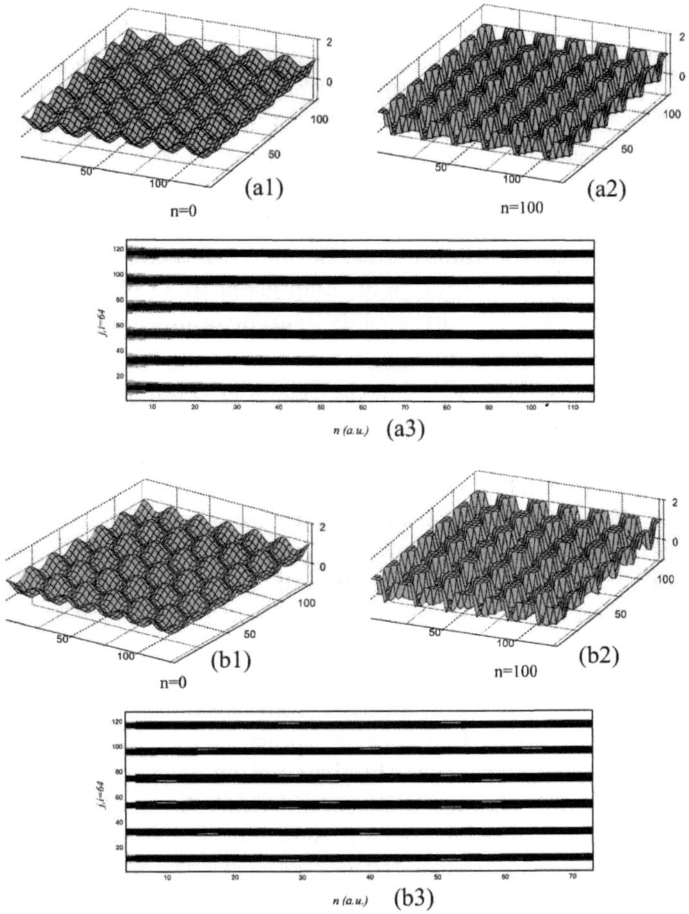

Fig. 7.15. Spatially inhomogeneous patterns in a single layer of the CML. (**a1,b1**) Initial distribution; (**a2,b2**) Terminal state. (**a3,b3**) Space–time evolution of the cross-section at $j = 64$. (**a**) Appearance of a regular pattern steady in time. Parameter values: $a = 0.4, \alpha = 0.7, d = 0.001$. (**b**) Regular pattern with oscillations near the fixed point O_2. Parameter values: $a = 0.4, \alpha = 4.5, d = 0.01$

Then, from (7.55) using (7.86) we have

$$\begin{aligned} x_{j,k}(n) = d\big(\Delta x(n)\big)_j + px_{j,k}(n) - \alpha x_{j,k}(n)G_{j,k}\big(x_{j,k}(n), y_{j,k}(n)\big) \\ +d\big(\Delta x(n)\big)_k, \quad j,k = 1,2,\ldots,N, \end{aligned} \tag{7.87}$$

with

$$p \equiv 1 + \frac{\alpha(1-a+a^2)}{3} - 4d - 2h, \quad G(x,y) \equiv F(y) + \frac{x^2}{4},$$

$$F(y) \equiv \frac{3}{4}\left(y - \frac{2(1+a)}{3}\right)^2.$$

There is a manifold of interlayer synchronization, $M = x_{j,k}(n), (j,k = 1,2,\ldots,N)$, in the phase space. Let us obtain the conditions for M to attract all motions of (7.55). To do this we consider

$$V(n) = \sum_{j,k=1}^{N} \frac{x_{j,k}^2(n)}{2}.$$

Changing $V(n)$ with respect to (7.87) yields

$$\Delta V = V(n+1) - V(n) = -\sum_{j,k=1}^{N} (K_{j,k} + G_{j,k}P_{j,k}), \qquad (7.88)$$

with

$$\begin{aligned} K_{j,k} \equiv & -d^2\left(\Delta x(n)\right)_j^2 - 2dpx_{j,k}(n)\left(\Delta x(n)\right)_j - 2d^2\left(\Delta x(n)\right)_j\left(\Delta x(n)\right)_k \\ & +(1-p^2)x_{j,k}^2(n) - 2dpx_{j,k}(n)\left(\Delta x(n)\right)_k - d^2\left(\Delta x(n)\right)_k^2, \\ P_{j,k} \equiv & \; 2\alpha dx_{j,k}(n)\left(\Delta x(n)\right)_j + 2\alpha px_{j,k}^2(n) - \alpha^2x_{j,k}^2(n)G_{j,k}\left(x_{j,k}(n), y_{j,k}(n)\right) \\ & +2\alpha dx_{j,k}(n)\left(\Delta x(n)\right)_k. \end{aligned}$$

Let us find the condition such that all functions $K_{j,k}$ and $P_{j,k}$ are positive definite. Let us consider (7.55) with parameters ensuring the existence of the invariant domain Q. In the (h,α) plane this parameter region is located below the curve H_Q (Fig. 7.16). The function $P_{j,k}$ can be rewritten as follows:

$$P_{j,k} = \alpha^2x_{j,k}^2(n)\left[k - G_{j,k}\left(x_{j,k}(n), y_{j,k}(n)\right)\right] + \tilde{P}_{j,k},$$

with $k > 0$ and

$$\tilde{P}_{j,k} = 2\alpha dx_{j,k}(n)\left(\left(\Delta x(n)\right)_j + (2\alpha p - k\alpha^2)x_{j,k}^2(n) + 2\alpha dx_{j,k}(n)\left(\Delta x(n)\right)_k.$$

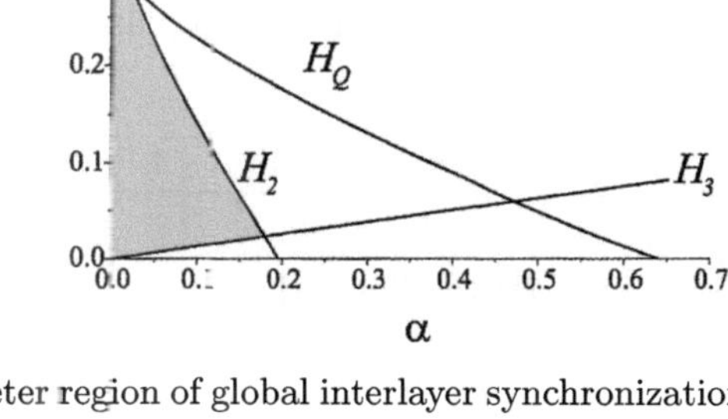

Fig. 7.16. Parameter region of global interlayer synchronization. Parameter values: $a = 0.4$, $d = 0.001$

We choose k such that

$$k - G_{j,k}\big(x_{j,k}(n), y_{j,k}(n)\big) \geq 0 \,. \tag{7.89}$$

Taking into account (7.86) and the condition $\{u_{j,k}(n), v_{j,k}(n) \in Q\}$ we find that (7.89) is satisfied if

$$k > 3r^2 + 2(1+a)r_1 + \frac{(1+a)^2}{3} \,. \tag{7.90}$$

Consider now the function

$$\tilde{P} = \sum_{j,k}^{N} \tilde{P}_{j,k} \,, \tag{7.91}$$

and introduce the vector $\mathbf{z} = (z_1, z_2, \ldots, z_{N^2})$, where $z_1 = x_{11}$, $z_2 = x_{12}$, $\ldots, z_{N^2} = x_{NN}$. Using the components of the vector $\mathbf{z}$ the function $\tilde{P}$ takes the form

$$\tilde{P} = \sum_{i,j=1}^{N^2} b_{ij} z_i z_j \equiv \mathbf{z}^T B \mathbf{z} \,,$$

with $b_{ij} = b_{ji}$. The superscript T denotes the transpose, and $B = (b_{ij})$ is a $N^2 \times N^2$ matrix. The quadratic form P will be positive definite if the eigenvalues of the symmetric matrix B are all positive. By applying Gershgorin theorem (Appendix G) to B, we find that all its eigenvalues are positive if

$$k < \frac{2(p-4d)}{\alpha} \,. \tag{7.92}$$

This inequality ensures that $\tilde{P}$ is positive definite. From the other side the parameter k must satisfy (7.90). The inequalities (7.90) and (7.92) have a solution if the parameters of (7.55) satisfy

$$8d + 2h < 1 + \frac{\alpha(a^2 - 4a + 1)}{6} \,, \tag{7.93}$$

which defines some parameter region in the (h, α) plane (Fig. 7.16, curve H_1). Furthermore,

$$r_1 < r_1^* \equiv \frac{-(1+a) + \sqrt{6(p-4d)/\alpha}}{3} \,. \tag{7.94}$$

We require $r_1^* > r_A$ (Fig. 7.13). Then the system of inequalities (7.90) and (7.92) has a solution inside the triangle ABC which determines the "size" of the invariant domain Q. It follows that the function $\sum_{j,k}^{N} P_{j,k}$ is positive definite if the following inequalities are satisfied:

$$\alpha < \alpha_2(a)\,, \quad \frac{-(1+a)+\sqrt{6(p-4d)/\alpha}}{3} > \frac{\varphi(z_{max})}{1-8d}\,, \tag{7.95}$$

or

$$\alpha_2(a) < \alpha < \alpha_3(a)\,, \quad \frac{-(1+a)+\sqrt{6(p-4d)/\alpha}}{3} > z_{max}\,. \tag{7.96}$$

They define the parameter region in the (h, α) plane located below curve H_2 (Fig. 7.16). Analogously, we obtain that the function $K = \sum_{j,k}^{N} K_{j,k}$ is positive definite if

$$h > \frac{\alpha(1-a+a^2)}{6}\,. \tag{7.97}$$

The corresponding parameter region is located above curve H_3. Finally, in the parameter region defined by inequalities (7.66–67), (7.93) and (7.95–97) the difference term ΔV is such that $\Delta V \leq 0$, and, hence, V is a Lyapunov function and the manifold of interlayer synchronization is globally asymptotically stable. This region is shown in Fig. 7.16.

Examples of Interlayer Synchronization.

Pattern Synchronization. Let us choose the parameter values in the region of global interlayer synchronization (Fig. 7.16). Note that by choosing the intralayer coupling small enough the existence of stationary patterns in each single layer taken separately is ensured (Sect. 7.5.2). Let the initial conditions in the two layers be two different patterns in the form of stripes with different orientations (Fig. 7.17a). The interlayer synchronization leads to identical offsprings. All elements tend to the main diagonal and group within the invariant domains described in Sect. 7.5.2. Lower pictures in Fig. 7.17a illustrate initial and terminal arrangement of the elements relative to the diagonal.

Pattern Replication. As earlier described (Chap. 6) multilayer lattices with continuous time allow dynamical replication of form or dynamical copying. This occurs when the interlayer interaction allows one of the patterns (stimulus) to dominate the other one, which may be in a spatially chaotic state. Here we have extended this result to CML. Let one of the layers initially contain a regular pattern called the stimulus and the elements of the second layer exhibit a spatially irregular distribution (Fig. 7.17b). The result of the interlayer synchronization provides two copies of the stimulus. Note that in the region of global interlayer synchronization (Fig. 7.16) the layered CML keeps all the characteristic features of replication in lattices with continuous time such as quality factor, spatial resolution, contour selection property, etc. (see Sect. 6.3).

For example, the contour of the original image may appear as the difference between the initial and synchronized patterns (Fig. 7.18). It preserves the key features of the image with only a few spatial points (about 2%).

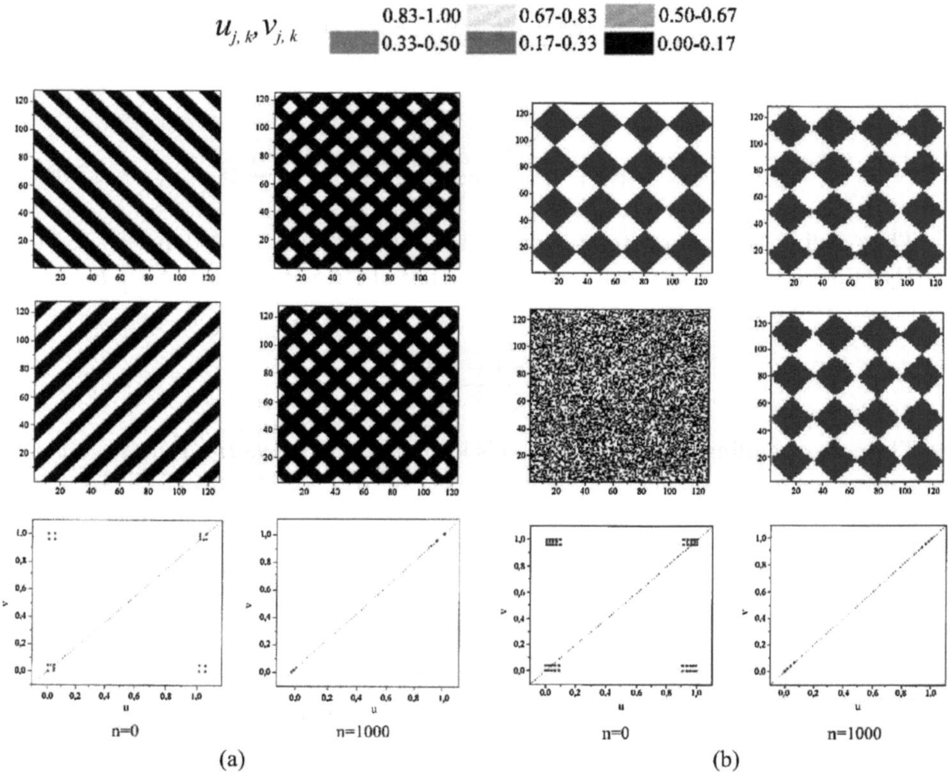

Fig. 7.17. Interlayer synchronization of stationary patterns. *Left column*: initial conditions. *Right column*: terminal patterns. *Lower row*: dynamics of the lattice elements in the (u, v) plane. (**a**) Synchronization of regular patterns leading to a new spatial form. Parameter values: $a = 0.6, \alpha = 0.1, h = 0.1, d = 0.001$. (**b**) Pattern replication or replication of form. Parameter values: $a = 0.5, \alpha = 0.1, h = 0.1, d = 0.001$

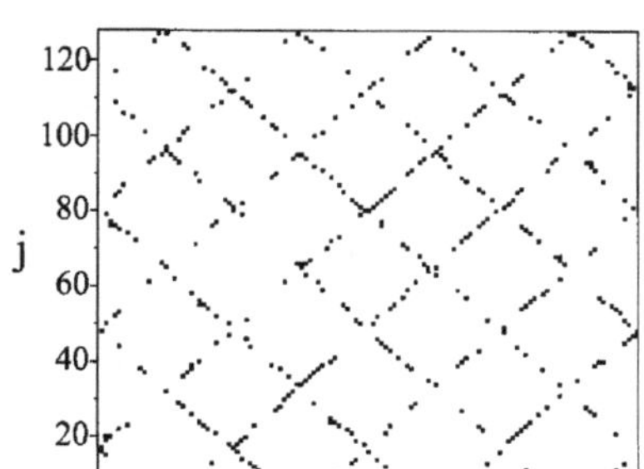

Fig. 7.18. Contour selection property of CML. Differential image obtained as the misprint signal between replicated patterns and the original image (Fig. 7.17b)

In fact, we have a 50 : 1 ratio in the data compression as the result of the interlayer synchronization in CML.

At variance with the continuous case, CML have an upper boundary of synchronization. This boundary has been theoretically estimated in Sect. 7.5.3, and it is shown in Fig. 7.16. Exceeding some threshold an instability of the synchronization manifold appears, which may lead to the appearance of new off-diagonal attractors.

7.7 Instability of the Synchronization Manifold

An important issue now is the possible destruction of the synchronization mode. This instability is associated with the so-called *riddling* and a *blowout* bifurcation of the attractor on the diagonal. In the first case some fixed points or cycles with a positive transverse Lyapunov exponent (transverse instability) appear at the synchronization manifold. If the trajectory of the attractor is slightly perturbed, for example, by noise, and comes near an unstable point, it is expelled from the diagonal. Then it may go to some other attractor or it may return to the neighborhood of the synchronization manifold. After the blowout bifurcation almost all trajectories in the diagonal subspace become transversely unstable and the system goes to a new off-diagonal attractor. However, if the manifold is weakly unstable near the bifurcation the *on–off* intermittency behavior may occur when trajectories are distributed mostly in a small neighborhood of the diagonal.

Let us illustrate how the manifold of *interlayer* synchronization in (7.55) loses its stability. It has been shown in Sect. 7.6.1 that above the curves C_0 and C_2 (Fig. 7.11), defined by (7.60), one of the multipliers of the homogeneous fixed points O_0 and O_2 makes them unstable, $\mu_{1,1}^{(2)} < -1$. Thus, at least one fixed point with one unstable direction appears on the diagonal. Then, for the local instability of the interlayer synchronization mode it suffices to have

$$h > \min \left\{ \begin{array}{c} \frac{1}{2}\left[2 - \alpha a - 4d\left(1 + \cos\frac{\pi}{N}\right)\right], \\ \frac{1}{2}\left[2 - \alpha(1-a) - 4d\left(1 + \cos\frac{\pi}{N}\right)\right] \end{array} \right\} . \tag{7.98}$$

When (7.98) is satisfied the diagonal may contain, however, some synchronized solutions with riddled basins (inhomogeneous fixed points, limit cycles, etc.) which are locally stable in the transverse direction.

7.7.1 Instability of the Synchronized Fixed Points

Let us choose the parameters of (7.55) providing mutual synchronization of stationary patterns (Sect. 7.5.2–3). It implies some restrictions on the values of d and h (Figs. 7.14 and 7.15). Then all 2^{N^2} attractors, or stable fixed points,

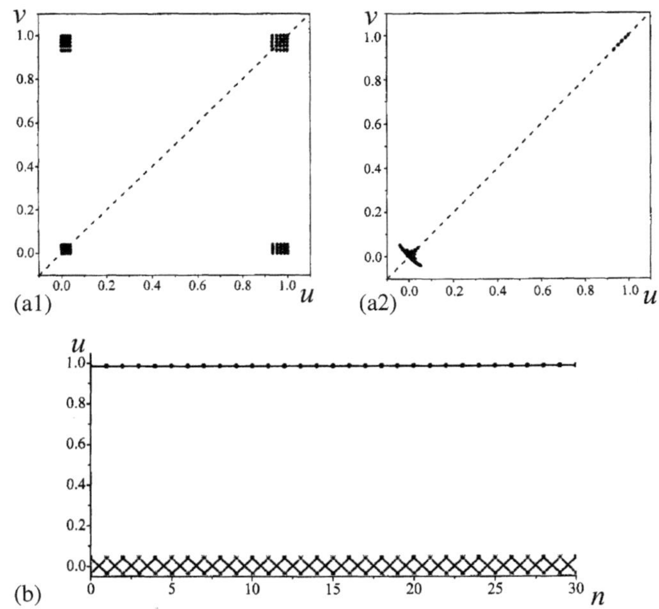

Fig. 7.19. Instability of synchronized stationary patterns. (**a1**) Initial conditions ($n = 0$) and (**a2**) terminal patterns ($n = 10^5$) in the (u, v) plane. (**b**) Time-series elements $u_{64,64}, v_{64,64}$ and $u_{64,65}, v_{64,65}$ taken from different clusters. Parameter values: $a = 0.4, \alpha = 0.7, h = 0.78, d = 0.004$

are located along the diagonal in the phase space of (7.55). Let us increase the interlayer coupling, h, to satisfy (7.98). For illustration we take initial conditions in the form of two different spatially chaotic patterns. Figure 7.19 shows the result of the interaction in the (u, v) plane. Each point here represents a pair of elements taken at the same spatial site (j, k) of the two layers. At the terminal state, there are points located near but outside the synchronization manifold. At the same site the elements taken from the two layers exhibit antiphase oscillations in the neighborhood of O_2 and a steady signal near O_0.

In the $2N^2$-dimensional phase space the synchronized patterns represent the inhomogeneous fixed point. To assess the transverse instability, among N^2 transversal Lyapunov exponents we have calculated the largest:

$$\lambda^{tr} = \max_{i=1,N} \log |\mu_i^{tr}| , \tag{7.99}$$

where μ_i^{tr} are the N^2 eigenvalues of the Jacobi matrix of (7.55) whose eigenvectors are transversal to the diagonal subspace. The dependence of λ^{tr} on h for different values of d is shown in Fig. 7.20a. There is a critical value of h above which the pattern loses its transverse stability. Note that higher values of d yield higher λ_{tr}. For any other of the 2^{N^2} possible spatial configurations the instability curves have similar shape.

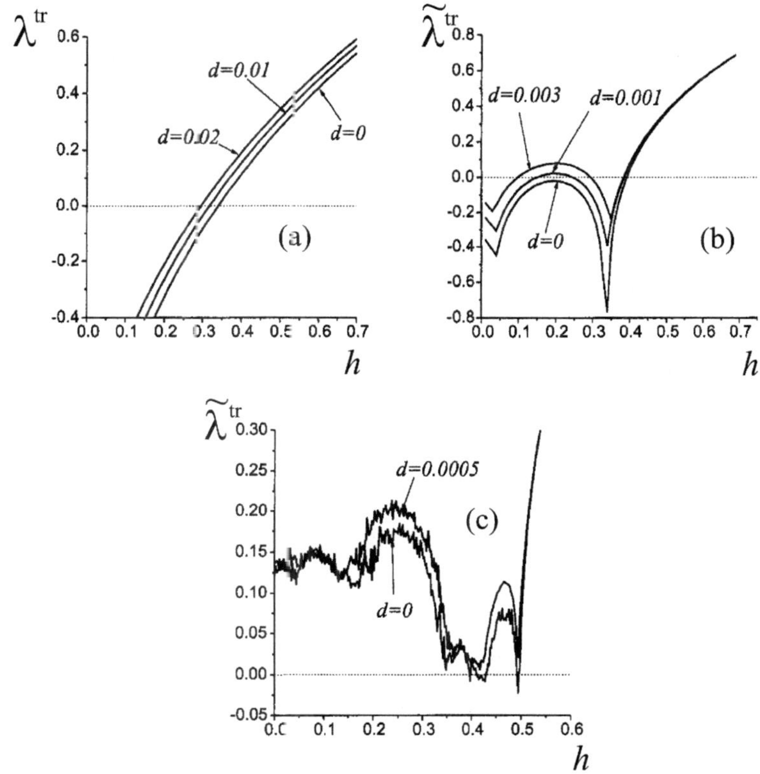

Fig. 7.20. (**a**) Largest transverse Lyapunov exponent, λ^{tr}, for various values of the intralayer coupling, d. Parameter values: $N = 10, a = 0.5, \alpha = 2.64$. Initial conditions are randomly distributed synchronized steady patterns. (**b**) Instability factor, $\tilde{\lambda^{tr}}$, for a period-four periodic orbit attractor on the diagonal. Parameter values: $N = 30, a = 0.1, \alpha = 2.8$. (**c**) Instability factor, $\tilde{\lambda^{tr}}$, for a chaotic attractor on the diagonal. Parameter values: $N = 30, a = 0.1, \alpha = 2.921$

7.7.2 Instability of Synchronized Attractors and On–Off Intermittency

Let us now increase α to have in each element a limit cycle or a chaotic attractor (Fig. 7.10). Then, for small values of d in the CML we may obtain an attractor in the N^2-dimensional diagonal subspace. To assess its transversal stability we have calculated the quantity

$$\tilde{\lambda^{tr}} = \max_{j,k=1,N} \lim_{n\to\infty} \frac{1}{n} \log |\mu^{tr}_{j,k}(n)| , \tag{7.100}$$

where

$$\mu^{tr}_{j,k}(n) = 1 + f'(u_{j,k}(n)) - 2h$$

is the transverse multiplier of the trajectory $u_{j,k}(n) = v_{j,k}(n)$ at the n-iteration of the (j,k) pair of units taken with $d = 0$. Then, $\tilde{\lambda}^{tr}$ gives the transverse Lyapunov exponent for $d = 0$. When $d \neq 0$ but is small enough, the trajectory of each local unit becomes slightly perturbed, changing the transverse multiplier at each iteration of the map. Then, we obtain a spectrum of Lyapunov exponents of the perturbed trajectory of the lattice units. Accordingly, positiveness of $\tilde{\lambda}^{tr}$ indicates possible transverse instability of a unit, and hence the synchronized attractor of the CML.

We have analyzed the behavior of the instability factor, $\tilde{\lambda}^{tr}$, for a periodic orbit of period four and a chaotic attractor (Fig. 7.20b,c). In the limit cycle case there is a critical value of h above which the cycle becomes transversally unstable. The increase of d (Fig. 7.20b) brings positive $\tilde{\lambda}^{tr}$ for a given value of h. For the chaotic attractor the instability factor has positive values, but it oscillates near zero for some region of h. This indicates the weak instability of the diagonal subspace, and hence may give rise to on–off intermittency. Figure 7.21a provides a snapshot of the resulting attractor of the CML in the (u,v) plane. To characterize the attractor we introduce the quantity

$$\rho(x) = \lim_{n\to\infty} \frac{1}{n} \lim_{\delta x\to 0} \frac{N_p(x,\delta x,n)}{N^2}, \tag{7.101}$$

where $N_p(x,\delta x,n)$ is the number of points lying in the stripe $(x, x+\delta x)$ of the (u,v) plane, $x = u-v$, at the n-iteration in the CML. This quantity represents an average density of points relative to the diagonal and characterizes the probability of the points to be at distance x from the diagonal. Then, if all points are in the diagonal subspace $\rho(0) = 1, \rho(x) = 0, \forall x \neq 0$. For the off-diagonal attractor (Fig. 7.21a) the behavior of $\rho(x)$ for various values of d is shown in Fig. 7.21b. For vanishing d the function $\rho(x)$ has a high maximum at $x = 0$, indicating that for a long time the CML trajectory stays in the vicinity of the diagonal with a typical on–off intermittency behavior. The space–time plot in Fig. 7.21c illustrates the evolution of the difference variables $x_{j,k=15}(n)$. The on-diagonal stages (white color) are intermitted with off-diagonal excursions (black color) of the unit trajectories. For higher values of d the function $\rho(x)$ tends to a constant in some region of x, indicating that the attractor is almost uniformly distributed. Note that for CML we deal not only with intermittency in time but also with intermittency in lattice space and hence space–time intermittency.

Consider the case of pattern instability (Figs. 7.19 and 7.20a). If the diagonal is weakly unstable the points are distributed in the neighborhood of the diagonal (Fig. 7.19b). Let us increase h and hence λ^{tr} (Fig. 7.20a). The snapshot of the attractor in this case is shown in Fig. 7.22a. The time evolution of each pair of units occurs in the following way: In the vicinity of the unstable synchronized fixed point the trajectory is strongly expelled from the diagonal subspace. Then it comes to the region where it is attracted back to the diagonal and evolves in its neighborhood until it reaches the vicinity of an unstable fixed point, completing the on–off diagonal loop. The

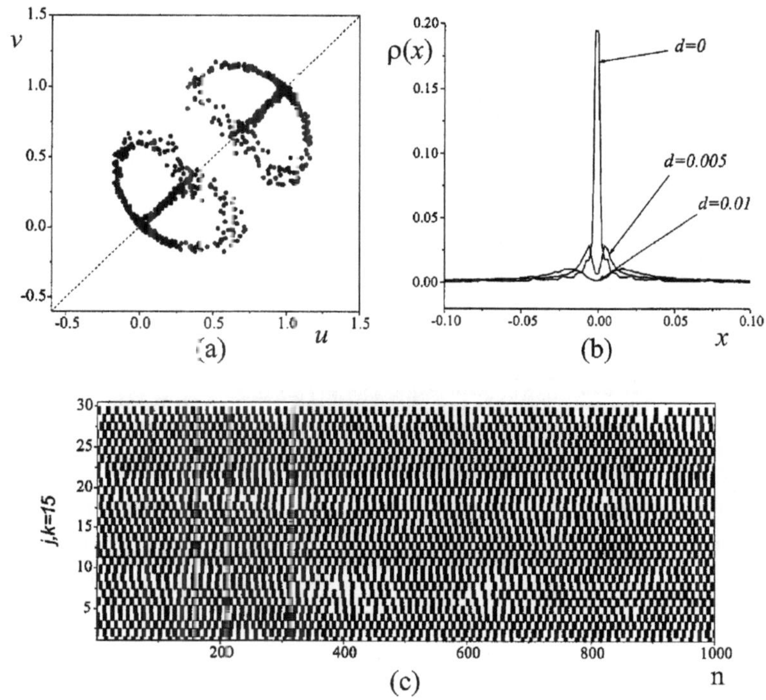

Fig. 7.21. Instability of a chaotic attractor and on–off intermittency. Parameter values: $N = 30, a = 0.1, \alpha = 2.921, h = 0.370, d = 0.001$. (**a**) A snapshot in the (u, v) plane. (**b**) Density function, $\rho(x)$, for various values of d. (**c**) Space–time evolution of difference variables $x_{j,k=15}$. White color shows the units in the vicinity of the diagonal for which $x_{j,k}(n)$ are distributed in the range $\in [0; 0.01)$; black color is for $x_{j,k}(n) \geq 0.01$

probability distribution for such on–off behavior is shown in Fig. 7.22b. The difference with the intermittent behavior in Fig. 7.22 is that by increasing d two maxima appear, indicating that the trajectory in the "on-diagonal stage" is mostly distributed within a small distance from the diagonal. The space–time behavior of the cross-section, $x_{j,k=15}(n)$, shown in Fig. 7.22c is seemingly more regular or ordered in space and time relative to Fig. 7.21c as it emerges out of a regular steady pattern distribution. We see that in this case, in spite of the synchronization failure, all elements are well separated and form two clusters located near O_0 and O_2. Hence, the distribution of "species" in the layers still has a bistable character. It allows the replication of form discussed in the previous section. For appropriate initial conditions we can obtain self-replicated patterns much the same as those shown in Fig. 7.17b. However, in this case the information must be decoded or extracted from the terminal state of the layers which is not steady in time. Then the CML preserves all features of pattern selection (Fig. 7.17a) and pattern replication (Fig. 7.17b, 7.18) that were found for systems with continuous time.

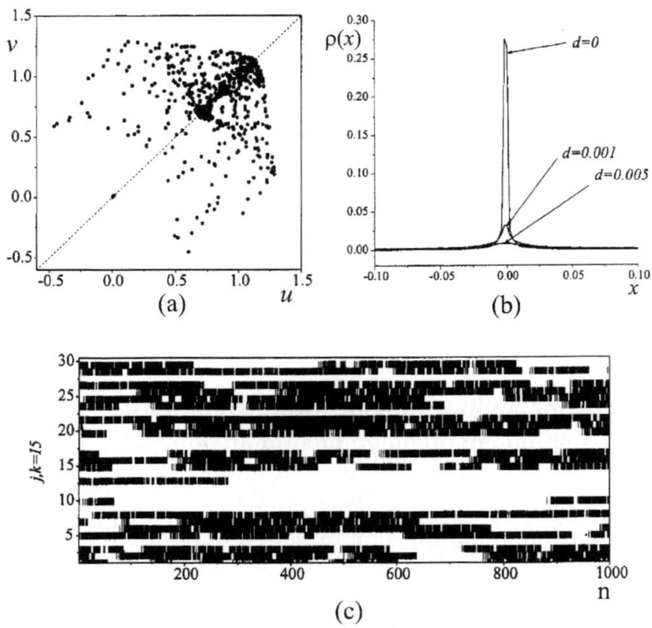

Fig. 7.22. Instability of a synchronized steady pattern and on–off loop attractor. Parameter values: $N = 30, a = 0.5, \alpha = 2.64, h = 0.750, d = 0.005$. (**a**) A snapshot in the (u, v) plane. (**b**) Density function $\rho(x)$ for various values of d. (**c**) Space–time evolution of the difference variables $x_{j,k=15}$. Black and white as in Fig. 7.21c

For the loss of synchronization it is important to note the significant role of nonvanishing values of d, however small, in the dynamics of the lattice. When $d = 0$, there are independent pairs of interacting cubic maps (7.57). For each pair the influence of a small d plays the role of noise, which perturbs a trajectory, expelling it from the synchronization manifold.

7.8 Salient Features of Coupled Map Lattices

In this chapter we have presented a constructive method for finding enough conditions for the stability of steady and dynamical structures in CML of infinite dimensions. The essence of the method consists in the following: The infinite matrix realizing the linearized operator on some fixed basis is approximated by a sequence of finite-dimensional matrices whose dimension exhibits unlimited growth. This method allows us to reduce the investigation of stable solutions in the unbounded CML to the study of the spectrum characteristics of some finite-dimensional matrices corresponding to CML composed of a finite number of units.

We have carried out the investigation of spatial chaos in the 1D diffusive FNS equation. We have determined the parameter set such that, depending on the initial conditions, a steady structure (pattern) is a solution of this equation. The profile of the structure is determined by an infinite sequence (1D case) or a matrix (2D case) consisting of two symbols and is given beforehand.

Our investigation has shown that the 1D FNS equation has an infinite set of stationary waves moving with the velocity $c = 1$. The profiles of these waves are determined by randomly chosen infinite sequences of two symbols. Each of these waves is linearly stable to perturbations moving with the same velocity. Hence, chaotic traveling waves occur in the diffusive FNS equation.

We have investigated the phenomena of pattern formation and synchronization in multilayer bistable CML. In the corresponding phase space we have constructed a set of invariant domains containing all fixed points of the CML. Thus, we have obtained the existence conditions for the possible 2^{N^2} stationary patterns in a single layer. The distribution of "species" has a bistable character, and hence for large enough N the CML provides a "memory" tool capable of storing binary data, black-and-white images, etc.

Studying the interaction between layers we have obtained the conditions of global interlayer synchronization and hence pattern synchronization and replication of form. This pattern replication allows duplication of an original image (stimulus), selecting in the copies its key features. From the viewpoint of the applications, the multilayer CML has potential for information processing, stored in the layers as binary patterns.

We have obtained sufficient conditions for the interlayer synchronization mode to become unstable. In the neighborhood of the blow-out bifurcation, when the synchronization manifold is weakly unstable, the CML may display space–time, on–off intermittency. The intralayer coupling, d, tends to increase the instability growth factor. When synchronized steady patterns become strongly unstable, we have illustrated the possibility of an on–off loop attractor. Such an attractor does not break their bistable spatial configuration or *clustering* while the units oscillate in a complex way within the two clusters. Then, in spite of complex dynamics in time, the spatial distribution of CML variables stays well organized, i.e. it may retain a binary pattern encoded with the two clusters.

The list of references on the CML dynamics given is, of course, far from complete. Readers interested in other results on CML dynamics may find the book by Kaneko and Tsuda [7.31] and the special issue of *Physica D* edited by Chaté and Courbage [7.16] worth studying.

8. Conclusions and Perspective

Hopefully, the reader has found that the results presented in our book give a fair view of self-organization processes together with a great deal of methodology for a very diverse class of active spatially extended systems. The contents of the book may also appear divided into two parts. One deals with self-organization in single systems and the other with processes in interconnected active 1D (chain) and 2D lattices, thus leading to a 3D architecture.

We have shown how one can qualitatively, and quantitatively characterize typical self-organization phenomena for single systems including spatio-temporal patterns of different degrees of complexity (homogeneous, periodic, chaotic, etc.) and traveling waves (wave fronts or kinks, pulses, solitons, soliton-bound states, spiral waves and chaotic waves). We have studied how from their underlying dynamics these processes depend on control parameters and what allows the dynamic properties of the systems to change effectively.

A characteristic feature of the systems considered is, indeed, the appearance of self-organizing, synergetic, dissipative forms with a complex spatial configuration. Take, for instance, chaotic patterns, soliton-bound states, wave pulse trains, oscillating fronts etc. These structures are long lived, hence evolutionarily stable, for a wide class of initial distributions, and their complex dynamic behavior is not exotic but a typical property of active extended systems. Another interesting feature of active lattice systems is the relationship and mutual influence of their spatial and temporal dynamics. For example, in a 1D chain composed of auto-oscillating bistable elements we have obtained the conditions for which the system exhibits chaos of oscillation amplitudes (chaotic dynamics in space) while the time evolution stays regular. Moreover, we have shown that the oscillations may become mutually synchronized on the background of amplitude disorder. We have also shown that several levels of synchronization are possible, including in-phase and splay-phase modes, phase synchronization and synchronization in the mean. These modes appear through the formation of phase and frequency clusters.

The interconnected lattice systems considered in our book may be interpreted as systems of coupled lattices or layers. We call such systems multilayer lattices. They possess a three-dimensional geometry, anatomy or architecture with both intra- and interlayer connections (bonds or couplings). We have described typical phenomena for such systems, such as interlayer synchronization and desynchronization through space–time intermittency, excitation reentry and circulation, destruction and damping of dynamic patterns and dynamic copying or replication of form. We think that the latter is very interesting from both the viewpoint of applications and its fundamental aspects. Replicating a form means that an image (stimulus) such as some dynamic structure (pattern or wave) initially existing in a layer can be transferred to other layers that were initially in, say, spatially chaotic and/or unrelated states. The latter may be considered as raw material readily available for the replicating process. The replication dynamics is based on the mechanisms of competition (albeit not a true natural selection) and synchronization. We have found that using these mechanisms the system may reproduce, in a number of copies, the key features of an original *dynamic* pattern. Hence it provides one of the most important functions of natural systems: to replicate many times a necessary form according to a given template. The imperfection of copies relative to the original, which always occurs, allows us to treat the misprints as a possible degree of freedom for the appearance of new features or new information in the process of multistep replication. Clearly, this replication process is not merely like that of a printing press.

We have also shown that, in general, a most adequate approach for studying self-organization in active lattice systems is using the qualitative methods of dynamical system theory, bifurcation theory and synergetics, to which we added, when available, numerical experimental evidence to support our findings. These methods naturally contain some basics for self-organization notions and mechanisms. Take, for instance, attractors (regular and chaotic), bifurcations, multistability, homoclinic and heteroclinic orbits, etc., allowing us to explain and predict such typical self-organization phenomena as the existence of spatio-temporal structures and their changing number, the development of instabilities and so on.

We think that the results presented in this book have interest from two points of view. On the one hand, we hope that the self-organization forms which we have described in a constructive way are quite universal and typical for many other systems well beyond those treated here. For the models considered in the book the salient features are the dynamical properties of elements and the type of their connections but not the origin (physical, chemical, biological, etc.) and particular details or specificity of the systems described. On the other hand, we can say that our results may be useful for the design of new information processing devices using neurodynamical principles involving parallel processing. Neurocomputers must have, indeed, a lattice-like 3D architecture. An elementary unit of information processing in neurocomputers must be a spatio-temporal structure of activity, i.e. a device with extended memory. At present, there is intensive research in this direction. Hopes here are connected with molecular electronic technology. Modern engineering allows us to built devices with single elements having polymer molecular size. In particular, some interesting results have been obtained in the design of artificial reaction–diffusion systems, providing a foundation for future massively parallel molecular computers.

Before concluding we would like to point out some interesting and important problems closely related to the contents of our book and demanding further work. We have considered self-organization in lattice systems with mainly linear and local diffusive interactions or bonds. It would be very interesting to consider the cases of nonlinear coupling, coupling with delay and nonlocal couplings and to investigate the influence of regular and noisy external and intrinsic perturbations on the self-organization processes. Another problem is the consideration of the dynamics of multilayer lattice systems composed of elements or units with really sophisticated and complex dynamics, thus mimicking the properties of real neurons, like Haken and others have recently put forward or we have proposed in collaboration with neurobiologist R.R. Llinas. This opens new possibilities, for instance, for the design of really autonomous, robots (with no external, additional computer) based on ideas and methods of the theory of synergetic systems and nonlinear dynamics. (This approach should allow a distributed "intelligence" rather than having such a capacity concentrated in a single "brain" center).

Let us conclude by further commenting on the fascinating evolutionary problem of replication of form and/or function, to which, indeed, we have devoted part of this book. In evolution from periodic to higher levels, replication of form and function are basic processes. F. Crick [8.2, 8.3] noted that "at first sight it would seem a very difficult task to make an exact copy of the intact three-dimensional structure of a protein in its well-organized native fold. One could conceive making a molecular cast of the surface, as one might for a piece of sculpture, but how would one copy the inside of the molecule? Nature has solved this difficulty with a neat trick. The polypeptide chain is synthesized as one extended, rather one-dimensional structure and then folds itself up ... The molecule explores the constant opportunities offered by thermal movement until, by trial and error, the best fold is discovered ... To produce this miracle of molecular construction all the cell needs to do is to string together the amino acids in the correct order ... Life is an infinitely rare event, and yet we see it teeming all around us. How can such rare things be so common? ... We need to carry the considerable amount of information as instructions to form the complexity which characterizes life, and unless this information is copied with reasonable accuracy the mechanism will decay under accumulated weight of errors. Perfect accuracy is not a requirement. Many of the copying errors will be a handicap but a few are likely to be an improvement ... We need these for natural selection to operate on. Thus, we need mutations, as these genetic errors are called." Replication and mutation (and/or imperfect fidelity when copying) are necessary requirements for evolution.

Replication of form seems necessary for the onset of function that belongs to an emergent synergetic level. However, at least in the very early stages of evolution, a clearcut separation may not have existed between these two processes. Discussion persists, for instance, about which came first, protein sequence or structure. On the other hand, biochemical pathways did not evolve by the sequential addition of steps to pathways that become functional only at the end. Instead, they have been rigged up with pieces brought from other pathways, hence using nearby accessible patterned or functionally operating material. This is the tinkering process advocated by F. Jacob [8.5]. A minimum of function (and its replication) may be necessary to have replication of form. But without a significant portion of the latter a minimal function may not emerge, hence form and function may have grown by successive steps of disparate time duration and degree of sophistication. If there is a trial-and-error search, its efficiency may be drastically improved both in time and exploration of landscape. In recent books, Haken [8.4] and Scott [8.7] have beautifully illustrated this tinkering process when discussing brain dynamics and other neurobioprocesses.

The (bio)evolutionary problem is indeed quite complex. Take again proteins. The great majority of sequences have multiple steady states and hence may fold into different structures. Here faithful or high-fidelity replication of 3D form is indeed crucial. Failure to do so may lead to a new species, with functional inactivity or misfunction. For the simpler case of prions, prion diseases such as Creutzfeldt–Jacob disease in humans or the related BSE in (mad) cows are thought to result from a conformational change of a normal isoform of a prion to a pathogenic form. The cause would be a transition between two possible steady states of the system exhibiting, for example, bistability properties [8.6]. Yet, we feel that faithful replication of form in 3D may not be such a rare and difficult event, at an early stage before the (complex) specific function of the 3D architecture is acquired. One can approach say the very early stage of bioevolution in a metaphoric sense. As said by various authors, in science, metaphor spawns theory, the ultimate value of which is judged by comparison with experiment. Metaphor also plays another role: as a vehicle for carrying powerful concepts and images from one area of science to another.

The reader can easily see that there is indeed ample territory for research in this area, for there is yet no overwhelming evidence for any one theory of replication of form in complex systems. Nonetheless, there is sufficiently strong argument for the scenario of tinkering in the realm of unsteady, metastable, long-lasting structures to push for its development. Indeed, the life of those who bear, say, genes may be long but it is certainly transient. Even genes do not last for ever. What (bio)evolution seems to demand is opportunities (including a long enough lifetime) with available appropriate raw materials to replicate on. The Darwinian survival of the "fittest", when only steady states, permanent forms in "infinite" time, are considered, appears to be an attitude that is too conservative [8.1]. Long-lasting metastable forms, having significant function, may be enough to win, and so on.

Appendices

A. Integral Manifolds of Stationary Points

Consider (4.36) within $-1 \le z \le 1$:

$$\begin{cases} \mu\dot{x} = y - x - (1+a)x\,, \\ \delta\dot{y} = x - y + z\,, \\ \dot{z} = -\beta y - \gamma z\,. \end{cases} \tag{A.1}$$

The system (A.1) has at its origin the single stationary point O. Its corresponding characteristic equation is

$$\lambda^3 + \left(\gamma + \frac{1}{\delta} + \frac{a+1}{\mu}\right)\lambda^2 + \left(\frac{\gamma+\beta}{\delta} + \frac{\gamma(a+1)}{\delta} + \frac{a}{\mu\delta}\right)\lambda + \frac{\beta(a+1)}{\mu\delta} + \frac{\gamma a}{\mu\delta} = 0\,. \tag{A.2}$$

Within the G_u parameter range (4.38), (A.2) has one positive root $\lambda = \lambda_a$ and a pair of complex-conjugate roots. Therefore, the stationary point O is a saddle-focus with a 1D unstable manifold $W^u_{1,2}$ and a stable separatrix plane $W^s_\mu(O)$. Let us derive the equations which describe these manifolds. To define $W^s_\mu(O)$ we change variables from (x, y, z) to (u_1, u_2, u_3),

$$\begin{bmatrix} x \\ y \\ z \end{bmatrix} = \begin{bmatrix} 1 & -k_1^a & -k_2^a \\ 0 & 1 & 0 \\ 0 & 0 & 1 \end{bmatrix} \begin{bmatrix} u_1 \\ u_2 \\ u_3 \end{bmatrix}, \tag{A.3}$$

with

$$\begin{aligned} k_1^a &= \delta(\lambda_a + \tfrac{a+1}{\mu})\,, \\ k_2^a &= 1/\beta\left[-\tfrac{a}{\mu} - \left(1 + \tfrac{\delta(a+1)}{\mu}\right)\lambda_a - \delta\lambda_a^2\right]. \end{aligned} \tag{A.4}$$

Then (A.1) becomes

$$\begin{cases} \dot{u}_1 = \lambda_a u_1\,, \\ \dot{u}_2 = \frac{1}{\delta}u_1 - \frac{k_1^a+1}{\delta}u_2 - \frac{k_2^a-1}{\delta}u_3\,, \\ \dot{u}_3 = -\beta u_2 - \gamma u_3\,. \end{cases} \tag{A.5}$$

It follows from the first equation of (A.5) that, in the new phase variables (u_1, u_2, u_3), $W^s_\mu(O)$ is given by $u_1 = 0$. In terms of the original variables (x, y, z) it is

$$W^s_\mu(O): \left\{x + \delta\left(\lambda_a + \tfrac{a+1}{\mu}\right) y + \tfrac{1}{\beta}\left[-\tfrac{a}{\mu} - \left(1 + \tfrac{\delta(a+1)}{\mu}\right)\lambda_a - \delta\lambda_a^2\right] z = 0\right\}. \tag{A.6}$$

Thus all motions in $W^s_\mu(O)$ are governed by the 2D system

$$\begin{cases} \dot{y} = -\dfrac{k_1^a + 1}{\delta} y - \dfrac{k_2^a - 1}{\delta} z\,, \\ \dot{z} = -\beta y - \gamma z\,. \end{cases} \tag{A.7}$$

To derive the equations of $W^u_{1,2}$, we now change variables in the form

$$\begin{bmatrix} x \\ y \\ z \end{bmatrix} = \begin{bmatrix} -k_4^a & 0 & 1 \\ 1 & 0 & 0 \\ 1 + k_4^a - k_3^a & 1 & -1 \end{bmatrix} \begin{bmatrix} v_1 \\ v_2 \\ v_3 \end{bmatrix}, \tag{A.8}$$

with

$$\begin{aligned} k_3^a &= -\lambda_a \delta\,, \\ k_4^a &= -1/\lambda_a\mu + a + 1\,. \end{aligned} \tag{A.9}$$

Then (A.1) becomes

$$\begin{cases} \dot{v_1} = \lambda_a v_1 + \frac{1}{\delta} v_2\,, \\ \dot{v_2} = \left(-\gamma - \frac{1+\lambda_a\delta}{\delta}\right) v_2 + \left(\gamma - \frac{a+1}{\mu}\right) v_3\,, \\ \dot{u_3} = -\frac{1}{\delta(\lambda_a\mu + a + 1)} v_2 - \frac{(a+1)}{\mu} v_3\,. \end{cases} \tag{A.10}$$

$W^u_{1,2}$ is given by $v_2 = 0$ and $v_3 = 0$. In terms of the original variables (x, y, z) it is

$$W^u_\mu(O): \left\{-\frac{x}{k_4^a} = \frac{y}{1} = \frac{z}{1 + k_4^a - k_3^a}\right\}. \tag{A.11}$$

A similar approach permits us to derive the equations of the integral manifolds of the stationary points $P^\pm$. Their corresponding characteristic equations can be written in the form (A.2) using the substitution $a \to b$. Both stationary points $P^\pm$ have one negative eigenvalue $\lambda = \lambda_b$ and two complex-conjugate eigenvalues $\lambda = h_b \pm i\omega_b$. In the parameter region G_u (Fig. 4.18) h_b is positive, and in the region G_s, h_b is negative. Therefore, these stationary points are either asymptotically stable (in G_s), or saddle-foci with dim $W^s_\mu(P^\pm) = 1$ and dim $W^u_\mu(P^\pm) = 2$, respectively. The equations describing these manifolds are

$$W_\mu^u(P^\pm) : \left\{x \mp x_0 + \delta\left(\lambda_b + \tfrac{b+1}{\mu}\right)(y \mp y_0) \right.$$
$$\left. +1/\beta\left[-\tfrac{\delta}{\mu} - \left(1 + \tfrac{\delta(b+1)}{\mu}\right)\lambda_b - \delta\lambda_b^2\right](z \mp z_0) = 0\right\}, \qquad \text{(A.12)}$$

$$W_\mu^s(P^\pm) : \left\{-\frac{x \mp x_0}{k_4^b} = \frac{y \mp y_0}{1} = \frac{z \mp z_0}{1 + k_4^b - k_3^b}\right\}, \qquad \text{(A.13)}$$

where $k_1^b, k_2^b, k_3^b, k_4^b$ are defined by (A.4) and (A.9) using the substitution $\lambda_a \to \lambda_b$.

B. Relative Location of the Manifolds $W_\mu^s(O)$ and $W_\mu^u(P^+)$

It follows from (A.6–7) that the lines at the intersections of the manifolds $W_\mu^s(O)$ and $W_\mu^u(P_+)$ with the plane U_{+1} are, respectively,

$$\begin{aligned} l_\mu^s &: \{k_1^a y + k_2^a z + 1 = 0\}, \\ l_\mu^u &: \{k_1^b(y - y_0) + k_2^b(z - z_0) + 1 - x_0 = 0\}. \end{aligned} \qquad \text{(B.1)}$$

Note that for $\mu = 0$ these lines become a single one $l_0 : \{y = a + 1\}$. Since $k_1^a/k_1^a \neq k_1^b/k_1^b$, l_μ^s and l_μ^u intersect each other at some point $L(y_l, z_l)$. Therefore, in the phase space of (4.36) there exists a trajectory Γ_l which contains this point and, simultaneously, belongs to $W_\mu^s(O)$ and $W_\mu^u(P^+)$. As $\xi \to +\infty$, Γ_l tends to O and with $\xi \to -\infty$ to P^+. Using (B.1) the coordinates of the point L can be written as

$$y_l = \frac{\Delta_1}{\Delta}, \quad z_l = \frac{\Delta_2}{\Delta},$$

with

$$\begin{aligned} \Delta &\equiv k_1^a k_2^b - k_1^b k_2^a, \\ \Delta_1 &\equiv -k_2^b - k_2^a(x_0 - 1 + k_1^b y_0 + k_2^b z_0), \\ \Delta_2 &\equiv k_1^b + k_1^a(x_0 - 1 + k_1^b y_0 + k_2^b z_0). \end{aligned} \qquad \text{(B.2)}$$

Note that in (B.2) the coordinates y_l and z_l are given by the parameters of (4.36), (A.1) and by the eigenvalues λ_a and λ_b. To make y_l and z_l depend only on the parameters of (4.36), we use an asymptotic representation of the eigenvalues λ_a and λ_b,

$$\lambda_q = -\frac{q+1}{\mu} - \frac{1}{\delta(q+1)} + \sum_{k=1}^{4} \Psi_k^q \mu^q + O_q(\mu^2), \qquad \text{(B.3)}$$

where the index q stands for either a or b and the values Ψ_k^q are given by the following expressions:

$$\Psi_1^q \equiv -\frac{q}{\delta^2(q+1)^3}\,, \quad \Psi_2^q \equiv -\frac{q(q-1)}{\delta^3(q+1)^5} + \frac{\beta}{\delta^2(q+1)^3}\,,$$

$$\Psi_3^q \equiv -\frac{q(1-3\,q+q^2)}{\delta^4(q+1)^7} + \frac{\gamma\beta}{\delta^2(q+1)^4} - \frac{2\beta(1-q)}{\delta^3(q+1)^5}\,,$$

$$\Psi_4^q \equiv \frac{q(1-q)(1-5q+q^2)}{\delta^5(q+1)^9} - \frac{\gamma\beta(3-2q)}{\delta^3(q+1)^6} + \frac{3\beta(1-3q+q^2)}{\delta^4(q+1)^7} + \frac{\gamma^2\beta}{\delta^2(q+1)^5} - \frac{\beta^2}{\delta^3(q+1)^5}\,.$$

Using (A.4) and (B.2–3) we obtain the coordinates of L:

$$\begin{aligned} y_l &= (a+1)\left(1 - \frac{\gamma a + \beta(a+1)}{a(a+1)^2(b+1)}\mu^2 + O(\mu^3)\right)\,, \\ z_l &= a\left[1 - \frac{[\gamma a + \beta(a+1)](a+b+2)}{a(a+1)(b+1)}\mu \right. \\ &\quad \left. + \frac{\gamma a + \beta(a+1)}{a(a+1)(b+1)}\left(\gamma + \frac{(a+1)^3+(b+1)^3}{\delta(a+1)^2-(b+1)^2}\right)\mu^2 + O(\mu^3)\right]\,, \end{aligned} \tag{B.4}$$

where, as the values Δ, Δ_1 and Δ_2 in (B.2) are of order μ, we have used the asymptotic representation of λ_a and λ_b with an accuracy up to μ^4 in order to obtain y_l and z_l with an accuracy up to μ^2.

C. Flow Trajectories on the Manifolds $W_\mu^s(O)$ and $W_\mu^u(P^+)$

Let us obtain the coordinates of the points M and K which are located on the lines l_μ^s and l_μ^u, respectively. These points are used as the boundaries dividing the flow located on the planes $W_\mu^s(O)$ and $W_\mu^u(P^+)$ into incoming and outgoing trajectories from these planes.

Let us start with the analysis of the behavior of trajectories on $W_\mu^s(O)$ described by (A.7). In order to study the orientation of the vector field (A.7) on $l_\mu^s = W_\mu^s(O) \cap U_1$, we consider the function

$$w = k_1^a y + k_2^a z + 1$$

and its derivative with respect to (A.7). Then we apply the condition

$$\dot{w}\mid_{w=0} = 0\,,$$

which defines the coordinates of $M(y_m, z_m)$:

$$
\begin{aligned}
y_m &= (a+1) + O(\mu^3)\,,\\
z_m &= a - \frac{\beta(a+1)+\gamma a}{(a+1)}\mu + \frac{\beta(a+1)+\gamma a}{\delta(a+1)^3}\mu^2 + O(\mu^3)\,.
\end{aligned} \tag{C.1}
$$

This point belongs to l_μ^s and divides the flow located on $W_\mu^s(O)$ into incoming and outgoing trajectories from $W_\mu^s(O)$. Hence, the vector field orientation of (4.36) along l_μ^s is the following:

- in the interval $z > z_m$, it is directed to decreasing values of the x-coordinate;
- in the interval $z < z_m$, it is directed to increasing values of the x-coordinate;
- at the point $z = z_m$, it is tangent to the plane U_1.

Using a similar technique we can obtain the coordinates of K which belongs to l_μ^u. At K, the flow located on $W_\mu^u(P^+)$ is divided into incoming and outgoing trajectories. The coordinates of K are

$$
\begin{aligned}
y_k &= (a+1) + O(\mu^3)\,,\\
z_k &= a - \frac{\beta(b+1)+\gamma b}{(b+1)}\mu + \frac{\beta(b+1)+\gamma b}{\delta(b+1)^3}\mu^2 + O(\mu^3)\,.
\end{aligned} \tag{C.2}
$$

Let us see the properties of the trajectories lying on $W_\mu^u(P^+)$. It follows from (A.7), (A.9) and (B.3) that motions on $W_\mu^u(P^+)$ in the region $x \geq 1$ are governed by the differential equations

$$
\begin{cases}
\dot{y} = \dfrac{B_\mu}{\delta}y + \dfrac{I_\mu}{\delta}z + \dfrac{D_\mu}{\delta}\,,\\
\dot{z} = -\beta y - \gamma z\,,
\end{cases} \tag{C.3}
$$

with

$$
\begin{aligned}
B_\mu &\equiv B\left[1 - \frac{\mu}{\delta(b+1)^2} + \left(\frac{1-b}{\delta^2(b+1)^4} + \frac{\beta}{\delta b(b+1)^2}\right)\mu^2 + O(\mu^3)\right],\\
I_\mu &\equiv 1 - \frac{\mu}{\delta(b+1)^2} - \left(\frac{\gamma}{\delta(b+1)^3} - \frac{(1-b)}{\delta^2(b+1)^4}\right)\mu^2 + O(\mu^3)\,,\\
D_\mu &\equiv D\left(1 - \frac{\mu}{\delta(b+1)^2} + \frac{(1-b)}{\delta^2(b+1)^4}\mu^2 + O(\mu^3)\right).
\end{aligned}
$$

The system (C.3) has a stationary point (y_0, z_0) (4.39). Its corresponding eigenvalues are

$$
\lambda_{1,2} = -h_b \pm \mathrm{i}\omega_b\,, \tag{C.4}
$$

with

$$
h_b \equiv \frac{\gamma - \dfrac{B_\mu}{\delta}}{2}\,,
$$

$$\omega_b \equiv \sqrt{\frac{I_\mu \beta}{\delta} - \frac{\left(\gamma + \frac{B_\mu}{\delta}\right)^2}{4}}\,.$$

Thus in the parameter range G_u (Fig. 4.18) the stationary point (x_0, y_0) is an unstable focus ($h_b < 0$), but within the region G_s it is a stable focus ($h_b > 0$).

D. Instability of Spatially Homogeneous States

As earlier mentioned the instability of the spatially homogeneous solution associated with the homoclinic orbit is a criterion for instability of the solitary waves. Here we illustrate the instability of the spatially homogeneous solution of (4.33) using the following boundary conditions:

$$\begin{aligned} y_{N+1} &= y_N\,, \\ w_0 &= w_1\,. \end{aligned} \tag{D.1}$$

Consider the stationary points of (4.33) and (D.1) contained in the region $|x_j| \leq 1$, $(j = 1, 2, \ldots, N)$. The stationary states of the array correspond to these stationary points. In the region $|x_j| \leq 1$, (4.33) and (D.1) have a one-parameter family of stationary states

$$\{x_j = y_j = z_j = 0,\; w_j = w_0 = \text{const.}\}\,,$$
$$j = 1, 2, \ldots, N\,.$$

To examine the stability of these states, we analyze the characteristic determinant $Q_N = \det(Q - \lambda I)$, where I is the $4N \times 4N$ identity matrix and Q is the $N \times N$ block matrix

$$Q = \begin{pmatrix} \Sigma_0 & P & 0 & & \cdots & 0 \\ E & \Sigma & P & & \cdots & 0 \\ 0 & E & \Sigma & & \cdots & 0 \\ \vdots & & & \ddots & & \\ \vdots & & & & \Sigma & P \\ 0 & \cdots & \cdots & & E & \Sigma_N \end{pmatrix},$$

with

$$\Sigma_0 = \begin{pmatrix} \sigma & \alpha & 0 & 0 \\ 1 & -1 & 1 & 0 \\ 0 & -\beta & -\gamma & 0 \\ 0 & d & 0 & 0 \end{pmatrix}, \quad \Sigma = \begin{pmatrix} \sigma & \alpha & 0 & 0 \\ 1 & -1 & 1 & -1 \\ 0 & -\beta & -\gamma & 0 \\ 0 & d & 0 & 0 \end{pmatrix},$$

$$
\Sigma_N = \begin{pmatrix} \sigma & \alpha & 0 & 0 \\ 1 & -1 & 0 & -1 \\ 0 & -\beta & -\gamma & 0 \\ 0 & 0 & 0 & 0 \end{pmatrix}, \quad P = \begin{pmatrix} 0 & 0 & 0 & 0 \\ 0 & 0 & 0 & 0 \\ 0 & 0 & 0 & 0 \\ 0 & -d & 0 & 0 \end{pmatrix},
$$

$$
E = \begin{pmatrix} 0 & 0 & 0 & 0 \\ 0 & 0 & 0 & 1 \\ 0 & 0 & 0 & 0 \\ 0 & 0 & 0 & 0 \end{pmatrix},
$$

and $\sigma \equiv -\alpha(1+a)$.

Expanding Q_N along the first four rows, we rewrite it in the form

$$
Q_N = q_1 A_{N-1} + q_2 B_{N-1}\,, \tag{D.2}
$$

where A_{N-1} and B_{N-1} are given determinants and

$$
q_1 \equiv \lambda^4 + [1+\gamma+\alpha(1+a)]\,\lambda^3 + [\alpha a + \alpha(1+a)\gamma + \gamma + \beta]\,\lambda^2 + \alpha a \gamma \lambda\,,
$$

$$
q_2 \equiv \frac{dq_1}{\lambda}\,.
$$

Next we expand the determinant A_{N-1} along the first four rows to obtain

$$
A_{N-1} = q_3 A_{N-2} + q_2 B_{N-2}\,, \tag{D.3}
$$

with

$$
\begin{aligned} q_3 &\equiv q_1 - \epsilon\,, \\ \epsilon &\equiv d(\sigma - \lambda)(\gamma + \lambda)\,. \end{aligned}
$$

Then, expanding the determinant A_{N-1} along the last four rows, we obtain

$$
A_{N-1} = q_1 S_{N-2} - \lambda \frac{\epsilon}{d} C_{N-2}\,, \tag{D.4}
$$

where the determinants S_{N-2} and C_{N-2} are

$$
\begin{aligned} S_{N-2} &= \det(S - \lambda I)\,, \\ C_{N-2} &= \det(C - \lambda I)\,, \end{aligned} \tag{D.5}
$$

where I is the $4(N-2) \times 4(N-2)$ identity matrix and S and C are block matrices of size $(N-2) \times (N-2)$:

$$
S = \begin{pmatrix} \Sigma & P & 0 & & \cdots & 0 \\ E & \Sigma & P & & \cdots & 0 \\ 0 & E & \Sigma & & \cdots & 0 \\ \vdots & & & \ddots & & \\ \vdots & & & & \Sigma & P \\ 0 & \cdots & \cdots & & E & \Sigma \end{pmatrix}, \quad C = \begin{pmatrix} \Sigma & P & 0 & & \cdots & 0 \\ E & \Sigma & P & & \cdots & 0 \\ 0 & E & \Sigma & & \cdots & 0 \\ \vdots & & & \ddots & & \\ \vdots & & & & \Sigma & P \\ 0 & \cdots & \cdots & & E & \Sigma_{N-2} \end{pmatrix},
$$

with

$$\Sigma_{N-2} = \begin{pmatrix} \sigma & \alpha & 0 & 0 \\ 1 & -1 & 1 & 0 \\ 0 & -\beta & -\gamma & 0 \\ 0 & d & 0 & -d \end{pmatrix} .$$

Then we expand the determinants S_{N-2} and C_{N-2} along the last four rows and obtain

$$\begin{aligned} S_{N-2} &= q_3 S_{N-3} - \lambda\epsilon/d C_{N-3} , \\ C_{N-2} &= q_2 S_{N-3} - \epsilon C_{N-3} . \end{aligned} \tag{D.6}$$

It follows from (D.6) that

$$C_{N-2}\lambda = d(q_1 - q_3) S_{N-3} + d S_{N-2} . \tag{D.7}$$

From (D.4) and (D.7) we obtain

$$A_{N-1} = (q_1 - \epsilon) S_{N-2} - \epsilon^2 S_{N-3} . \tag{D.8}$$

Finally, from (D.2–3) and (D.8) we find

$$Q_N = (q_1 - \epsilon) S_{N-1} + \epsilon(q_1 - 2\epsilon) S_{N-2} - \epsilon^3 S_{N-3} . \tag{D.9}$$

The Lyapunov characteristic eigenvalues corresponding to the stationary homogeneous states are given by

$$(q_1 - \epsilon) S_{N-1} + \epsilon(q_1 - 2\epsilon) S_{N-2} - \epsilon^3 S_{N-3} = 0 . \tag{D.10}$$

On the other hand, from (D.7–8) we obtain the following recurrent relation:

$$S_{N-1} = 2z\epsilon S_{N-2} - \epsilon^3 S_{N-3} , \tag{D.11}$$

with

$$2z\epsilon \equiv q_1 - 2\epsilon .$$

Now, as done in Sect. 4.2.2, we treat the recurrent relation (D.11) as a 2D mapping with the following initial conditions:

$$\begin{aligned} S_1 &= \epsilon(2z + 1) , \\ S_2 &= \epsilon^2(4z^2 + 2z - 1) . \end{aligned} \tag{D.12}$$

Solving (D.11–12) we obtain

$$S_{N-1} = \epsilon^{N-1} [U_{N-1}(z) + U_{N-2}(z)] , \tag{D.13}$$

where $U_{N-2}(z)$ is the Chebyshev polynomial of the second kind, i.e.

$$U_m(z) \equiv \frac{(z+\sqrt{z^2-1})^{m+1} - (z-\sqrt{z^2-1})^{m+1}}{2\sqrt{z^2-1}} .$$

Substituting (D.13) to (D.10) and using a property of the Chebyshev polynomials we obtain the following equation:

$$(z+1)[U_{N-1}(z) + U_{N-2}(z)] = 0 , \tag{D.14}$$

which is equivalent to (D.10).

Equation (D.14) has a root $z = -1$, hence $q_1 = 0$, i.e.

$$\lambda^4 + [1+\gamma+\alpha(a-1)]\,\lambda^3 + [\alpha a + \alpha(a+1)\gamma + \gamma]\,\lambda^2 + \alpha a \gamma \lambda = 0 . \tag{D.15}$$

To find the other roots we introduce

$$v = z + \sqrt{z^2-1} . \tag{D.16}$$

Substituting (D.16) into (D.14) we obtain

$$(v)^{2N-1} = 1 . \tag{D.17}$$

It follows from (D.16–17) that the other roots of (D.14) are

$$z = z_s , \quad z_s \equiv \cos\frac{2\pi s}{2N-1} , \tag{D.18}$$

$$s = 0, 1, \ldots, N-2 .$$

These roots provide $q_1 = 2\epsilon(1+z_s)$, which is equivalent to

$$\begin{aligned} &\lambda^4 + [1+\gamma+\alpha(a+1)]\,\lambda^3 + [\alpha a + \alpha(a+1)\gamma + \gamma + 2d(1+z_s)]\,\lambda^2 \\ &+\{\alpha a\gamma + 2d[\gamma+\alpha(a+1)](1+z_s)\}\,\lambda + 2d\alpha(1+a)\gamma(1+z_s) = 0 , \end{aligned} \tag{D.19}$$

$$s = 0, 1, 2, \ldots, N-2 .$$

Therefore, the Lyapunov characteristic eigenvalues associated with the stationary homogeneous state are determined by (D.15) and (D.19). It follows from (D.15) that the set of these eigenvalues contains a root with a positive value. Hence, any homogeneous stationary state of the array is unstable. Note that in the set of eigenvalues there is one zero root associated with the existence in (4.33) and (D.1) of a one-parameter family of stationary states.

E. Topological Entropy and Lyapunov Exponent

To describe the chaotic sets of the map f we use two quantities: the *topological entropy* and the *Lyapunov exponent*. For completeness we recall their definitions:

Topological Entropy

Consider the map $f : I \to I$ of the interval I. Let A be an f-invariant set. Denote by $\Gamma = \{f^k x\}_{k=0}^{n}$ the segments of the map trajectory of time duration or length n. Two segments $\Gamma_1 = \{f^k x\}_{k=0}^{n-1}$ and $\Gamma_2 = \{f^k y\}_{k=0}^{n-1}$ of the trajectories of time length n are (n, ϵ)-separated if there exists k_0, $0 \leq k_0 \leq n-1$ such that

$$|f^{k_0}x - f^{k_0}y| \geq \epsilon\,.$$

Let $A_{n,\epsilon}$ be the set of the map trajectories of time length n which obeys the following conditions:

- every two segments of the trajectories from this set are (n, ϵ)-separated;
- in A there exists no segment of the trajectory of time length n which is (n, ϵ)-separated from any element of $A_{n,\epsilon}$.

Let $C_{n,\epsilon}$ be the number of elements of $A_{n,\epsilon}$. The topological entropy of the map f, $h_{top}(f)$, is

$$h_{top}(f) = \lim_{\epsilon \to 0} \overline{\lim_{n \to \infty}} \frac{\ln C_{n,\epsilon}}{n}\,.$$

The topological entropy characterizes the *instability* of the trajectories and the complexity of the structure of A. If $h_{top}(f) > 0$, then A is chaotic.

Lyapunov Exponent

Let x_0 be a point on the interval I. The Lyapunov exponent is defined by

$$\lambda(x_0) = \lim_{n \to \infty} \frac{1}{n} \ln \left| \frac{df^n(x_0)}{dx_0} \right|.$$

The Lyapunov exponent characterizes the rate of the *exponential separation* of two close trajectories or linear instability in phase space.

Note that the definition of chaos, $h_{top}(f) > 0$, does not distinguish between observable or attracting behavior and cases when the chaotic set is unstable. Together with the unstable chaotic set, a map can have attracting periodic points of different periods. An attracting chaotic interval of the map or *chaotic attractor* is characterized by a positive Lyapunov exponent $\lambda(x)$ for $x \in I$. Thus, *a map f of the interval I describes a chaotic attractor* if

- the map is chaotic $h_{top}(f) > 0$;
- in I there are no stable periodic points, i.e. $\lambda(x) > 0$, for all $x \in I$.

F. Multipliers of the Fixed Point of the Coupled Map Lattice (7.55)

Let $\{u_{j,k}(n) = u^*_{j,k},\ v_{j,k}(n) = v^*_{j,k}\}$ be a fixed point of (7.55). We set

$$u_{j,k}(n) = u^*_{j,k}(n) + \xi_{j,k}(n)\,, \quad v_{j,k}(n) = v^*_{j,k}(n) + \eta_{j,k}(n)\,,$$
$$j, k = 1, 2, \ldots, N\,,$$

where $\xi_{j,k}(0)$ and $\eta_{j,k}(0)$ are arbitrary small enough initial perturbations. Linearizing (7.55) in the neighborhood of the fixed point we obtain

$$\begin{cases} \xi_{j,k}(n+1) = d[\xi_{j+1,k}(n) + \xi_{j-1,k}(n)] + \sigma_1(j,k)\xi_{j,k}(n) \\ \qquad +d[\xi_{j,k+1}(n) + \xi_{j,k-1}(n)] + h\eta_{j,k}(n)\,, \\ \eta_{j,k}(n+1) = d[\eta_{j+1,k}(n) + \eta_{j-1,k}(n)] + \sigma_2(j,k)\eta_{j,k}(n) \\ \qquad +d[\eta_{j,k+1}(n) + \eta_{j,k-1}(n)] + h\xi_{j,k}(n)\,, \end{cases} \tag{F.1}$$

with

$$\begin{aligned} &\sigma_1(j,k) \equiv 1 + \alpha f'(u^*_{j,k}) - h - 4d\,, \\ &\sigma_2(j,k) \equiv 1 + \alpha f'(v^*_{j,k}) - h - 4d\,, \\ &\xi_{0,k}(n) = \xi_{1,k}\,, \quad \xi_{j,0}(n) = \xi_{j,1}(n)\,, \\ &\xi_{j,N+1}(n) = \xi_{j,N}(n)\,, \quad \xi_{N+1,k}(n) = \xi_{N,k}(n)\,, \\ &\eta_{0,k}(n) = \eta_{1,k}\,, \quad \eta_{j,0}(n) = \eta_{j,1}(n)\,, \\ &\eta_{j,N+1}(n) = \eta_{j,N}(n)\,, \quad \eta_{N+1,k}(n) = \eta_{N,k}(n)\,. \end{aligned} \tag{F.2}$$

Multipliers of Homogeneous Fixed Points

The coordinates of homogeneous fixed points do not depend on j and k. Let $\{u_{j,k}(n) = u^*,\ v_{j,k}(n) = v^*\}$ be such a fixed point. Then the solution of (7.55) is

$$\xi_{j,k}(n) = A_j B_k(\mu)^n\,, \quad \eta_{j,k}(n) = qA_j B_k(\mu)^n\,, \tag{F.3}$$

where the variables A_j and B_k are time-independent and q is a constant. Substituting (F.3) into (7.55) we find that A_j and B_k satisfy

$$d(A_{j+1} + A_{j-1})B_k + (\sigma_1 - \mu + hq)A_j B_k + d(B_{k+1} + B_{k-1})A_j = 0\,, \tag{F.4}$$
$$d(A_{j+1} + A_{j-1})B_k + (\sigma_2 - \mu + h/q)A_j B_k + d(B_{k+1} + B_{k-1})A_j = 0\,. \tag{F.5}$$

and hence

$$\sigma_1 - \mu + hq = \sigma_2 - \mu + \frac{h}{q}\,, \tag{F.6}$$

with

$$q = \frac{-(\sigma_1 - \sigma_2) \pm \sqrt{(\sigma_1 - \sigma_2)^2 + 4h^2}}{2h}\,. \tag{F.7}$$

Therefore, A_j and B_k satisfy

$$d(A_{j+1} + A_{j-1})B_k + \sigma A_j B_k + d(B_{k+1} + B_{k-1})A_j = 0\,, \tag{F.8}$$

$$j, k = 1, 2, \ldots, N\,,$$

with $\sigma \equiv \sigma_1 - \mu + hq$. Each equation from (F.8) can be rewritten in the form

$$\frac{A_{j+1} + A_{j-1} + \dfrac{\sigma - \mu}{d} A_j}{A_j} = \frac{B_{k+1} + B_{k-1}}{B_k}\,. \tag{F.9}$$

Then, (F.9) is satisfied if the left- and right-hand parts are equal to the same constant independent of j and k. We denote this constant by γ. From (F.2) and (F.9) it follows that

$$\begin{aligned} &B_{k+1} + \gamma B_k + B_{k-1} = 0, k = 1, 2, \ldots, N\,, \\ &B_0 = B_1\,, \quad B_{N+1} = B_N\,, \end{aligned} \tag{F.10}$$

which is a homogeneous linear equation system. It has nontrivial solution if its determinant, D, vanishes. The determinant is

$$D = \begin{vmatrix} \gamma+1 & 1 & 0 & 0 & \ldots & 0 & 0 \\ 1 & \gamma & 1 & 0 & \ldots & 0 & 0 \\ 0 & 1 & \gamma & 1 & \ldots & 0 & 0 \\ & & & \ddots & & & \\ 0 & 0 \ldots & & 1 & \gamma & 1 & 0 \\ 0 & 0 \ldots & & 0 & 1 & \gamma & 1 \\ 0 & 0 \ldots & & 0 & 0 & 1 & \gamma+1 \end{vmatrix}.$$

Using the technique described in Chap. 4 we find that

$$D = \left(\frac{\gamma}{2} + 1\right) U_{N-1}\left(\frac{\gamma}{2}\right)\,, \tag{F.11}$$

where $U_{N-1}(\gamma/2)$ denotes Chebyshev polynomials of the second type whose zeros correspond to

$$\gamma = 2\cos\frac{s\pi}{N}\,, \quad s = 1, 2, \ldots, N-1\,. \tag{F.12}$$

Therefore, the roots of $D = 0$ are

$$\gamma = \gamma_s\,, \quad \text{with} \quad \gamma_s \equiv 2\cos\frac{s\pi}{N}\,, \quad s = 1, 2, \ldots, N-1\,. \tag{F.13}$$

Using (F.13), from (F.9) we obtain

$$\begin{aligned} &A_{j+1} + \left(\frac{\sigma - \mu}{d} - \gamma_s\right) A_j + A_{j-1} = 0\,, \quad j = 1, 2, \ldots, N\,, \\ &A_0 = A_1\,, \quad A_{N+1} = A_N\,. \end{aligned} \tag{F.14}$$

The determinant of the homogeneous linear system (F.14) is equal to zero if

$$\frac{\sigma-\mu}{d}\gamma_s=\gamma_l\,,\quad \text{with}\quad \gamma_l\equiv 2\cos\frac{l\pi}{N}\,,\quad l=1,2,\,\ldots,N-1\,,\quad \text{(F.15)}$$

and hence the multipliers of the homogeneous fixed point are

$$\mu_{s,l}^{(1),(2)}=\frac{\sigma_1+\sigma_2-2d(\gamma_s+\gamma_l)\pm\sqrt{(\sigma_1+\sigma_2)^2+4h^2}}{2}\,,$$
$$s,l=1,2,\,\ldots,N\,. \quad \text{(F.16)}$$

Multipliers of Inhomogeneous Fixed Points

For such fixed points the multipliers cannot be written explicitly. However, applying the Gershgorin theorem (see Appendix G) we can locate all the multipliers on the complex plane. Let us rewrite (7.55) in the coordinate form:

$$\begin{cases}\xi_{j,k}(n+1)=\sum_{l=j-1}^{j+1}\sum_{s=k-1}^{k+1}a_{jkls}\xi_{j,k}(n)+h\eta_{j,k}(n)\,,\\ \eta_{j,k}(n+1)=\sum_{l=j-1}^{j+1}\sum_{s=k-1}^{k+1}b_{jkls}\eta_{j,k}(n)+h\xi_{j,k}(n)\,,\end{cases}\quad \text{(F.17)}$$

where

$$\begin{cases}a_{j,k,j,k}=\sigma_1(j,k)\,,\quad b_{j,k,j,k}=\sigma_2(j,k)\,,\\ a_{j,k,j-1,k}=a_{j,k,j,k-1}=a_{j,k,j+1,k}=a_{j,k,j,k+1}=d\,,\\ b_{j,k,j-1,k}=b_{j,k,j,k-1}=b_{j,k,j+1,k}=b_{j,k,j,k+1}=d\,,\end{cases}$$

and all other coefficients in (F.17) equal to zero. Applying the Gershgorin theorem to (F.17) we obtain that the spectrum of the linear operator given by (F.17) lies within the union of $2N^2$ disks D_{jk}:

$$\{\mu\in\mathbf{C}\;:\;|1+\alpha f'(u^*_{j,k})-h-4d-\mu|\le 4d+h\}\,,$$
$$\{\mu\in\mathbf{C}\;:\;|1+\alpha f'(v^*_{j,k})-h-4d-\mu|\le 4d+h\}\,.$$

It follows that the multipliers $\left(\mu_{j,k}^{(1)},\mu_{j,k}^{(2)}\right)$ of each inhomogeneous fixed point satisfy the condition

$$\begin{aligned}1+\alpha f'(u^*_{j,k})-2h-8d\le\mu_{j,k}^{(1)}\le 1+\alpha f'(u^*_{j,k})\,,\\ 1+\alpha f'(v^*_{j,k})-2h-8d\le\mu_{j,k}^{(2)}\le 1+\alpha f'(v^*_{j,k})\,.\end{aligned}\quad \text{(F.18)}$$

G. Gershgorin Theorem

Let us consider a matrix $A=[a_{ij}]\in M_n(C)$, where $M_n(C)$ is a complex set of $n\times n$ matrices. The location of the eigenvalues of the matrix A on the complex plane can be estimated using the following theorem [G.1]:

Theorem (Gershgorin). *Let* $A = [a_{ij}] \in M_n$, *and let*

$$R'_i(A) \equiv \sum_{j=1, j \neq i}^{n} |a_{ij}|, \qquad 1 \leq i \leq n$$

denote the deleted absolute row sums of A. *Then all the eigenvalues of* A *are located in the union of* n *disks*

$$\bigcup_{i=1}^{n} \{z \in \mathbf{C} : |z - a_{ii}| \leq R'_i(A)\} \equiv G(A)$$

Furthermore, if a union of k *of these* n *disks forms a connected region that is disjoint from all the remaining* $n - k$ *disks, then there are precisely* k *eigenvalues of* A *in this region.*

Note that the disks in $G(A)$ taken separately are usually called Gershgorin disks.

Since A and A^T (A^T is the transposed matrix) have equivalent eigenvalues, then the Gershgorin theorem yields the following corollary:

Corollary. *If* $A = [a_{ij}] \in M_n(C)$, *and*

$$C'_j \in \sum_{i=1, i \neq j}^{n} |a_{ij}|,$$

then, all the eigenvalues of the matrix A *are located in the union of* n *disks*

$$\bigcup_{i=1}^{n} \{z \in \mathbf{C} : |z - a_{ii}| \leq C'_i(A)\} \equiv G(A^T)$$

Furthermore, if a union of k *of these disks forms a connected region that is disjoint from all the remaining* $n - k$ *disks, then there are precisely* k *eigenvalues of the matrix* A *in this region.*

References

Preface

Listed here are books, book chapters, review and feature articles of general interest providing reference, motivation and context.

Afraimovich, V. S., Nekorkin, V. I., Osipov, G. V. and Shalfeev, V. D., *Stability, Structures and Chaos in Nonlinear Synchronization Networks* (World Scientific, Singapore, 1995).

Andronov, A. A., Leontovich, E. A., Gordon, I. I. and Maier, A. G., *Theory of Bifurcations of Dynamic Systems on a Plane* (Wiley, New York, 1973).

Andronov, A. A., Vitt, A. A. and Chaikin, S. E., *Theory of Oscillations* (Pergamon, New York, 1966).

Arecchi, F. T. and Farini, A., *Lexicon of Complexity* (Studio Editoriale Fiorentino, Firenze, 1996).

Bishop, A. R., Krumhansl, J. A. and Trullinger, S. E., "Solitons in condensed matter: A paradigm", *Physica D* **1** (1980) 1–44.

Boussinesq, J. V., "Essai sur la théorie des eaux courantes", *Mém. présentés par divers savants à l'Acad. Sci. Inst. France (Paris)* **23** (1877) 1–680. Additions et éclaircissements au mémoire intitulé (supra), *ibidem* **24**, No. 2 (1878) 1–64.

Changeux, J. P., *L'homme neuronal* (Fayard, Paris, 1983) (in French).

Chaté, H. and Courbage, M. (Editors), "Lattice dynamics", *Physica* **D 1-4** (1997).

Christov, C. I., Maugin, G. A. and Velarde, M. G., "Well-posed Boussinesq paradigm with purely spatial higher-order derivatives", *Phys. Rev. E* **54** (1996) 3621–3638.

Chua, L. O., *CNN: A Paradigm for Complexity* (World Scientific, Singapore, 1998).

Colinet, P., Legros, J. C. and Velarde, M. G., *Nonlinear Dynamics of Surface-Tension-Driven Instabilities* (Wiley-VCH, New York, 2001).

Cronin, J., *Mathematical Aspects of Hodgkin–Huxley Neural Theory* (University Press, Cambridge, 1987).

Drazin, P. G. and Johnson, R. S., *Solitons: An Introduction* (University Press, Cambridge, 1989).

Feistel, R. and Ebeling, W., *Evolution of Complex Systems. Selforganization, Entropy and Development* (Kluwer, Dordrecht, 1989).

Fife, P. C., *Mathematical Aspects of Reacting and Diffusing Systems* (Springer-Verlag, Berlin, 1979).

Gaponov-Grekhov, A. V. and Rabinovich, M. I., *Nonlinearities in Action. Oscillations, Chaos, Order, Fractals* (Springer-Verlag, Berlin, 1992)

Haken, H., "Cooperative phenomena in systems far from thermal equilibrium and in nonphysical systems", *Rev. Modern Phys.* **47** (1975) 67–121.

Haken, H., *Synergetics. An Introduction. Nonequilibrium Phase Transitions and Self-Organization in Physics, Chemistry, and Biology*, 3rd Edition (Springer-Verlag, Berlin, 1983a).

Haken, H., *Advanced Synergetics. Instability Hierarchies of Self-Organizing Systems and Devices* (Springer-Verlag, Berlin, 1983b).

Haken, H., *Principles of Brain Functioning. A Synergetic Approach to Brain Activity, Behavior and Cognition* (Springer-Verlag, Berlin, 1996).

Haken, H., *Information and Self-Organization. A Macroscopic Approach to Complex Systems*, 2nd Edition (Springer-Verlag, Berlin, 2000).

Hirsch, M. W. and Smale, S., *Differential Equations, Dynamical Systems and Linear Algebra* (Academic Press, New York, 1974).

Hoppensteadt, F. C. *An Introduction to the Mathematics of Neurons* (University Press, Cambridge, 1986).

Izhikevich, E. M., "Neural excitability, spiking, and bursting", *Int. J. Bifurcation Chaos* **10** (2000) 1171–1266.

Kaneko, K. (Editor), *Theory and Applications of Coupled Map Lattices* (Wiley, Chichester, 1993)

Kaneko, K. and Tsuda, I., *Complex Systems: Chaos and Beyond* (Springer-Verlag, Berlin, 2001).

Korteweg, D. J. and de Vries G., "On the change of form of long waves advancing in a rectangular channel, and on a new type of long stationary waves", *Phil. Mag.* **39**[5] (1895) 422–443.

Koschmieder, E. L., *Bénard Cells and Taylor Vortices* (University Press, Cambridge, 1993).

Llinás, R. R. (Editor), *The Biology of the Brain. From Neurons to Networks* (Freeman, New York, 1988).

Llinás, R. R., *I of the Vortex. From Neurons to Self* (M. I. T. Press, Cambridge, Mass., 2001).

Madan, R. N. (Editor), *Chua's Circuit: A Paradigm for Chaos* (World Scientific, Singapore, 1993).

Manganaro, G., Arena, P. and Fortuna, L., *Cellular Neural Networks. Chaos, Complexity and VLSI Processing* (Springer-Verlag, Berlin, 1999).

Milton, J., *Dynamics of Small Neural Populations* (AMS, Providence, 1996).

Mira, C., "Chua's circuit and the qualitative theory of dynamical systems", *Int. J. Bifurcation Chaos* **7** (1997) 1910–1916.

Nepomnyashchy, A. A., Velarde, M. G. and Colinet, P., *Interfacial Phenomena and Convection* (Chapman and Hall/CRC, New York, 2002).

Nicolis, G., *Introduction to Nonlinear Science* (University Press, Cambridge, 1995).

Nicolis, G. & Prigogine, I., *Self-Organization in Non-equilibrium Systems. From Dissipative Structures to Order through Fluctuations* (Wiley, New York, 1977).

Rabinovich, M.I., Ezersky, A. B. and Weidman, P. D., *The Dynamics of Patterns* (World Scientific, Singapore, 2000)

Remoissenet, M., *Waves Called Solitons. Concepts and Experiments*, 3rd Edition (Springer-Verlag, Berlin, 2001).

Scott, A. C., Chu, F. Y. F. and McLaughlin, D. W., "The Soliton: A new Concept in Applied Science", *Procs. IEEE* **61** (1973) 1443–1483.

Scott, A., *Nonlinear Science: Emergence and Dynamics of Coherent Structures* (University Press, Oxford, 1999)

Scott, A. C., *Stairway to the Mind* (Copernicus-Springer, New York, 1995).

Shepherd, G. M., *Foundations of the Neuron Doctrine* (University Press, Oxford, 1991).

Shepherd, G. M. (Editor), *The Synaptic Organization of the Brain* (University Press, Oxford, 1998).

Shilnikov, L. P., "Mathematical problems of nonlinear dynamics: A tutorial", *Int. J. Bifurcation Chaos* **7** (1997) 1953–2001.

Shilnikov, L. P., Shilnikov, A. L., Turaev, D. V. and Chua, L. O., *Methods of Qualitative Theory in Nonlinear Dynamics* (World Scientific, Singapore; Part I, 1998; Part II, 2001)

Tass, P. A., *Phase Resetting in Medicine and Biology. Stochastic Modelling and Data Analysis* (Springer-Verlag, Berlin, 1999).

Thompson, J. M. T. and Stewart, H. B., *Nonlinear Dynamics and Chaos. Geometrical Methods for Engineers and Scientists*, 2nd Edition (Wiley, New York, 2001).

Ustinov, A. V., "Solitons in Josephson junctions", *Physica D* **123** (1998) 315–329.

Velarde, M. G. and Normand, Ch., "Convection", *Sci. Amer.* **243** [1] (1980) 92–108.

Velarde, M. G., Nepomnyashchy, A. A. and Hennenberg, M., "Onset of oscillatory interfacial instability and wave motions in Bénard layers", *Adv. Appl. Mech.* **37** (2000) 167–238.

Chapter 1

[1.1] Afraimovich, V. S. Nekorkin, V. I., Osipov, G. V. and Shalfeev, V. D., *Stability, Structures and Chaos in Nonlinear Synchronization Networks* (World Scientific, Singapore, 1995).

[1.2] Andronov, A. A., Leontovich, E. A., Gordon, I. I. and Maier, A. G., *Theory of Bifurcations of Dynamic Systems on a Plane* (Wiley, New York, 1973).

[1.3] Andronov, A. A., Vitt, A. A. and Chaikin, S. E., *Theory of Oscillations* (Pergamon, New York, 1966).

[1.4] Bénard, H., "Les tourbillons cellulaires dans une nappe liquide", *Rev. Gén. Sci. Pures Appl.* **11** (1900) 1261–1271.

[1.5] Bénard, H., "Les tourbillons cellulaires dans une nappe liquide transportant de la chaleur par convection en régime permanent", *Ann. Chim. Phys.* **23** (1901) 62–143.

[1.6] Christov, C. I. and Velarde, M. G., "Dissipative solitons", *Physica D* **86** (1995) 323–347.

[1.7] Colinet, P., Legros, J. C. and Velarde, M. G., *Nonlinear Dynamics of Surface-Tension-Driven Instabilities* (Wiley-VCH, New York, 2001).

[1.8] Courant, R. and Friedrichs, K. O., *Supersonic Flow and Shock Waves* (Wiley-Interscience, New York, 1948).

[1.9] Domb, C., *The Critical Point. A historical Introduction To The Modern Theory of Critical Phenomena* (Taylor & Francis, London, 1996)

[1.10] Fife, P. C., *Mathematical Aspects of Reacting and Diffusing Systems* (Springer-Verlag, Berlin, 1979).

[1.11] Haken, H., *Synergetics. An Introduction. Nonequilibrium Phase Transitions and Self-Organization in Physics, Chemistry, and Biology*, 3rd Edition (Springer-Verlag, Berlin, 1983a).

[1.12] Haken, H., *Advanced Synergetics. Instability Hierarchies of Self-Organizing Systems and Devices* (Springer-Verlag, Berlin, 1983b).

[1.13] Hoppensteadt, F. C. *An Introduction to the Mathematics of Neurons* (University Press, Cambridge, 1986).

[1.14] Kaneko, K. and Tsuda, I., *Complex Systems: Chaos and Beyond* (Springer-Verlag, Berlin, 2001).

[1.15] Koschmieder, E. L., *Bénard Cells and Taylor Vortices* (University Press, Cambridge, 1993).

[1.16] Mach, E. and Wosyka, J., "Über einige mechanische Wirkungen des elektrischen Funkens", Sitzungsber. *Akad. Wiss. Wien* **72**(II) (1875) 44–52.

[1.17] Nekorkin, V. I. and Velarde, M. G., "Solitary waves, soliton bound states and chaos in a dissipative Korteweg–de Vries equation", *Int J. Bifurcation Chaos* **4** (1994) 1135–1146.

[1.18] Nicolis, G., *Introduction to Nonlinear Science* (University Press, Cambridge, 1995).

[1.19] Nicolis, G. and Prigogine, I., *Self-Organization in Non-equilibrium Systems. From Dissipative Structures to Order through Fluctuations* (Wiley, New York, 1977).

[1.20] Russell, J.S., "Report on waves", *Rep. 14th Meet. British Ass. Adv. Sci., York*, 311–390, (J. Murray, London, 1844).

[1.21] Russell, J. S., *The Wave of Translation in the Oceans of Water, Air and Ether*, with *Appendix* (Trübner, London, 1885).

[1.22] Ustinov, A. V., "Solitons in Josephson junctions", *Physica D* **123** (1998) 315–329.

[1.23] Velarde, M. G. and Normand, Ch., "Convection", *Sci. Amer.* **243**[1] (1980) 92–108.

Chapter 2

[2.1] Bazin, H., "Recherches experimentales sur la propagation des ondes", *Mém. présentés par divers savants à l'Acad. Sci. Inst. France* (Paris) **19** (1865) 495–644.

[2.2] Bishop, A. R., Krumhansl, J. A. and Trullinger, S. E., "Solitons in condensed matter: A paradigm", *Physica D* **1** (1980) 1–44.

[2.3] Bouasse, H., *Houle, Rides, Seiches et Marées* (Delagrave, Paris, 1924), pp. 291–292.

[2.4] Boussinesq, J. V., "Théorie de l'intumescence liquide appelée onde solitaire ou de translation se propageant dans un canal rectangulaire", *C. R. Hebd. Séances Acad. Sci. (Paris)* **72** (1871) 755–759.

[2.5] Boussinesq, J. V., "Théorie des ondes et des remous qui se propagent le long d'un canal rectangulaire horizontal en communiquant au liquide contenu dans ce canal des vitesses sensiblement pareilles de la surface au fond", *J. Math. Pures Appl.* **17** (1872) 55–108.

[2.6] Boussinesq, J. V., "Essai sur la théorie des eaux courantes", *Mém. présentés par divers savants à l'Acad. Sci. Inst. France (Paris)* **23** (1877) 1–680. Additions et éclaircissements au mémoire intitulé (supra), *ibidem* **24**[2] (1878) 1–64.

[2.7] Christov, C. I. and Velarde, M. G., "Dissipative solitons", *Physica D* **86** (1995) 323–347.

[2.8] Christov, C. I., Maugin, G. A. and Velarde, M. G., "Well-posed Boussinesq paradigm with purely spatial higher-order derivatives", *Phys. Rev. E* **54** (1996) 3621–3638.

[2.9] Dodd, R. K., Eilbeck, J. C., Gibbon, J. D. and Morris, H. C., *Solitons and Nonlinear Wave Equations* (Academic Press, Inc., 1984).
[2.10] Drazin, P. G. and Johnson, R. S., *Solitons: An Introduction* (University Press, Cambridge, 1989).
[2.11] Garazo, A. N. and Velarde, M. G., "Dissipative Korteweg–de Vries description of Marangoni–Bénard oscillatory convection", *Phys. Fluids A* **3** (1991) 2295–2300.
[2.12] Gardner, P. L., Greene, J. M., Kruskal, M. D. and Miura, R. M., "Method for solving Korteweg–de Vries equation", *Phys. Rev. Lett.* **19** (1967) 1095–1097.
[2.13] Kliakhandler, I. L., Porubov, A. V. and Velarde, M. G., "Localized finite-amplitude disiturbances and selection of solitary waves", *Phys. Rev. E* **62** (2000) 4959–4962.
[2.14] Korteweg, D. J. and de Vries G., "On the change of form of long waves advancing in a rectangular channel, and on a new type of long stationary waves", *Phil. Mag.* **39**[5] (1895) 422–443.
[2.15] Krehl, P. and van der Geest, M., "The discovery of the Mach reflection effect and its demonstration in an auditorium", *Shock Waves* **1** (1991) 3–15.
[2.16] Linde, H., Chu, X.-L. and Velarde, M. G., "Oblique and head-on collisions of solitary waves in Marangoni–Bénard convection", *Phys. Fluids A* **5** (1993) 1068–1070.
[2.17] Linde, H., Chu, X.-L., Velarde, M. G. and Waldhelm, W., "Wall reflections of solitary waves in Marangoni–Bénard convection", *Phys. Fluids A* **5** (1993) 3162–3166.
[2.18] Mach, E. and Wosyka, J., "Über einige mechanische Wirkungen des elektrischen Funkens", Sitzungsber. *Akad. Wiss. Wien* **72**[II] (1875) 44–52.
[2.19] Nekorkin, V. I. and Velarde, M. G., "Solitary waves, soliton bound states and chaos in a dissipative Korteweg–de Vries equation", *Int J. Bifurcation Chaos* **4** (1994) 1135–1146.
[2.20] Nepomnyashchy, A. A., Velarde, M. G. and Colinet, P., *Interfacial Phenomena and Convection* (Chapman & Hall/CRC, New York, 2002).
[2.21] Newell, A. C., *Solitons in Mathematics and Physics* (SIAM, Philadelphia, 1985).
[2.22] Rayleigh, Lord, "On waves", *Phil. Mag.* **1** (1876) 257–279.
[2.23] Rednikov, A. Y., Velarde, M. G., Ryazantsev, Yu. S., Nepomnyashchy, A. A. and Kurdyumov, V. N., "Cnoidal Wave Trains ans Solitary Waves in a Dissipation-Modified Korteweg–de Vries Equation", *Acta Appl. Math.* **39** (1995) 457–475.
[2.24] Remoissenet, M., *Waves Called Solitons. Concepts and Experiments*, 3rd Edition (Springer-Verlag, Berlin, 2001).
[2.25] Russell, J.S., "Report on waves", *Rep. 14th Meet. British Ass. Adv. Sci., York*, 311–390, (J. Murray, London, 1844).
[2.26] Russell, J. S., *The Wave of Translation in the Oceans of Water, Air and Ether*, with *Appendix* (Trübner, London, 1885).
[2.27] Ursell, F., "The long-wave paradox in the theory of gravity waves", *Proc. Cambridge Phil. Soc.* **49** (1953) 685–694.
[2.28] Velarde, M. G., Nekorkin, V. I. and Maksimov, A. G., "Further results on the evolution of solitary waves and their bound states of a dissipative Korteweg–de Vries equation", *Int. J. Bifurcation Chaos* **5** (1995) 831–839.
[2.29] Velarde, M. G., Nepomnyashchy, A. A. and Hennenberg, M., "Onset of oscillatory interfacial instability and wave motions in Bénard layers", *Adv. Appl. Mech.* **37** (2000) 167–238.
[2.30] Zabusky, N. and Kruskal, M. D., "Interaction of 'solitons' in a collisionless plasma and the recurrence of initial states", *Phys. Rev. Lett.* **15** (1965) 57–62.

Chapter 3

[3.1] Barone, A. and Paterno, G., *Physics and Applications of the Josephson Effect* (Wiley, New York, 1982).
[3.2] Likharev, K. K., *Dynamics of Josephson Junctions and Circuits* (Gordon and Beach, New York 1986).
[3.3] Lonngren, K. and Scott, A., *Solitons in Action* (Academic Press, New York, 1978).
[3.4] Maksimov, A. G. and Nekorkin, V. I., "Fronts in extended Josephson junctions", *Izvestiya Vyssh. Uchebn. Zaved. Radiofizika* **34** (1991) 956–965 (in Russian).
[3.5] Maksimov, A. G., Nekorkin, V. I. and Rabinovich, M. I., "Soliton trains and $I-V$ characteristics of long Josephson junctions", *Int. J. Bifurcation Chaos* **5** (1995) 491–505.
[3.6] Maksimov, A. G., Pederson, N. F., Christiansen, P. L., Molkov, Ya. I. and Nekorkin, V. I., "On kink-dynamics of the perturbed sine-Gordon equation", *Wave Motion* **23** (1996) 203–213.
[3.7] Ustinov, A. V., "Solitons in Josephson junctions", *Physica D* **123** (1998) 315–329.

Chapter 4

[4.1] Arnéodo, A., Coullet, P. and Tresser, C., "Possible new strange attractors with spiral structure", *Commun. Math. Phys.* **79** (1981) 573–579.
[4.2] Chua, L. O., *CNN: A paradigm for Complexity* (World Scientific, Singapore, 1998).
[4.3] Feigenbaum, M. J., "Quantitative Universality for a Class of Nonlinear Transformations", *J. Stat. Phys.* **19** (1978) 25–52.
[4.4] Horn, R. A. and Johnson, C. R., *Matrix Analysis* (Cambridge University press, Cambridge, 1985).
[4.5] Kazantsev, V. B., Nekorkin, V. I. and Velarde, M. G., "Pulses, fronts and chaotic wave trains in a one-dimensional Chua's lattice", *Int. J. Bifurcation Chaos* **7** (1997) 1775–1790.
[4.6] Madan, R. N. (Editor), *Chua's Circuit: A Paradigm for Chaos* (World Scientific, Singapore, 1993).
[4.7] Mira, C., "Chua's circuit and the qualitative theory of dynamical systems", *Int. J. Bifurcation Chaos* **7** (1997) 1910–1916.
[4.8] Nekorkin, V. I. and Chua, L. O., "Spatial disorder and wave fronts in a chain of coupled Chua's circuits", *Int. J. Bifurcation Chaos* **3** (1993) 1281–1291.
[4.9] Nekorkin, V. I., Kazantsev, V. B. and Chua, L. O., "Chaotic attractors and waves in a one-dimensional array of modified Chua's circuits", *Int. J. Bifurcation Chaos* **6** (1996) 1295–1317.
[4.10] Nekorkin, V. I., Kazantsev, V. B. and Velarde, M. G., "Travelling waves in a circular array of Chua's circuits", *Int. J. Bifurcation Chaos* **6** (1996) 473–484.
[4.11] Nekorkin, V. I., Kazantsev, V. B., Rulkov, N. F., Velarde, M. G. and Chua, L. O., "Homoclinic orbits and solitary waves in a one-dimensional array of Chua's circuits", *I.E.E.E. Trans. Circuits and Systems* **42** (1995) 785–801.
[4.12] Shilnikov, L. P., "Chua's circuit: rigorous results and future problems", *Int. J. Bifurcation Chaos* **4** (1994) 489–519.

Chapter 5

[5.1] Cronin, J., *Mathematical Aspects of Hodgkin–Huxley Neural Theory* (University Press, Cambridge, 1987).

[5.2] Defontaines, A.-D., Pomeau, Y. and Rostand, B., "Chain of coupled bistable oscillators: A model", *Physica D* **46** (1990) 201–216.

[5.3] Haken, H., *Principles of Brain Functioning. A Synergetic Approach to Brain Activity, Behavior and Cognition* (Springer-Verlag, Berlin, 1996).

[5.4] Hoppensteadt, F. C. *An Introduction to the Mathematics of Neurons* (University Press, Cambridge, 1986).

[5.5] Izhikevich, E. M., "Neural excitability, spiking, and bursting", *Int. J. Bifurcation Chaos* **10** (2000) 1171–1266.

[5.6] Llinás, R. R. (Editor), *The Biology of the Brain. From Neurons to Networks* (Freeman, New York, 1988).

[5.7] Llinás, R. R., *I of the Vortex. From Neurons to Self* (M. I. T. Press, Cambridge, Mass., 2001).

[5.8] MacKay, R. S. and Sepulchre, J.-A., "Multistability in networks of weakly coupled bistable units", *Physica D* **82** (1995) 243–254.

[5.9] McNell, K., "Bifurcations in ring arrays of phase-bistable systems", *Int. J. Bifurcation Chaos* **1** (1999) 107–117.

[5.10] Milton, J., *Dynamics of Small Neural Populations* (AMS, Providence, 1996).

[5.11] Nekorkin, V. I., Kazantsev, V. B., Velarde, M. G. and Chua, L. O., "Pattern interaction and spiral waves in a two-layer system of excitable units", *Phys. Rev. E* **58** (1998) 1764–1773.

[5.12] Nekorkin, V. I. and Makarov, V. A., "Spatial chaos in a chain of coupled bistable oscillators", *Phys Rev. Lett.* **74** (1995) 4819–4822.

[5.13] Nekorkin, V. I., Makarov, V. A. and Velarde, M. G., "Spatial disorder and waves in a ring chain of bistable oscillators", *Int. J. Bifurcation Chaos* **6** (1996) 1845–1858.

[5.14] Nekorkin, V. I., Makarov, V. A. and Velarde, M. G., "Clustering and phase resetting in a chain of bistable nonisochronous oscillators", *Phys. Rev. E* **58** (1998) 5742–5747.

[5.15] Nekorkin, V. I., Makarov, V. A., Kazantsev, V. B. and Velarde, M. G., "Spatial disorder and pattern formation in lattices of coupled bistable elements", *Physica D* **100** (1997) 330–342.

[5.16] Nekorkin, V. I., Voronin, M. L. and Velarde, M. G., "Clusters in an assembly of globally coupled bistable oscillators", *Eur. Phys. J. B* **9** (1999) 533–543.

[5.17] Pikovsky, A., Rosenblum, M. and Kurths, J., "Phase synchronization in regular and chaotic systems", *Int. J. Bifurcation Chaos* **10** (2000) 2291–2305.

[5.18] Scott, A. C., *Stairway to the Mind* (Copernicus-Springer, New York, 1995).

[5.19] Shepherd, G. M., *Foundations of the Neuron Doctrine* (University Press, Oxford, 1991).

[5.20] Shepherd, G. M. (Editor), *The Synaptic Organization of the Brain* (University Press, Oxford, 1998).

[5.21] Tass, P. A., *Phase Resetting in Medicine and Biology. Stochastic Modelling and Data Analysis* (Springer-Verlag, Berlin, 1999).

[5.22] Thiran, P., *Dynamics and Self-Organization of Locally Coupled Neural Networks* (Presses Politechniques et Universitaires Romandes, Lausanne, 1997).

[5.23] Winfree, A. T., *The Geometry of Biological Time*, 2nd Edition (Springer-Verlag, Berlin, 1990).

[5.24] Zhabotinsky, A. M., *Concentration Auto-oscillations* (Nauka, Moscow, 1974) (in Russian).

[5.25] Zykov, V. S., *Modelling of Wave Processes in Excitable Media* (University Press, Manchester, 1988).

Chapter 6

[6.1] Babloyantz, A. and Lourenço, C., "Computation with chaos: A paradigm for cortical activity", *Proc. Natl. Acad. Sci. USA* **91** (1994) 9027–9031.

[6.2] Binczak, S., Elibeck, J. C. and Scott, A. C., "Ephatic coupling of myelinated fibers", *Physica D* **148** (2001) 159–179.

[6.3] Bose, A., "Symmetric and antisymmetric phases in parallel coupled nerve fibers", *SIAM J. Appl. Math.* **55** (1995) 1650–1674.

[6.4] Brindley, J., Holden, A. V. and Palmer, A., *A numerical model for reentry in weakly coupled parallel excitable fibres*, in *Nonlinear Wave Processes in Excitable Media*, A. V. Holden, M. Markus and H. G. Othmer (Editors), (Plenum Press, New York, 1991), pp. 123–126.

[6.5] Haken, H., *Principles of Brain Functioning. A Synergetic Approach to Brain Activity, Behavior and Cognition* (Springer-Verlag, Berlin, 1996).

[6.6] Heagy, J. F., Carroll, T. L. and Pecora, L. M., "Synchronous chaos in coupled oscillator systems", *Phys Rev. E* **50** (1994) 1874–1885.

[6.7] Heagy, J. F., Carroll, T. L. and Pecora, L. M., "Experimental and mumerical evidence for riddled basins in coupled chaotic systems", *Phys. Rev. Lett.* **73** (1994) 3528–3531.

[6.8] Heagy, J. F., Carroll, T. L. and Pecora, L. M., "Desynchronization by periodic orbits", *Phys. Rev. E* **52** (1995) 1253–1256.

[6.9] Heagy, J. F., Pecora, L. M. and Carroll, T. L., "Short wavelength bifurcations and size instabilities in coupled oscillator systems", *Phys. Rev. Lett.* **74** (1995) 4185–4188.

[6.10] Hopfield, J. J., "Neural networks and physical systems with emergent collective computational abilities", *Proc. Natl. Acad. Sci. USA* **79** (1982) 2554–2558.

[6.11] Hopfield, J. J., "Pattern recognition computation using action potential timing for stimulus representation", *Nature* **376** (1995) 33–36.

[6.12] Hoppensteadt, F. C. and Izhikevich, E. M., "Synchronization of laser oscillators, associative memory, and optical neurocomputing", *Phys. Rev. E* **62** (2000) 4010–4013.

[6.13] Hoppensteadt, F. C. and Izhikevich, E. M., "Oscillatory neurocomputers with dynamic connectivity", *Phys. Rev. Lett.* **82** (1999) 2983–2986.

[6.14] Keener, J. P., "Homogenization and propagation in the bistable equation", *Physica D* **136** (2000) 1–17.

[6.15] Kladko, K., Mitkov, I. and Bishop, A. R., "Universal scaling of wave propagation failure in arrays of coupled nonlinear cells", *Phys. Rev. Lett.* **84** (2000) 4505–4508.

[6.16] Marquie, P., Comte, J. C. and Bilbault, J. M., "Contour detection using a two-dimensional diffusive nonlinear electrical network", *Proc. 2000 Int. Symposium On Nonlinear Theory and Its Applications* (NOLTA 2000, Dresden, Germany, 2000), pp. 331–334.

[6.17] Nekorkin, V. I., Kazantsev, V. B. and Velarde, M. G., "Image transfer in multilayered assemblies of lattices of bistable oscillators", *Phys. Rev. E* **59** (1999) 4515–4522.

[6.18] Nekorkin, V. I., Kazantsev, V. B. and Velarde, M. G., "Mutual synchronization of two lattices of bistable elements", *Phys. Lett. A* **236** (1997) 505–512.

[6.19] Nekorkin, V. I., Kazantsev, V. B., Artyukhin, D. V. and Velarde, M. G., "Wave propagation along interacting fiber-like lattices", *Eur. Phys. J. B* **11** (1999) 677–685.

[6.20] Nekorkin, V. I., Kazantsev, V. B., Rabinovich, M. I. and Velarde, M. G., "Controlled disordered patterns and information transfer between coupled neural lattices with oscillatory states", *Phys. Rev. E* **57** (1998) 3344–3351.

[6.21] Nekorkin, V. I., Kazantsev, V. B., Velarde, M. G. and Chua, L. O., "Pattern interaction and spiral waves in a two-layer system of excitable units", *Phys. Rev. E* **58** (1998) 1764–1773.

[6.22] Palmer, A., Brindley, J. and Holden A. V., "Initiation and stability of reentry in two coupled excitable fibers", *Bull. Math. Biology* **54** (1992) 1039–1056.

[6.23] Panfilov, A. V. and Holden A. V., "Vortices in a system of two coupled excitable fibers", *Phys. Lett. A* **147** (1990) 463–466.

[6.24] Pecora, L. M. and Carroll, T. L., "Synchronization of chaotic systems", *Phys. Rev. Lett.* **64** (1990) 821–824.

[6.25] Velarde, M. G., Nekorkin, V. I., Kazantsev, V. B. and Ross, J., "The emergence of form by replication", *Proc. Natl. Acad. Sci. USA* **94** (1997) 5024–5027.

[6.26] Zinner, B., "Existence of traveling wavefront solutions for the discrete Nagumo equation", *SIAM J. Diff. Eqs.* **96** (1992) 1–27.

[6.27] Zinner, B., "Stability of traveling wavefronts for the discrete Nagumo equation", *SIAM J. Math. Anal.* **22** (1991) 1016–1020.

Chapter 7

[7.1] Afraimovich, V. S. and Bunimovich, L. A., "Simplest structures in coupled map lattices and their stability", *Int. J. Random Comput. Dyn.* **1** (1993) 423–444.

[7.2] Afraimovich, V. S. and Bunimovich, L. A., "Density of defects and spatial entropy in extended systems", *Physica D* **80** (1995) 277–288.

[7.3] Afraimovich, V. S. and Chow, S.-N., "Topological spatial chaos and homoclinic points of Z^d-actions in lattice dynamical systems", *Jpn. J. Indust. Appl. Math.* **12** (1995) 367–383.

[7.4] Afraimovich, V. S., Glebsky, L. Yu. and Nekorkin, V. I., "Stability of stationary states and topological spatial chaos in multidimensional lattice dynamical systems", *Int J. Random Comput. Dyn.* **2** (1994) 287–303.

[7.5] Afraimovich, V. S. and Nekorkin, V. I., "Chaos of travelling waves in a discrete chain of diffusively coupled maps", *Int. J. Bifurcation Chaos* **4** (1994) 631–637.

[7.6] Afraimovich, V. S. and Pesin, Ya., "Traveling waves in lattice models of multi-dimensional and multicomponent media: I. General hyperbolic properties", *Nonlinearity* **6** (1993) 429–455.

[7.7] Afraimovich, V. S. and Pesin, Ya., "Traveling waves in lattice models of multi-dimensional and multicomponent media: II. Ergodic properties and dimension", *Chaos* **3** (1993) 233–241.

[7.8] Arnéodo, A., Coullet, P. and Tresser, C., "Possible new strange attractors with spiral structure", *Commun. Math. Phys.* **79** (1981) 573–579.

[7.9] Ashwin, P., Buescu, J. and Stewart, I., "Bubbling of attractors and synchronization of chaotic oscillators", *Phys. Lett. A* **193** (1994) 126–139.

[7.10] Ashwin, P., Buescu, J. and Stewart, I., "From attractor to chaotic saddle: Tale of transverse instability", *Nonlinearity* **9** (1996) 703–737.

[7.11] Ashwin, P. and Terry, J., "On riddling and weak attractors", *Physica D* **142** (2000) 87–100.

[7.12] Bunimovich, L. A. and Sinai, Ya. G., *Statistical Mechanics of Coupled Map Lattices*, in *Coupled Map Lattices: Theory and Applications*, K. Kaneko (Editor), (Wiley, New York 1992).

[7.13] Bunimovich, L. A., Livi, R., Martinez-Mekler, G. and Rutto, S., "Coupled trivial maps", *Chaos* **2** (1992) 283–291.
[7.14] Bunimovich, L. A. and Sinai, Ya. G., "Space-time chaos in coupled map lattices", *Nonlinearity* **1** (1998) 491–516.
[7.15] Carretero-Gonzalez, R., Arrowsmith, D. K. and Vivaldi, F., "One-dimensional dynamics for traveling fronts in coupled map lattices", *Phys. Rev. E* **61** (2000) 1329–1336.
[7.16] Chaté, H. and Courbage, M. (Editors), "Lattice dynamics", *Physica D* **1-4** (1997).
[7.17] Coutinho, R. and Fernandez, B., "On the global orbits in a bistable CML", *Chaos* **7** (1997) 301–310.
[7.18] Coutinho, R. and Fernandez, B., "Extended symbolic dynamics in bistable CML: Existence and stability of fronts", *Physica D* **108** (1997) 60–80.
[7.19] Feigenbaum, M. J., "Quantitative Universality for a Class of Nonlinear Transformations", *J. Stat. Phys.* **19** (1978) 25–52.
[7.20] Fernandez, B., "Kink dynamics in one-dimensional coupled map lattices", *Chaos* **5** (1995) 602–608.
[7.21] Fernandez, B., "Existence and stability of steady fronts in bistable CML", *J. Stat. Phys.* **82** (1996) 931–950.
[7.22] Glendinning, P., "Transitivity and blowout bifurcation in a class of globally coupled maps", *Phys. Lett. A* **264** (1999) 303–310.
[7.23] Haken, H., *Advanced Synergetics. Instability Hierarchies of Self-Organizing Systems and Devices* (Springer-Verlag, Berlin, 1983b).
[7.24] Heagy, J. F., Carroll, T. L. and Pecora, L. M., "Desynchronization by periodic orbits", *Phys. Rev. E* **52** (1995) 1253–1256.
[7.25] Johnston, M. E., "Bifurcations of coupled bistable maps", *Phys. Lett. A* **229** (1997) 156–164.
[7.26] Kaneko, K., "Pattern dynamics in spatiotemporal chaos", *Physica D* **34** (1989) 1–41.
[7.27] Kaneko, K., "Spatiotemporal chaos in one-spatio and two-dimensional coupled map lattices", *Physica D* **37** (1989) 60–82.
[7.28] Kaneko, K., *Simulating Physics with Coupled Map Lattices-Pattern Dynamics, Information Flow and Thermodynamics of the Spatio-temporal Chaos*, in *Formation, Dynamics and Statistics of Patterns*, K. Kawasaki, A. Onuki and M. Suzuki (Editors), (World Scientific, Singapore, 1990), pp. 1–50.
[7.29] Kaneko, K, "Overview of Coupled map lattices", *Chaos* **2** (1992) 279–282.
[7.30] Kaneko, K., "Chaotic traveling waves in a CML", *Physica D* **68** (1993) 299–317.
[7.31] Kaneko, K. and Tsuda, I., *Complex Systems: Chaos and Beyond* (Springer-Verlag, Berlin, 2001).
[7.32] Nekorkin, V. I., "Spatial chaos in a discrete model of radiotechnical medium", *Radiotekh. Elektron.* **37** (1992) 651–660 (in Russian).
[7.33] Nekorkin, V. I., Kazantsev, V. B. and Velarde, M. G., "Synchronization in two-layer bistable coupled map lattices", *Physica D* **151** (2001) 1–26.

Chapter 8

[8.1] Cairns-Smith, A. G., *Genetic Takeover and the Mineral Origins of Life* (Cambridge University Press, Cambridge, 1982).
[8.2] Crick, F., *Life Itself: Its origins and Nature* (Simon and Shuster, New York, 1981).
[8.3] Crick, F., *What Mad Pursuit* (Basic Books, New York, 1988).

[8.4] Haken, H., *Principles of Brain Functioning. A Synergetic Approach to Brain Activity, Behavior and Cognition* (Springer-Verlag, Berlin, 1996).
[8.5] Jacob, F., *The possible and the Actual* (Pantheon Books, New York, 1982)
[8.6] Laurent, M., "Prion diseases and the *protein only* hypothesis: a theoretical dynamic study", *Biochem. J.* **318** (1996) 35–38.
[8.7] Scott, A. C., *Stairway to the Mind* (Copernicus-Springer, New York, 1995).

Appendix G

[G.1] Horn, R. A. and Johnson, C. R., *Matrix Analysis* (Cambridge University press, Cambridge, 1985).

Index

MIX
Papier aus verantwortungsvollen Quellen
Paper from responsible sources
FSC® C105338

If you have any concerns about our products,
you can contact us on
ProductSafety@springernature.com

In case Publisher is established outside the EU,
the EU authorized representative is:
Springer Nature Customer Service Center GmbH
Europaplatz 3, 69115 Heidelberg, Germany

Printed by Libri Plureos GmbH
in Hamburg, Germany